EXERCISES IN INVERTEBRATE PALEONTOLOGY

Exercises in invertebrate paleontology

Frank K. McKinney

Department of Geology
Appalachian State University
Boone, North Carolina 28608
and Field Museum of
Natural History (Chicago)

Illustrated by Marjorie J. McKinney

BOSTON
BLACKWELL SCIENTIFIC PUBLICATIONS
OXFORD LONDON EDINBURGH
MELBOURNE PARIS BERLIN VIENNA

Editorial offices:
3 Cambridge Center, Cambridge
Massachusetts 02142, USA
Osney Mead, Oxford OX2 0EL, England
25 John Street, London WC1N 2BL
England
23 Ainslie Place, Edinburgh EH3 6AJ
Scotland
54 University Street, Carlton
Victoria 3053, Australia

Other editorial offices:

Arnette SA
2, rue Casimir-Delavigne
75006 Paris
France

Blackwell Wissenschaft
Meinekestrasse 4
D–1000 Berlin 15
Germany

Blackwell MZV
Feldgasse 13
A–1238 Wien
Austria

First published 1991

Set by Setrite Typesetters, Hong Kong
Printed and bound in America by
BookCrafters, Ann Arbor, Michigan

91 92 93 94 5 4 3 2 1

DISTRIBUTORS

USA
Blackwell Scientific Publications, Inc.
3 Cambridge Center
Cambridge, MA 02142
(*Orders*: Tel: 800 759-6102)

Canada
Oxford University Press
70 Wynford Drive
Don Mills
Ontario M3C 1J9
(*Orders*: Tel: 416 441-2941)

Australia
Blackwell Scientific Publications
(Australia) Pty Ltd
54 University Street
Carlton, Victoria 3053
(*Orders*: Tel: 03 347-0300)

Outside North America and Australia
Marston Book Services Ltd
PO Box 87
Oxford OX2 0DT
(*Orders*: Tel: 0865 791155
Fax: 0865 791927
Telex: 837515)

Library of Congress
Cataloging-in-Publication Data

McKinney, Frank K. (Frank Kenneth)
Exercises in invertebrate paleontol
Frank K. McKinney
Illustrated by Marjorie J. McKinney.
p. cm.
Includes bibliographical references
(p.) and index.
ISBN 0–86542–074–2
1. Invertebrates, Fossil—Laborato
manuals. 2. Paleontology—
Laboratory manuals. I. McKinney,
Marjorie J. II. Title.
OE770.M32 1991
562—dc20

Contents

Preface

Introductory invertebrate paleontology laboratories are generally designed to give students an understanding of and some familiarity with the more common and widely encountered groups of fossil invertebrates, through study of the fossils themselves. This is not an end in itself, but is a first step in making fossils useful in studies of evolution, biostratigraphy, analysis of paleoenvironments, paleoceanography, and several other areas of historical geology and biology.

A series of laboratory exercises devoted entirely to taxonomy and morphology ought to give students enough familiarity with fossils to use them for various problems in historical geology and paleobiology. However, few students would come through such a course with any enthusiasm for the relicts of the history of life. Therefore, although it is necessary to understand basic taxonomy and morphology for the intelligent use of fossils in geology and biology, the exercises in this manual include portions in which the fossils are put to use, analyzed, or interpreted from their occurrence in the rock record or from examination of their modern relatives.

The exercises in this manual were chosen to form an introduction to the major groups of invertebrate fossils, plus some of the protists that are among the most valuable microfossils for various geological applications (Exercise 3) or that were at one time considered to be invertebrate animals but possibly are vertebrates (Exercise 16). The omission of fossil annelids, insects, and malacostracan arthropods is no judgement that they have not been important in the history of life, but an acknowledgement that they are so uncommon in most university and college paleontology collections that they normally are not available for study in introductory paleontology laboratories.

Examples of the use of fossils for biostratigraphy, paleoenvironmental and paleogeographical reconstruction, functional analysis, and other geological and paleobiological studies in this manual, were chosen because they appear to serve well to introduce students to the varied applications and subfields of paleontology. The examples used are either based on works that I consider particularly elegant studies, or that take an approach that students and I have found interesting and instructive.

Classification of organisms is not firmly agreed upon for many groups of organisms, and different classifications may be found in current paleontology textbooks. Except where noted, I have followed the classifications used in *Fossil invertebrates*, edited by Boardman, Cheetham & Rowell (1987). For the arthropods, two superficially similar but conceptually very different classifications are presented. Two methods of dividing trilobites are also presented. Some instructors may choose classifications other than those used here, without much change in the text or assignments.

This manual is intended to be broad-based in its coverage of taxonomy, morphology, and stratigraphic and paleobiologic uses of fossils. I hope that it is potentially useful in most paleontology courses regardless of emphasis or approach, and that it will complement any of the currently popular paleontology texts. The set of exercises is designed for flexibility. The sequence in which the exercises are given follows the order in which I teach them. However, there are more exercises than can be covered in a single semester course, if one exercise is covered each week.

The exercises will almost fill a two-quarter course consisting of 10 weeks per term. Also, if used along with an independent laboratory research project of a few weeks duration, the exercises would suffice for a two-semester sequence of laboratory work. However, it should also be useful in the large number of introductory invertebrate paleontology courses that are taught within one semester, having one laboratory period of 2–3 h per week. For this purpose, each exercise is designed to be self-contained and to be modular, so that the exercises may be covered in any order. Some of the longer exercises contain more material than can be covered within a normal laboratory assignment, but these and all the others are organized in modules. Therefore, specific components within each may be omitted, or parts of more than one exercise may be recombined in order to cover more than one fossil group during a single laboratory period. In some cases, conceptual parts of some exercises may be moved into other exercises on different groups of organisms.

Many exercises contain portions that are not dependent on fossils being immediately at hand. These parts can serve as homework exercises either for laboratory or for lecture.

Materials required of students for these laboratory exercises include the textbook, one or more pencils, and a metric ruler. It would be useful also to have a hand lens, and extra graph and drawing paper. Students should also own paperback dictionaries of geological and biological terms or have access to dictionaries in the laboratory.

Many people have critically evaluated earlier versions of these exercises. The most important critics have been students who for several years have asked for clarification and reasons as the exercises evolved (some longer than others) toward their present form. In addition, William Ausich, William Berry, Daniel Blake, Richard Boardman, Scott Brande, Thomas Broadhead, Karen Callahan, Alan Cheetham, David Clarke, Mike Foote, Jere Lipps, Patricia Kelley, Tom Klee, Michael LaBarbera, George Pemberton, David Raup, Keith Rigby, Richard Robison, Philip Signor, George Stanley, Steven Stanley, Karl Stock, Helen Tappan, Peter Ward, and Mark Wilson each have read and criticized one or more exercises. Several photomicrographs were made with the assistance of Ruth Dewel, in the Electron Microscope Laboratory, Appalachian State University. To all I give thanks, including any whom I have overlooked but whose name should have been on the list.

Undoubtedly, some errors, omissions, or warped emphases remain, despite the best efforts of the readers mentioned above to keep me from straying. I will appreciate them being pointed out to me by either students or instructors.

Finally, most line drawings in the manual were made by my wife, Marjorie. Without them, the manual would be much less than it is. Thanks, Marg.

Frank K. McKinney
Appalachian State University

EXERCISES IN INVERTEBRATE PALEONTOLOGY

EXERCISE 1

Taphonomy

1.1 INTRODUCTION

No organism has been preserved unchanged in its original life position. Instead, each fossil has been altered to some degree, ranging from dessication or freezing but with the majority of soft tissues still present to degradation of hard parts so that all that remains is a partial imprint or a residual grain too small to assign to any phylum. In addition, the majority of fossils are not entombed in life position or even in their position of death, but have generally at least rotated over so that their longest axis lies on the substratum. Many have been moved by currents, waves or mobile organisms for some unknown distance from the place at which they died.

The study of this history of individual fossils and of fossil assemblages, from death to observation or collection by a paleontologist or field geologist is known as **taphonomy**. Taphonomy has yielded a distinct body of knowledge and new understanding of the process of preservation. Many exciting new insights into carnivory and scavenging activity, utilization of skeletal remains by subsequent organisms, hydrodynamic processes that affect skeletal remains, sedimentary geochemistry, and other topics are currently developing in taphonomic studies. For many paleontologists, however, the changes that have afflicted fossils constitute a set of filters that diminish the amount of information available about the once-living organisms.

The purposes of this exercise are:

1 to consider as much as possible the history of changes that fossils in various states of preservation have experienced; and

2 to examine the probable effects of some taphonomic processes on the potential for preservation of a living marine community.

1.2 TAPHONOMIC CHANGES

The taphonomic history of most fossils may be divided into three intergrading stages: pre-burial, burial, and post-burial. (Fossils of burrowing organisms that were never exhumed from their enclosing sediment before they were collected have only a post-burial history.) Generally, burial is the briefest part of the history and post-burial is the most extended. Potential events in a fossil's taphonomic history are summarized below. Possible pre-burial alterations include the following:

1 Transportation. This redistributes the remains of organisms within the original habitat or mixed into other environments. Often breakage or abrasion reduces the size of transported skeletal fragments so that their identification becomes progressively more difficult. However, if the original skeletal microstructure is preserved, particularly if that is supplemented by retention of some original surface morphology, even rather small grains in thin sections can be identified for many marine invertebrates as indicated in Fig. 1.1. Over a century ago, Charles Darwin (1881, p. 251) used microstructure to recognize sand-sized fragments of "land mollusc" shells in castings made by earthworms.

(a) Currents. Currents cause sorting by size, shape, and density. Abrasion begins to erode surface morphology and causes rounding if there is extended current transportation. In addition, a preferential orientation of non-spherical skeletons is likely to develop, which gives an indication of the final direction of current movement before burial occurred.

(b) Shoreface waves. These waves, which stir up the sediment in the intertidal zone and just offshore along beaches, generate some fragmentation, but abrasion and sorting are especially characteristic. The effects of shoreface waves on fossil remains are difficult to distinguish from the effects of currents except through study of geometry, sedimentary structures, and lithologic associations of the enclosing sediments.

(c) Storm waves. These relatively brief but vigorous waves often result in more fragmentation than abrasion and sorting. They extend deeper into the water and so affect organic remains on parts of the sea or lake floor that are not touched by normal waves.

(d) Rafting and transportation by animals. Either of these may be hypothesized for exotic occurrences, but is difficult or impossible to demonstrate. Possible modes of transportation include ice, pumice, floating vegetation, and mobile organisms (such as sea gulls that break mollusks open by dropping them onto pavement). If unaffected by other types of transportation, abrasion and fragmentation are commonly slight.

2 Biological.

(a) Preservation by bioimmuration. Bioimmuration is the covering over of an organism by some other encrusting organism such that the shape and perhaps surface texture of the underlying organism is preserved on the underside of the encruster after the two become detached.

(b) Destruction. (i) Exogenous: these are external destructive forces, including fragmentation by such organisms as crabs and grazing fish. (ii) Endogenous: endogenous destruction occurs from within the organism or its remains, e.g. excavation of skeleton by boring organisms such as endolithic algae, sponges, barnacles, and drill holes made by carnivorous gastropods, or bacterial decomposition of organic portions of skeletons.

3 Chemical.

(a) Preservation. Skeletal cavities may be filled by minerals during early diagenesis (Fig. 1.2), which is the sequence of changes that occur as sediment is transformed to rock. This includes permineralization, which fills mineral material into tiny pore spaces within organisms or their skeletons.

(b) Destruction. Changes in acidity (pH) or oxidation potential (Eh) so that a critical value termed a chemical fence is crossed, results in the original material being dissolved or replaced by a different material (Fig. 1.3).

Figure 1.4 summarizes the common potential pre-burial taphonomic alterations that corals distributed across a reef may experience. Clearly, the types and severity of taphonomic alterations, even for this one group of organisms distributed

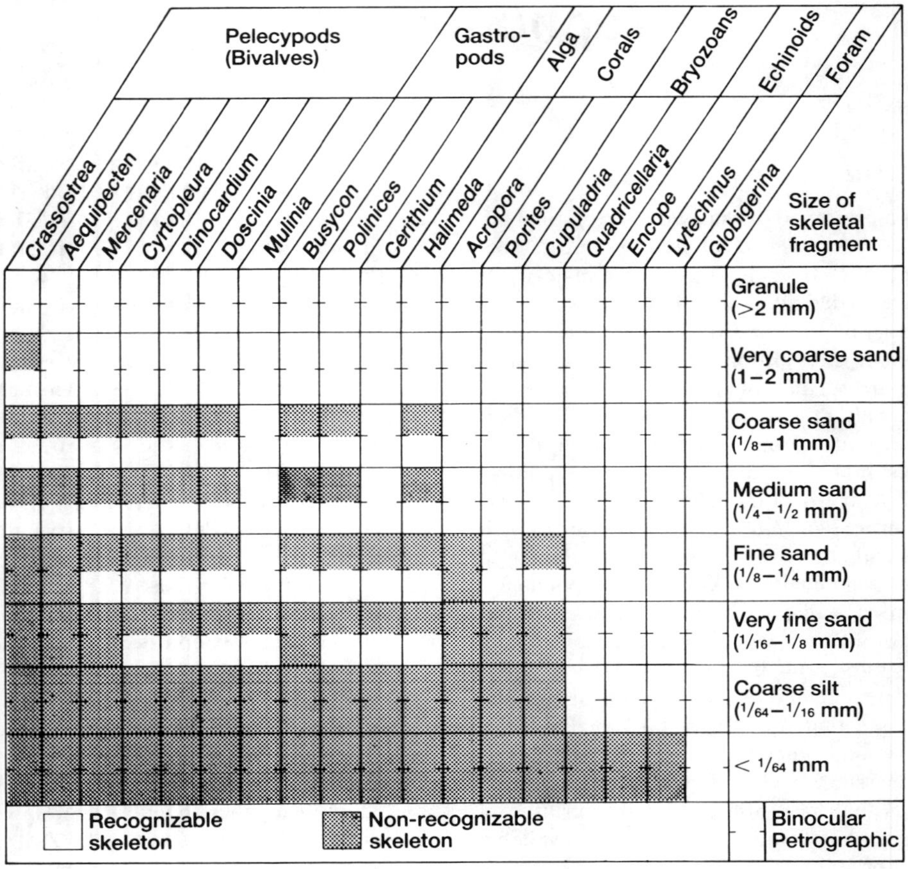

Fig. 1.1 Sizes at which fragments derived from skeletons of various marine organisms are recognizable when studied with a binocular microscope and when studied in thin sections with a petrographic microscope. From Feray *et al.* (1962) with permission from the American Association of Petroleum Geologists.

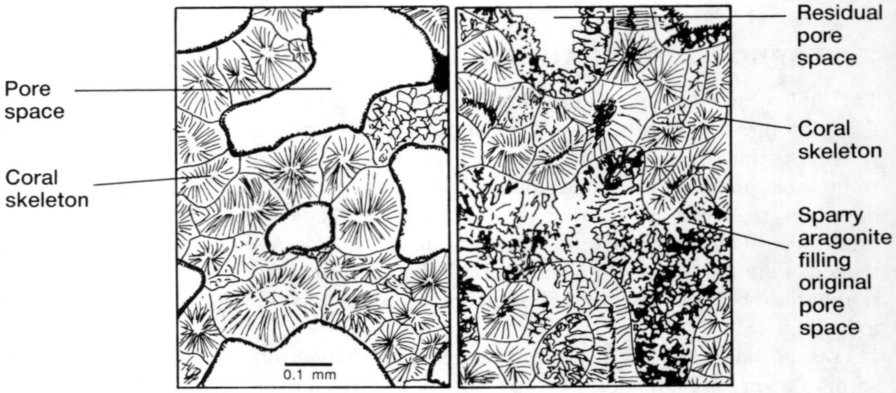

Fig. 1.2 Drawings from photomicrographs of: (a) a thin-section of skeleton cut from a living coral that has been diagenetically altered only by minute boring algae along the surface of the skeleton (producing the dark margin known as a micritic envelope); and (b) a coral, collected about 30 cm below the surface of a reef, in which aragonite cement has grown profusely into the interstices of the skeleton. From Friedman & Amiel (1974).

across subzones of a single environment, are basically different and characteristic for each specific type of occurrence. *(Questions 1 and 2)*

Burial situations include the following:

1 Rate.
(a) Catastrophic. This includes sudden influxes of sediment that cover over the fossil in its original habitat or move the fossil some distance as a sedimentary particle that ends up on the bottom or is buried within an accumulation of sediments.
(b) Slow. This is gradual burial of the fossil because of grain-by-grain or layer-by-layer accumulation.

2 Locale.
(a) *In situ*. Organisms may be buried at the place where they died. This is most common for benthic organisms that lived on fine-grained sediments. (i) In life position: preservation in life position generally is most common for

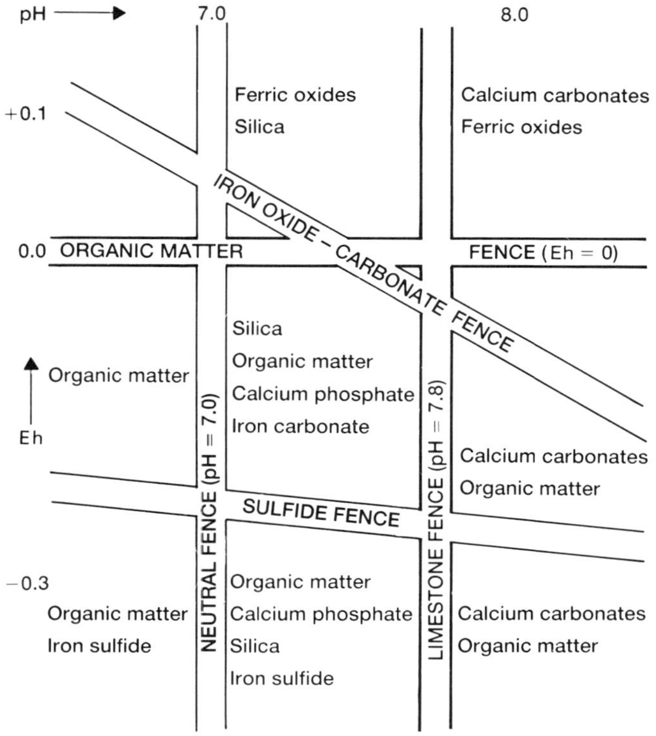

Fig. 1.3 Diagram of the common chemical fences that are pertinent in diagenesis of sedimentary rocks and in taphonomic alteration of fossils. Each fence separates two chemical fields characterized by contrasting occurrences of specific materials. For example, organic matter can exist below the organic matter fence (Eh=0), in conditions of negative Eh, but it does not exist above the organic matter fence, in conditions of positive Eh. From Towe in Boardman *et al.* (1987) after Krumbein & Garrels (1952).

burrowers or stationary organisms that live partially buried, less common for epifaunal sedentary organisms, and least common for mobile animals. (ii) Prone: most organisms topple over or collapse into a stable orientation on the substratum before they are buried.

(b) Transported. Transportation is indicated by abrasion and rounding and is suggested by preservation in coarse-grained sediment. It should be considered as a possibility where fossils in fine-grained sediments appear to rest on one another, forming the structural framework of the rock. *(Question 3)*

Possible post-burial alterations include the following:

1 Pre-lithification (before consolidation of sediment into rock).
(a) Compaction. Especially severe in dewatering of clays. It results in plastic-style flattening of flexible fossils and in flattening of rigid fossils by cracking into fragments that generally are contiguous.
(b) Chemical. (i) *Preservation*: this continues the filling of cavities during early diagenesis and includes permineralization. (ii) *Destruction*: fossils may be destroyed when they cross certain chemical fences as illustrated in Fig. 1.3.
(c) Biological. (i) *Exogenous*: this almost always consists of fragmentation by excavating or burrowing deposit feeders, but it includes shallowly buried shells trampled by heavy vertebrates such as dinosaurs. (ii) *Endogenous*: bacteria may decompose organic material after burial.

2 Post-lithification (after consolidation of sediment into rock).
(a) Physical. (i) *Structural deformation*. (ii) *Physical weathering*.
(b) Chemical. (i) *Recrystallization*. *(ii) Permineralization*. (iii) *Replacement*. (iv) *Distillation*: selective removal of oxygen and hydrogen from organic material results in the concentration of carbon. (v) *Dissolution*:

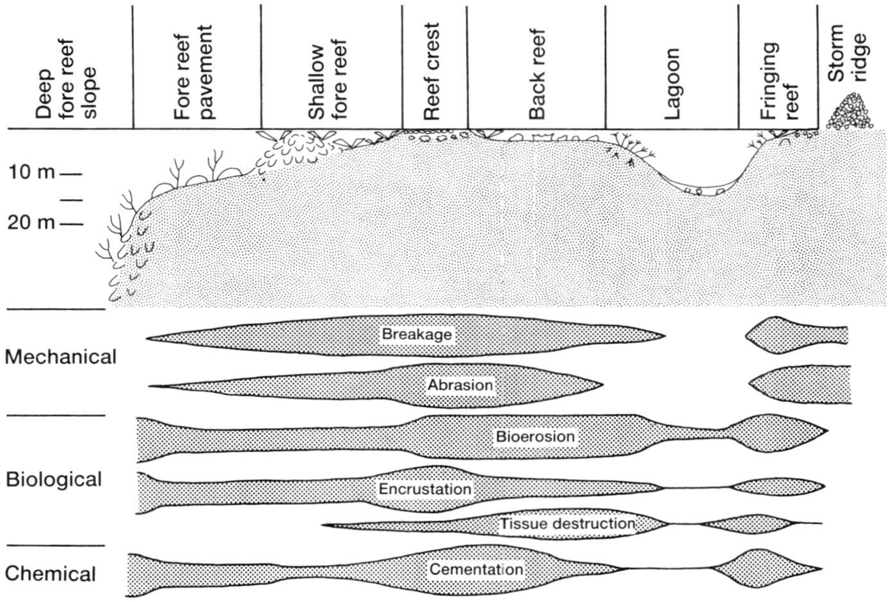

Fig. 1.4 Relative intensity of pre-burial taphonomic effects on corals across shore-parallel zones from offshore (left) to onshore (right). From MacIntyre (1984).

where skeletal material is dissolved from rock, *molds* of the exterior — and in some cases interior — surfaces are produced where the rock retains the impression of the fossil. (vi) *Casts*: casts are infillings (including plastics) of natural molds and replicate the original shape of the fossil material.
(Question 4)

1.3 LOSS OF INFORMATION FROM PRESERVED ASSEMBLAGES

Fossil assemblages, even those that are preserved *in situ*, lack large numbers of the original components of the living community. Certain size classes of the preserved species may be absent or under-represented, and many species have left no record at all. The similarity of the fossil assemblage to the original community depends on the original chemical composition of the species in the community and on the vicissitudes of preservation that have been summarized in Section 1.2.

Artificial or hypothetical degradation of modern communities can give useful insights into the types of organisms that may be lacking in fossil assemblages, particularly Cenozoic assemblages of fossils that have close living representatives.

The final question in this exercise will be a comparison of Oligocene oyster reefs from the Atlantic Coastal Plain of North Carolina with living oyster reefs in the estuaries and lagoons along the coast of North Carolina, based on work by Lawrence (1968). The Oligocene assemblage is dominated by densely packed, upright specimens of the oyster *Crassostrea gigantissima* which are associated with skeletons of several bryozoan species, one pelecypod (bivalve), one annelid, and one barnacle species. In addition, the oyster skeletons contain excavations made by sponges, bryozoans, an annelid, and a pelecypod; skeletal blisters that formed in response to a different annelid; and the imprint of a low-spired gastropod. In all, between 16 and 18 species are represented.

A total of 303 species has been identified in the *Crassostrea virginica* reefs in the vicinity of Beaufort, North Carolina (Wells, 1961), excluding monerans, copepods, ostracodes, nematodes, herring gulls, *Homo sapiens* and most protistans. Of the 303 species, 81 occur in 40% or more of the sites examined and will be the basis for comparison with the Oligocene reefs in the question below. The purpose of this part of the exercise is to see how comparison of a geologically

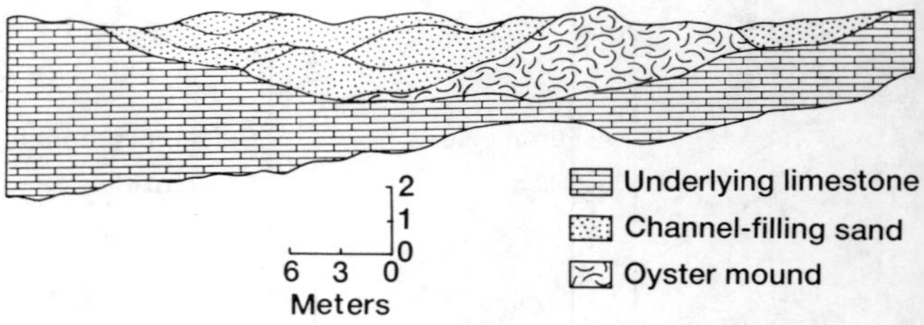

Fig. 1.5 A cross-section through an Oligocene oyster build-up and the channel deposits in which it is situated at Belgrade, North Carolina. From Lawrence (1968).

"degraded" recent community with a fossil analogue can be used to address such questions as probable similarity of communities and possible increase through time in species richness.
(Question 5)

The Belgrade oyster reef assemblage was preserved in porous terrigenous sediments deposited in a channel that was cut into underlying calcareous strata (Fig. 1.5). Movement of groundwater has strongly leached the deposit, allowing only calcitic fossils to be preserved as body fossils. Therefore, the hypothetical preservation state of the Beaufort oyster reefs in the column entitled "Calcite skeletons preserved, with species marked by r restored" should be most comparable to that of the Oligocene oyster reef assemblage at Belgrade.
(Question 6)

FURTHER READING

Allison, P. A., 1988. Konservat-Lagerstatten: cause and classification. *Paleobiology* **14**:331–344.

Behrensmeyer, A. K. & Kidwell, S. M., 1985. Taphonomy's contributions to paleobiology. *Paleobiology* **11**:105–119.

Durham, J. W., 1967. The incompleteness of our knowledge of the fossil record. *Journal of Paleontology* **41**:559–569.

Krumbein, W. C. & Garrels, R. M., 1952. Origin and classification of chemical sediments in terms of pH and oxidation-reduction potential. *Journal of Geology* **60**:1–33.

Thomas, R. D. K. (ed.), 1986. Taphonomy: ecology's loss is sedimentology's gain. *Palaios* **1(3)**. (An issue of the journal *Palaios* that consists of 10 papers on various aspects of taphonomy, written by various authors.)

EXERCISE 1

Worksheet

Name

QUESTION 1

Destruction due to transportation. Examine the specimens provided, which have been affected predominantly by one of the four categories of transportation as listed in the text of this exercise. Carefully note the range of effects that the specimens exhibit. In the spaces below, briefly (1) describe the effects, (2) determine the probable cause, and (3) explain why you answered (2) as you did.

1 Effects seen in specimens: ____________________

2 Probable type of transportation: ____________________

3 Explanation: ____________________

QUESTION 2

Endogenous biological destruction. Examine the specimens provided, which have been affected to varying degrees by boring organisms that are identified to phylum on the specimen label. (1) Describe the shape(s) of the cavities generated, (2) estimate the percent of original skeletal material that has been removed by the boring organisms in the most severely affected example, and (3) state how the specimen's ability to resist physical damage is affected by the presence of the borings.

1 Shape(s) of cavities: ____________________

2 Estimated % skeleton lost: ____________________

3 Effect on resistance to physical damage: ____________________

QUESTION 3

Examine the slabs provided. For each assess the probable burial situation and give the bases for your conclusions.

Burial situation for fossils in slab 1: ____________________

Bases for assessment: ____________________

____

Burial situation for fossils in slab 2: ____________________

Bases for assessment: ____________________

____

QUESTION 4

Labeled examples of several types of preservation are provided. For each give as complete a taphonomic history as possible, which consists of the taphonomic evidence exhibited and the implications or interpretation of each bit of evidence. Organize your answers in the following sequence: (1) chemical fences crossed and the direction they were crossed, (2) pre-burial events, (3) burial situation, (4) post-burial events. Each of the categories may not be pertinent for all specimens. Burial situation is impossible to determine for specimens that have been entirely removed from their original rock, unless they are conspicuously abraded. If a specimen provides no information for one of the parts (1)–(4), write NA on that line.

Specimen:

Type preservation:

Observation	*Interpretation*
1	
2	
3	
4	

Specimen:

Type preservation:

Observation	*Interpretation*
1	
2	
3	
4	

Specimen:

Type preservation:

Observation	*Interpretation*
1	
2	
3	
4	

Specimen:

Type preservation:

Observation	*Interpretation*
1 ______	______
2 ______	______
3 ______	______
4 ______	______

Specimen:

Type preservation:

Observation	*Interpretation*
1 ______	______
2 ______	______
3 ______	______
4 ______	______

Specimen:

Type preservation:

Observation	*Interpretation*
1 ______	______
2 ______	______
3 ______	______
4 ______	______

Specimen:

Type preservation:

Observation | *Interpretation*

1 ____________________ | ____________________

2 ____________________ | ____________________

3 ____________________ | ____________________

4 ____________________ | ____________________

Specimen:

Type preservation:

Observation | *Interpretation*

1 ____________________ | ____________________

2 ____________________ | ____________________

3 ____________________ | ____________________

4 ____________________ | ____________________

Specimen:

Type preservation:

Observation | *Interpretation*

1 ____________________ | ____________________

2 ____________________ | ____________________

3 ____________________ | ____________________

4 ____________________ | ____________________

Specimen:

Type preservation:

Observation	*Interpretation*
1 ______________________	______________________
2 ______________________	______________________
3 ______________________	______________________
4 ______________________	______________________

Specimen:

Type preservation:

Observation	*Interpretation*
1 ______________________	______________________
2 ______________________	______________________
3 ______________________	______________________
4 ______________________	______________________

QUESTION 5

In the following list of 81 species that are widespread in North Carolina oyster reefs, skeletal composition of each is recorded as A = aragonite or mixed aragonite and calcite, B = calcium phosphate (bone), C = calcite or predominantly calcite, O = largely or entirely organic, and Si = silica. In addition, organisms likely to generate preservable borings, imprints, or to stimulate identifiable responses in calcareous skeletons are indicated by an "r".

Complete the chart by filling an X in the boxes in each column if remains of the species should have survived to be represented at that stage of preservation. In column 1, score an X for all organisms that would still be represented if all well-mineralized skeletons are preserved and all organisms with organic skeletons have decayed and left no trace. In column 2, score only those organisms that would still be represented if only calcite or predominantly calcite skeletons — but *not* aragonite or mixed skeletons — are preserved. In column 3, *repeat* the scores from column 2 and *also* score those organisms that are likely to leave some identifiable boring, imprint, or response in a host. (Column 3 will have more species scored than are scored in column 2.)

Species	Skeleton	Column 1	Column 2	Column 3
Phylum Protozoa				
Folliculinid species	O			
Phylum Porifera				
Cliona cellata	Si, r			
Cliona spirilla	Si, r			
Cliona truitti	Si, r			
Haliclona permollis	Si			
Hymeniacidon heliophila	Si			
Streblospio benedicta	Si			
Phylum Coelenterata				
Aiplasia pallida	O			
Aiplasia eruplaurantia	O			
Diadumene leucoclena	O			
Diadumene luciae	O			
Leptogorgia virgulata	O			
Obelia sp.	O			
Phylum Platyhelmintha				
Stylochus ellipticus	O			
Phylum Nemertea				
Amphiporus ochraceus	O			
Tubulanus pellucidus	O			
Phylum Mollusca: Class Gastropoda				
Bittium varium	A			
Cantharus tinctus	A			
Crepidula fornicata	A			
Diodora cayenensis	A			
Fasciolaria hunteria	A			
Mitrella lunata	A			
Odostomia dianthophila	A			
Odostomia impressa	A			
Urosalpinx cinerea	A, r			
Phylum Mollusca: Class Bivalvia				
Anomia simplex	C			
Arca umbonata	A			
Brachidontes exustus	A			
Brachidontes recurvus	A			
Chione grus	A			
Crassostrea virginica	C			
Lithophaga bisulcata	A, r			
Martesia smithi	A			
Modiolus americanus	A			
Modiolus demissus	A			
Musculus lateralis	A			
Ostrea equestris	C			
Petricola pholadiformis	A, r			
Phylum Annelida				
Amphitrite ornata	O, r			
Capitella capitata	O			

Species	Skeleton	Column 1	Column 2	Column 3
Eteone heteropoda	O			
Eupomatus dianthus	O			
Haplosyllis spongicola	O			
Marphysa sanguinea	O			
Neanthes succinea	O			
Nereiphylla fragilis	O			
Nereis occidentalis	O			
Polydora websteri	O, r			
Sabella microphthalma	O			
Sabellaria vulgaris	O, r			
Phylum Arthropoda				
Balanus amphitrite	C			
Balanus eburneus	C			
Balanus improvisus	C			
Caprella acutifrons	O			
Caprella linearis	O			
Cassidiscus lunifrons	O			
Chthamalus fragilis	C			
Clibanarius vittatus	O			
Corophium cylindricum	O			
Dipteran larva	O			
Dyaneme perforata	O			
Eurypanopeus depressus	O			
Gammarus locusta	O			
Leptochelia savignyi	O			
Melita appendicula	O			
Melita dentata	O			
Neopanops texana sayi	O			
Panopeus herbati	O			
Pinnotheres ostreum	O			
Sphaeroma quadridentatum	O			
Phylum Bryozoa				
Alcyonidium polyoum	O			
Anguinella palmata	O			
Bugula neritina	O			
Electra crustulenta	C			
Membranipora tenuis	C			
Schizoporella errata	C			
Victorella pavida	O			
Phylum Pycnogonida				
Nymphon rubrum	O			
Phylum Chordata				
Bobiosoma bosci	B			
Molgula manhattensis	O			
Styela plicata	O			
Total for each column				

Complete the following chart by filling in the correct numbers from the information that you compiled above.

	Living oyster reefs	*All mineral skeletons preserved, but organic skeletons lost*	*Only calcite skeletons preserved*	*Calcite skeletons preserved, with species marked by r restored*	*Oligocene oyster reef assemblage at Belgrade, NC*
Phyla represented	________	________	________	________	5
Species present	________	________	________	________	16–18
Percentage of total community preserved	________	________	________	________	?
Percentage of species preserved only as r	________	________	________	________	~44

QUESTION 6

In the space provided below, describe the similarities and dissimilarities of the Belgrade Oligocene reef and the "degraded" modern oyster reef community from the North Carolina coast. What are the implications for original species richness (i.e. number of species) in the Oligocene community? For various purposes, some paleontologists have counted the number of taxa that are known for specific times in the past, or they have counted the number of taxa within specific habitats through time. There are many ways to go astray in making the counts and interpreting their meanings. Given the work you have done in preparation to answer Question 5, describe some possible problems with interpreting increase in diversity within habitats through time.

EXERCISE 2

Growth and simple statistical description of morphology

2.1 INTRODUCTION

Growth is a fundamental expression of life. It is more than simple increase in size, but involves various complexities such as changes in bodily proportions, and addition of organs and skeletal units so that relationships between parts of an organism change as growth progresses. These changes in measurable relationships may be due to such causes as structural necessity, attainment of some minimum size so that new functions can be engaged, or even so that certain relationships can be maintained uniformly.

Description and comparison of organisms can be quite cumbersome and subjective if words alone are available to accomplish the task. Even use of the most detailed possible drawings or photographs cannot adequately compensate for description, for each person must either still use words in thinking about the illustration or retreat even further into subjectivity in developing a nonverbal — though perhaps highly accurate — feeling or impression of an organism or how sets of organisms compare with one another. Use of numerical descriptions and comparisons is no panacea that necessarily develops completely objective thought and communication and that can be used arbitrarily or as a substitute for thought (Fig. 2.1). However, appropriate use of a combination of qualitative data, illustration, and numerical techniques can make thinking and communication much more successful and objective.

The purposes of this exercise are:

1 to introduce principles of numerical description of individual fossils and multiple-specimen samples drawn from populations of fossils;

2 to introduce some simple methods of estimating probability of differences between populations; and

3 to introduce some of the gross differences in growth patterns exhibited by invertebrate animals and examine why such radical differences exist.

Several repetitive but not difficult calculations are involved. They may be done by hand in a reasonable amount of time, or they may be done using a calculator or one of the numerous statistical programs available for computers and microcomputers.

2.2 DETERMINATE AND INDETERMINATE GROWTH

The shape that any organism has at any instant is basically the result of its growth history up to that point. (For some organisms, partial predation and loss of parts are also important shape determinants. However, this exercise focuses exclusively on growth.) Solitary organisms typically have a narrowly varying range of shapes and at full growth may reach some approximate maximum size, both of which are characteristic of the species. This is commonly termed **determinate growth**. In contrast, the shape and maximum size of many colonial organisms, which are built by budding additional functional units, are less constrained and are commonly determined more by the available space than by genetic programming. This is frequently designated **indeterminate growth**. Indeterminate growth is often an indication that one is dealing with a colonial organism or with non-colonial organisms that grow in dense aggregations in which crowding affects size and shape.
(Question 1)

2.3 NUMERICAL DESCRIPTION

Numerical description and comparison of organisms requires the scoring of some property (a **variable**) that is observed to differ or potentially differs among individuals. There are five types of variables:

1 counts of discrete structures such as number of teeth;

2 measures such as length, area, volume, etc.;

3 ranking along some scale such as from shortest to longest;

4 presence and category of some attribute such as a bump, widget, or gismo; and

5 ratios, such as length : volume.

This exercise focuses on description by measurement, which then may be used directly, or to establish rank, or to determine ratios between two different measures.

Linear measurements are typically made approximately along or transverse to the principal axis of growth, along the anterior–posterior and ventral–dorsal axes, or in some instances — especially where the principal axis of growth is

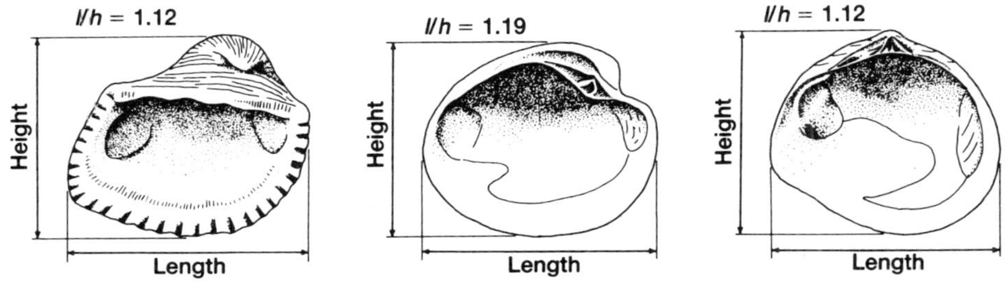

Fig. 2.1 Lateral views of three pelecypod shells that have approximately equal length : height proportions but that have conspicuous differences when other aspects of their morphology are considered. Redrawn after Clark (1904).

not inherently obvious — along the longest and shortest distances, i.e. the major and minor axes of length. Measurements of larger specimens not requiring magnification for study are made most accurately with calipers or by being laid directly on millimeter graph paper for measurement (Fig. 2.2). Commonly, such as for the pelecypods in Figs 2.1 and 2.2, length and height often are determined as straight-line measurements that at best are roughly parallel with and transverse to the principal growth axis. Figure 2.3 illustrates some principal linear measures of various types of fossils, some of which require magnification for measurement and study.

Many organisms add to their skeletons by **accretion**, which is direct addition of skeletal material onto the pre-existing part(s) so that the skeletal unit is extended or broadened, or both, and commonly thickened. Rate of accretion typically is variable so that along the growing margin, thickness, color, or some other feature varies and results in *growth lines* that record the history of the size and shape of the skeleton as it grew. Growth lines therefore allow an investigator to identify the earliest-formed part of the skeleton. Measurement can then be made exactly along the principal axis of growth by following a line perpendicular to the growth lines where their broadest widths indicate the most rapid accretion.
(Question 2)

2.3.1 Measures of central tendency

The set of measurements made for Question 2 do more than just describe 15 individual specimens. Collectively, they represent a sample drawn from an assemblage of fossils. They are thus thrice-removed from an original, living population of organisms:

1 even within a single bed there is generally at best mixing of remains from several successive populations that lived at the site of deposition;

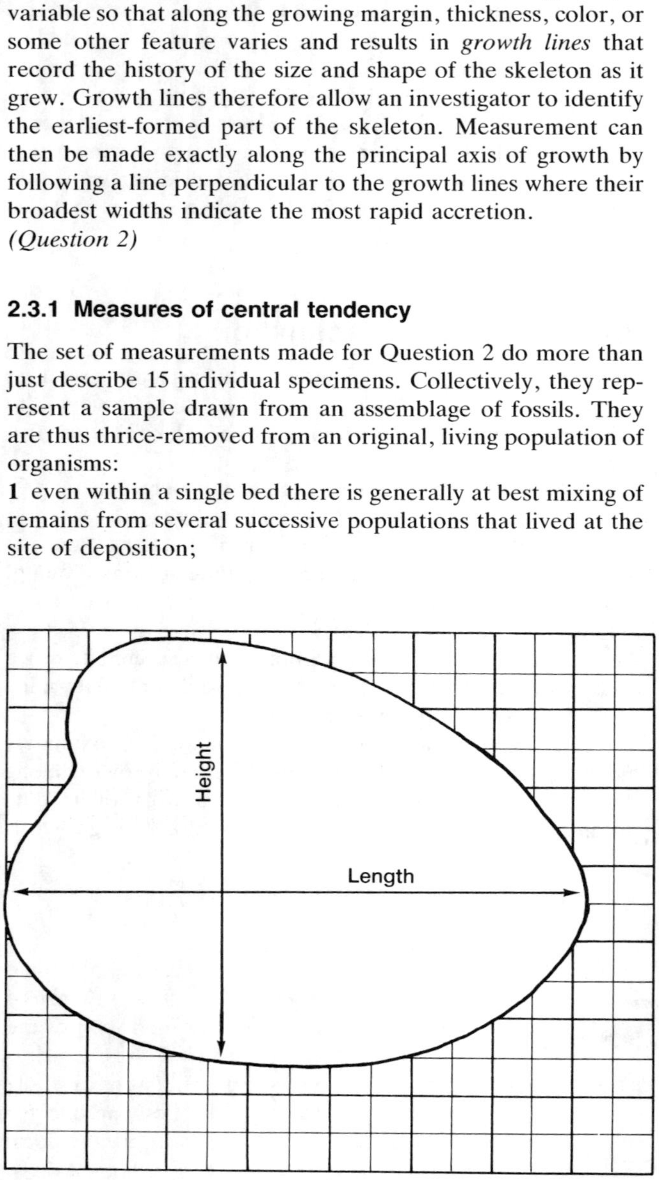

Fig. 2.2 Profile of pelecypod placed on graph paper for direct measurement (to nearest millimeter) of standard measures of length (anterior–posterior) and height (dorsal–ventral) for the group.

2 various size classes are selectively removed by taphonomic processes; and

3 the sample brought to the laboratory includes only part of the specimens preserved in the bed, and the collecting procedures can strongly influence the representativeness of the sample.

Among the various ways to determine "typical" values, the **arithmetic mean** ($\overline{X}$) is the most useful for general statistical analyses. The arithmetic mean is the sum of the values divided by the number of values in the set, generally designated N, or the number of observations:

$$\overline{X} = \frac{\Sigma X}{N}.$$

(Question 3)

2.3.2 Measures of dispersion

Among the various statistics that estimate scatter (dispersion), the **range** and the **standard deviation** are probably the commonest and most easily calculated. The range is the span of values and is given by listing the smallest and the largest values. It characteristically increases — progressively more slowly — as N increases. The standard deviation (s) is the square root of the variance (s^2):

$$s^2 = \frac{\Sigma(X - \overline{X})^2}{N - 1} \quad \left[= \frac{\Sigma X^2 - \frac{(\Sigma X)^2}{N}}{N - 1} \right].$$

Therefore,

$$s = \sqrt{\frac{\Sigma(X - \overline{X})^2}{N - 1}} \quad \left[= \sqrt{\frac{\Sigma X^2 - \frac{(\Sigma X)^2}{N}}{N - 1}} \right].$$

The denominator is $N-1$ rather than N, because statisticians have found that this compensates for consistent underestimation of variance and standard deviation for the population from which the sample is drawn. (If one wishes only to describe variance and standard deviation of the sample itself rather than to estimate them for the assemblage, then it would be accurate to use just N for the denominator.) Where two formulae are listed on a single line, the one on the right is enclosed in brackets and is an alternative formula for calculation by hand if no computer or calculator programs are available. The logic of the statistical processes is not as clear for the alternative formulae as for the basic formulae, but they make hand calculation easier.

Variation as expressed by s may not be immediately obvious and useful in comparison between different populations. For example, width in a sample of a species of brachiopods from a Devonian black shale may have an arithmetic mean of 5 mm and a standard deviation of 0.5 mm, whereas width in a sample from a different species of brachiopods from a coeval reef may have an arithmetic mean of 17 mm and a standard deviation of 1.2 mm. Which is more variable? The **coefficient of variation** (CV) is a way of assessing the variability relative to the average size and gives a clearer idea of how relatively variable the data actually are. It is calculated as

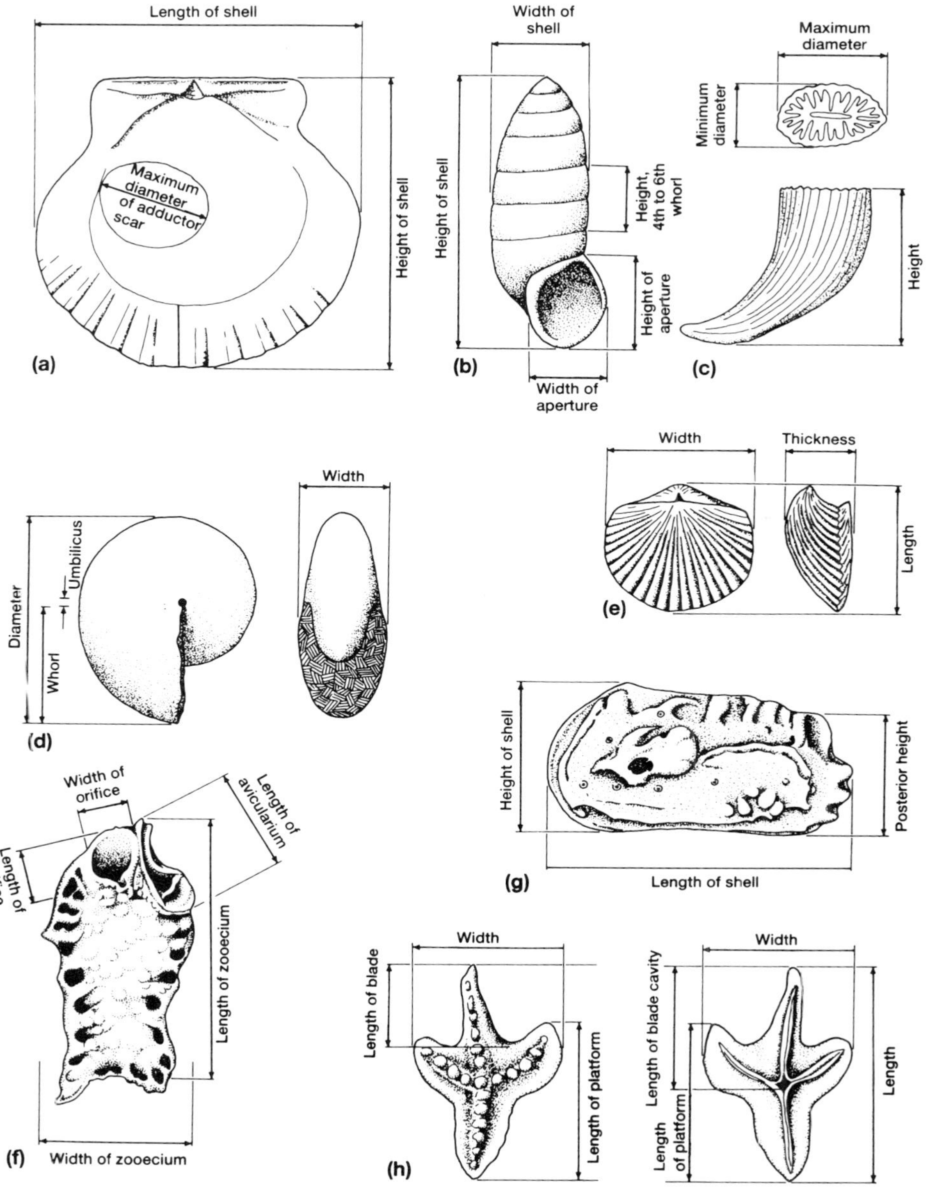

Fig. 2.3 Various invertebrates showing some of the principal measurements commonly made. (a) pelecypod mollusk, (b) gastropod mollusk, (c) scleractinian coral, (d) cephalopod mollusk, (e) brachiopod, (f) single zooecium from bryozoan colony, (g) ostracode arthropod, (h) conodont element. Diagrams (a)–(e) are approximately natural size, and (f)–(h) are magnified about 50×. After Cheetham (1968), Waller (1969), Petersen (1975), Cronin (1987), Gould & Woodruff (1987), Mori (1987), and Murphy & Cebeciogli (1987).

$$\mathrm{CV} = \frac{100\,s}{\bar{X}} \quad \left[= \frac{S}{\bar{X}} \times 100\% \right].$$

The coefficient of variation is a method of consistently comparing the relative size of s to $\bar{X}$. Multiplication by 100 is done simply to transform what would normally be a fractional value to a percentage given in whole numbers.
(Question 4)

Now that the arithmetic mean and standard deviation have been calculated for the lengths and widths that were recorded in Question 2, the arithmetic means for the original fossil assemblage in the outcrop can be estimated, presuming, of course, that the sample is an unbiased representation of the fossil assemblage. It is best to base estimates of the characteristics of the fossil assemblage on specimens that were carefully collected in a way to minimize bias. The best method for minimizing bias in field collections depends on the local conditions at the outcrop. Some of the possible methods include: bulk collection of sediment or *in situ* weathered residue of rock that is then screened and completely picked in a laboratory; marking off clearly defined areas from which every visible specimen is collected; and, in densely fossiliferous spots, marking off a gridwork that is the frame of reference for collecting at points specified by a sequence of random numbers. (There are tables of random numbers readily available, and many statistical computer packages can be used to generate a series of (almost) randomly generated numbers.)

Even with the most impeccably correct collecting, there is no guarantee that the mean of some variable in a sample of fossils would be exactly the same as the mean of a different sample drawn from the same fossil assemblage, or that either one would be the same as the true mean if we could determine it for the entire assemblage. In fact, it is almost certain that the sample's mean and the assemblage's mean values are not the same. Yet, short of collecting the entire assemblage, a sample is the only means available to estimate the mean values for the assemblage.

A mean value for some variable that is determined for a well-collected sample should not be too far off from the mean value for the fossil assemblage. But how far off might it be? It could be a little higher or a little lower. Therefore, if we take the sample's mean as the middle of a small range of values, it is more probable that the true mean for the assemblage is included than that the sample's mean value exactly equals whatever the mean value for the complete, uncollected fossil assemblage might be. The probability that the true mean for the fossil assemblage is included within the range of values (ranging a little above and a little below the sample mean) depends upon three things:

1 how wide the range is, because the wider the range of values the higher is the chance that the assemblage's true mean is included;

2 how variable the feature is that is measured, because the more variable it is, then the greater the chance that a different set of specimens would have given a substantially different mean value; and

3 the number of specimens in the sample, because an increasingly larger number of specimens used as a sample makes the mean value less affected by unusual specimens so that the sample's mean "settles" towards the assemblage mean.

The relationship between the three influences listed above are known, so that for each sample, it is easy to determine the size of the range of values around the sample's mean that must be included for a particular degree of confidence that the assemblage's mean is included. For example, if you want to be 68% confident (that is the same as eight chances out of 25 that you might be wrong) that the assemblage's mean is included, the formula is

68% **confidence interval** of mean $= \overline{X} \pm \dfrac{s}{\sqrt{N}}$.

If one is willing to accept only one chance in 20 that the true value is not included, that is referred to as the 95% confidence interval. The calculation is

95% confidence interval of mean $= \overline{X} \pm \dfrac{1.96\, s}{\sqrt{N}}$.

Generally, at least 95% confidence is required for statistical statements in science. Greater than 95% confidence that the mean of the statistical population (fossil assemblage) falls within the interval requires that s be multiplied by a number greater than 1.96, depending on the exact level of confidence required.

(Question 5)

2.3.3 Correlation and regression

As organisms grow, their increase in size can be expressed by more than one single measure. For example, greater length is generally accompanied by greater width, more ridges or segments, etc. Any two variables of an organism may increase relative to one another in very simple or in complex ways. The degree with which the variables co-vary, either within an individual as it grows or within a sample drawn from a population, is termed the **correlation** and may be expressed numerically by a number from −1 (a perfect inverse relationship) to zero (absolutely no relationship) to +1 (a perfectly related simultaneous increase or decrease in both).

(Question 6)

If the two characters are perfectly correlated, the points will fall along a single straight or curved line. Several measures of correlation exist, the most commonly used being the *Pearson coefficient of correlation*:

$$r = \frac{\Sigma(X-\overline{X})(Y-\overline{Y})}{\sqrt{\Sigma(X-\overline{X})^2\,\Sigma(Y-\overline{Y})^2}} = \frac{\Sigma(X-\overline{X})(Y-\overline{Y})}{(N-I)\,S_X S_Y}$$

$$\left[= \frac{\Sigma(XY) - \dfrac{(\Sigma X)(\Sigma Y)}{N}}{\sqrt{\left(\Sigma(X^2) - \dfrac{(\Sigma X)^2}{N}\right)\left(\Sigma(Y)^2 - \dfrac{(\Sigma Y)^2}{N}\right)}} \right]$$

where s_X and s_Y are the standard deviations for X and Y, respectively.

(Question 7)

The proportions that exist between two variables as their magnitudes change is referred to as **regression.** The simplest type is *linear regression*, in which the two variables are related as:

$$Y = a_Y + b_{YX} X^n,$$

in which a_Y is the intersection of the regression line with the Y axis, b_{YX} is the slope of the line, and n is one. The result is an equation of the straight line that relates Y to X. It expresses the relationship between Y and X, and how any X can be used to predict the corresponding value of Y once the intersection with the Y axis and slope are determined. These may be determined by the formulae,

$$b_{YX} = \frac{\Sigma(X-\overline{X})(Y-\overline{Y})}{\Sigma(X-\overline{X})^2} \quad \left[= \frac{\Sigma XY - \dfrac{\Sigma X \Sigma Y}{N}}{\Sigma X^2 - \dfrac{(\Sigma X)^2}{N}} \right]$$

and

$$a_Y = \overline{Y} - b_{YX}\,\overline{X}.$$

A moment's reflection on the formula for linear regression that is given above should indicate that a_Y is basically a value that helps to position the line $Y = b_{YX} X^n$ on the graph; it

does not enter into determining the slope of the line. Most regression lines should not be projected back toward the origin beyond the distribution of points on the graph, because there is no *a priori* reason to believe that the earlier growth phases maintained the same relative growth proportions as those recorded on the graph. In fact, the earliest growth stages often exhibit different proportions of growth, or their record may be entirely lost by metamorphosis from larval to adult stage. For example, some snails have early stages up to several millimeters in length that are differently shaped and grew differently before the accretion associated with adult growth began (Fig. 2.4). In such cases, it would make no biological sense to project a regression line to the graph's origin or to one of the axes based on relationships between variables measured from adult shells.
(Question 8)

2.4 TYPES OF GROWTH

If the hypothetical Y intercept for the population (a_Y is an estimate based on the sample) is zero, and the linear regression is a line that best fits the relationship between the two variables, then the growth is said to be **isometric**. In isometric growth, the proportion between the variables being compared does not change with increase in size.

Contrasting with isometric growth is **anisometric growth**, in which proportions do change with size increase. If the true Y intercept for the population is some value other than zero, anisometric growth occurred. If the exponent of X is some value other than 1, then **allometric growth** occurred, and the relationship is termed **allometry**. One of the most obvious types of allometric growth is the relationship between the radius of a circular encrusting patch and the area occupied by the patch, so that the regression formula is expressed as

$$Y = 0 + \pi X^2$$
$$= \pi X^2,$$

where Y is the area and X is the radius. Allometric growth is exceedingly common. It reflects changes in structural, physiological, reproductive, and other functions of organisms with growth.
(Question 9)

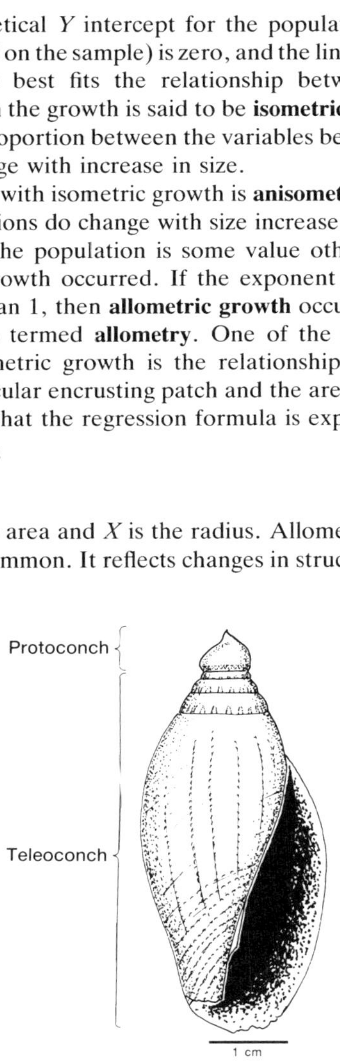

Fig. 2.4 The gastropod *Scaphella*, in which the earliest-formed part of the shell (**protoconch**) has a different expansion rate, cross-sectional shape of whorls, and orientation of axis of coiling from the later-formed part of the shell (**teleoconch**). The ratios of many morphological features differ between the two parts of the shell.

2.5 ESTIMATING SIMILARITY

Description is but one facet of the use of statistics in the study of organisms. One of the primary uses – perhaps *the* primary and most important use – of statistics is to assess the validity of a hypothesis. For example, collections made from two fossil assemblages of Cretaceous irregular echinoids may resemble one another in all visible ways except that one appears to have a slightly greater length : width ratio than the other, resulting in two possibilities: the original fossil assemblages may have essentially identical length : width ratios and the differences between the samples might be due to happenstance associated with random sampling, or the original assemblages might, in fact, have different length : width ratios. In order to estimate which is more likely, the hypothesis that no real difference exists (the **null hypothesis**) is usually tested. If it has very small probability of being true, then the alternative hypothesis (in this example that length : width differ between the assemblages) is accepted.

Probability of equivalence between two populations from which samples are drawn is commonly evaluated by *student's t-test*. The formula for this is

$$t = \frac{(\bar{X}_1 - \bar{X}_2)\sqrt{\dfrac{N_1 N_2}{N_1 + N_2}}}{\sqrt{\dfrac{(N_1 - 1)\,S_1^2 + (N_2 - 1)\,S_2^2}{N_1 + N_2 - 2}}}.$$

The size of the samples obviously affects the value of t, and therefore has to be taken into account when judging whether the two statistical populations sampled are similar or different. This is a concept known in statistics as the **degrees of freedom**, which in this case equals $N_1 + N_2 - 2$.
(Question 10)

The underlying assumption in the preceding part of this exercise was that the sample for Question 2 was drawn from a population for which the measured variables are *normally distributed* and can be adequately characterized and tested by **parametric** statistics. A **normal distribution** is a mathematically defined, symmetrical distribution that produces a bell-shaped curve when the frequencies of observations are plotted on a graph (Fig. 2.5). If the data are not normally distributed, then a different set of equations and tests apply that do not presume a normal distribution. They are termed **non-parametric** statistics and are generally no more difficult to understand or apply than the parametric statistics used up to this point as an introduction to the use of statistics in paleontology.

As an example, *Kendall's* τ is the nonparametric correlation test "equivalent" to Pearson's r, which was calculated in Question 7. Kendall's τ is based on sequential arrangement from smallest to largest of the observations of one of the characters being compared; the order of the specimens in this sequence is then compared with the order of the specimens if they had been sequentially arranged by magnitude of the other character. This test and other nonparametric tests avoid

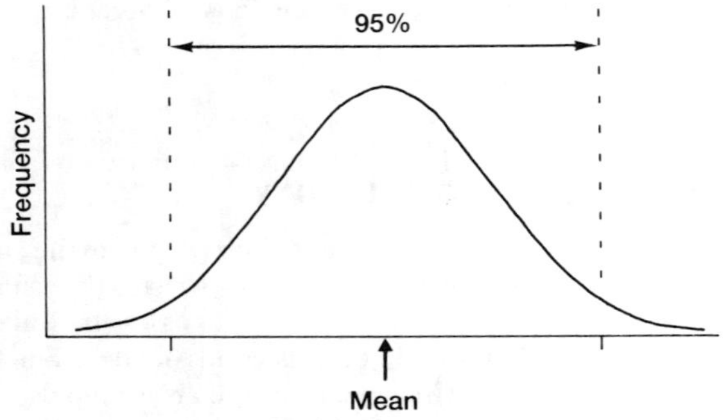

Fig. 2.5 Curve for normal distribution, with 95% of area of curve indicated.

comparison of individual observations with the sample's mean, which has somewhat limited usefulness for data that are not normally distributed.
(Question 11)

Kendall's τ has much more widespread application than does r. Because it pertains to normal and nonnormal distributions alike, no assumptions about form of distribution of the population need be made. Procedures exist for calculating confidence levels for each of the nonparametric tests and can be found in texts that cover nonparametric statistics.

One problem that is not addressed here is what happens if there is a "tie" in rank assignment. That is, if two observations have the same value, they are tied in rank: which should be listed first? The two different choices of sequence will affect the value of tau. There are procedures for correcting for ties, and this should be done. If you are using a computerized statistical package for calculations, use the "corrected for ties" value. If calculating by hand, you should realize that your value for τ is an approximation if there are tied cases within your data.

FURTHER READING

Cheeney, R. F., 1983. *Statistical methods in geology*. George Allen & Unwin: London.
Siegel, S., 1956. *Nonparametric statistics for the behavioral sciences*. McGraw-Hill: New York.
Sokal, R. R. & Rohlf, F. J., 1981. *Biometry*, 2nd edition. W. H. Freeman & Co.: New York.
Sprent, P., 1981. *Quick statistics, an introduction to nonparametric methods*. Penguin: London.

EXERCISE 2

Worksheet

Name

QUESTION 1

Several specimens of various invertebrate species are provided. For each species enter the name of the phylum and genus, whether the animal is solitary or colonial, and whether its growth is determinate or indeterminate.

	Phylum	*Species*	*Solitary/colonial*	*Determinate/ indeterminate growth*
1				
2				
3				
4				
5				

QUESTION 2

Measure the two characters specified by your instructor for the 15 specimens in the sample provided from an assemblage of a fossil species. Your instructor may wish to provide more than 15 specimens and explain how to use a randomizing process such as reading a table of random numbers to narrow the sample to 15. Enter the measurements in the appropriate columns in the Table. Note that this is a small sample that under many conditions may be too small to adequately characterize an assemblage of organisms that are highly variable. Record the measurements in millimeters.

Phylum:

Species:

Measured variable 1 (X):

Measured variable 2 (Y):

	X	X^2	*Rank (small to large)*	Y	Y^2	*Rank (small to large)*	(XY)
1							
2							
3							
4							
5							
6							
7							

8							
9							
10							
11							
12							
13							
14							
15							
SUM							

Now square the sum of the X column and the sum of the Y column:

$(\Sigma X)^2 =$

$(\Sigma Y)^2 =$

QUESTION 3

Determine the arithmetic mean of the two sets of measurements made in Question 2.

Arithmetic mean of X ($\overline{X}$):

Arithmetic mean of Y ($\overline{Y}$):

QUESTION 4

Record the range and compute the variance, standard deviation, and coefficient of variation for the measurements of the two characters made in Question 2.

Range of X:

Variance of X:

Standard deviation of X:

Coefficient of variation of X:

Range of Y:

Variance of Y:

Standard deviation of Y:

Coefficient of variation of Y:

QUESTION 5

For the two characters that were measured for Question 2, calculate the 95% confidence interval of the mean for the original fossil assemblage from which the sample in Question 2 was drawn.

95% confidence interval for X:

95% confidence interval for Y:

QUESTION 6

Label the axes of the graph below so that maximum and minimum values of variable 1 recorded in Question 2 are included on and nearly span the X axis, and maximum and minimum values of variable 2 in Question 2 are included on a nearly span the Y axis. Plot a scatter diagram of points representing the two characters of each specimen recorded in Question 2.

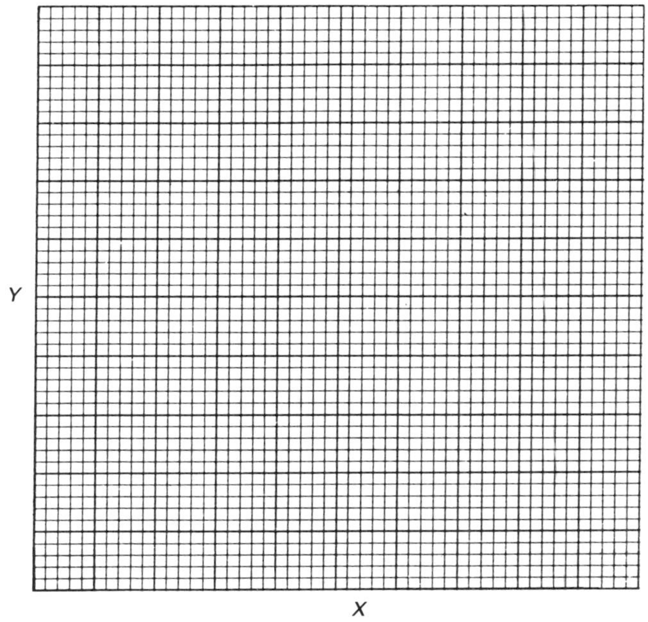

QUESTION 7

Determine the Pearson coefficient of correlation for the measurements determined for the sample in Question 2.

$r =$

QUESTION 8

Determine the hypothetical Y intercept and slope of the linear regression for the data in Question 2. Then write the resulting equation for linear regression for the two characters of the fossil sample measured for Question 2, calculate the Y value for some X value at the upper extreme of the data set, and the Y value for some X value at the lower extreme of the data set, plot the points on the scatter diagram in Question 7, and connect the two points by a straight line. This is the simple linear regression line for the two characters of the sample measured in Question 2 and, if it is correct, passes through the point defined by $\overline{X}$ and $\overline{Y}$.

$b_{YX} =$

$a_Y =$

Regression formula: $Y =$

Point 1. Low value chosen for X: Calculated value for Y:

Point 2. High value chosen for X: Calculated value for Y:

Plot the two points on the graph above and connect by a straight line.

QUESTION 9

Plot the data for shell height and ligament height for *Glycymeris glycymeris* (listed below), on graph A. Is the scatter of points linear? Would a linear regression line through the data accurately represent the relationship between the two variables? Now plot the same set of data on graph B, which has a semi-logarithmic scale. This graph transforms the relative spacing of the Y values to the spacings of their logarithms. How has this replotting transformed the scatter of points? Do you think that the fit of a linear regression of *X* against logarithm of *Y* would be better than *X* against *Y*?

Height of shell and height of ligament in 50 specimens of *Glycymeris glycymeris* from Port Erin, Isle of Man. Data courtesy of R. D. K. Thomas.

Shell height (mm)	*Ligament height (mm)*	*Shell height (mm)*	*Ligament height (mm)*
11.1	0.4	11.9	0.5
12.3	0.5	13.2	0.4
13.2	0.5	13.8	0.5
14.0	0.5	16.7	0.7
17.9	0.8	19.4	0.8
20.7	0.8	21.2	0.8
35.1	1.7	36.9	2.0
37.7	2.2	37.9	1.7
38.9	2.0	39.2	2.7
39.4	2.2	40.6	1.9
41.1	2.6	41.5	2.6
41.8	2.3	41.8	2.6
42.5	2.4	44.7	3.1
45.6	3.0	46.0	3.0
46.0	2.7	46.9	2.2
47.7	3.0	47.7	3.0
47.9	3.3	48.6	2.9
48.8	3.2	48.9	3.3
49.2	2.5	49.3	3.0
49.5	3.6	50.2	3.2
50.4	3.3	51.2	4.0
51.8	3.3	52.3	3.3
58.7	5.7	59.2	7.2
60.2	5.6	60.3	6.9
60.4	7.6	67.8	8.6

What is the transformation in shape of the distribution of points between Graph A and Graph B?

____

Which distribution would better fit a straight linear regression? ______________

The ligament lies adjacent to the hinge, on the outer side from the body cavity (see *Mercenaria* in Fig. 9.1). It functions to open the valves when the valve-closing muscles inside the body cavity relax, and it does this because the hinge serves as a fulcrum when the stretched ligament contracts to its neutral state. Keeping in mind the function of the ligament and changes in linear dimensions, areas, volumes, and centers of force of the ligament and shell as pelecypods grow, why is there an allometric relationship between ligament height and shell height in *Glycymeris glycymeris*?

____

____

____

GRAPH A

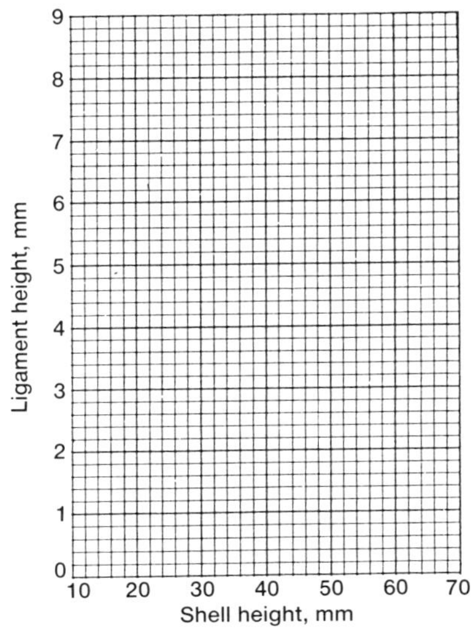

GRAPH B

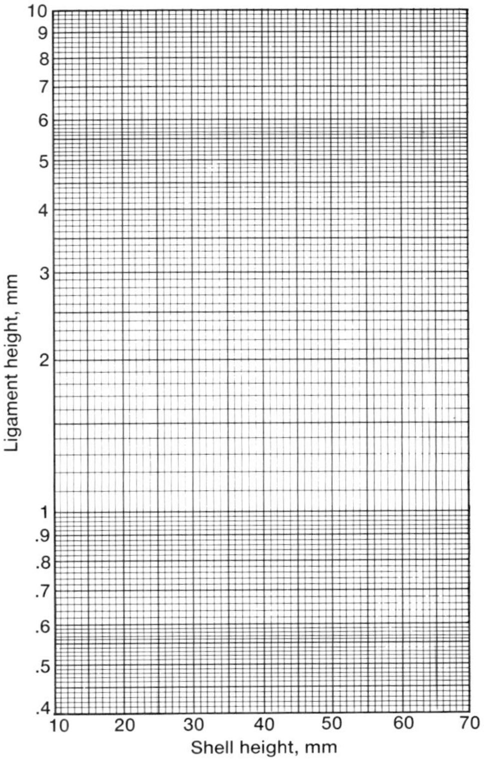

QUESTION 10

Measure the specified character of the 15 specimens in the sample provided, which is from a fossil population closely related to the species measured for Question 2. Determine the mean and standard deviation for the sample, then test the null hypothesis of no difference between the character in this sample's population and that for Question 2, by calculating student's t and checking the value against the following table. (The table lists probabilities for t, with 28 degrees of freedom. Degrees of freedom = $N_1 + N_2 - 2 = 15 + 15 - 2 = 28$.) If 0.05 is taken as the level at which to accept or reject the null hypothesis, then values of student's $t > 2.048$ would cause rejection of the null hypothesis.

t	*Probability of difference*	*Probability of no difference*
0.127	0.1 (10%)	0.9 (90%)
0.256	0.2	0.8
0.389	0.3	0.7
0.530	0.4	0.6
0.683	0.5	0.5
0.855	0.4	0.6
1.056	0.3	0.7
1.313	0.8	0.2
1.701	0.9	0.1
2.048	0.95	0.05
2.467	0.98	0.02
2.763	0.99	0.01
3.674	0.999	0.001

Variable (character) measured:

	X_2	$(X_2 - \overline{X}_2)$	$(X_2 - \overline{X}_2)^2$
1			
2			
3			
4			
5			
6			
7			
8			
9			
10			
11			
12			
13			
14			
15			
SUM			

$\overline{X}_2 =$

$s_2 =$

Student's $t =$

Approximate probability of difference in means:

QUESTION 11

Calculate Kendall's τ for the data recorded in Question 2 by following the directions given. First, in the table for Question 2 on the answer sheet, number the specimens from 1 to 15 from smallest to largest value of X, then do the same for each specimen for the values of Y. Transfer the ranks for Y to the table for Question 11. (The numbers already on the table are the ranks that you have just determined for X, but they have been rearranged from whatever order they have in Question 2, and are already listed in order from smallest to largest in the table for Question 11.)

Beginning with the number at the top of the Y column (Column 2), look at all the numbers located below it in the column, count the number that are larger, and enter the number in column 3. Continue the same procedure for the second number from the top in column 2, etc., ending with the bottom number in the column which – there being no numbers at all below it – will produce a zero in column 3. Now, begin again at the top of column 2, but this time count up the scores lower in the column that are smaller than the score in column 2, and enter the number in column 4. For each line, subtract the number in column 4 from the number in column 3 and enter the resulting number (including the + or – sign) in column 5. Get the total (S) for column 5 (subtracting the negative numbers and adding the positive numbers). Calculate Kendall's τ by

$\tau = 2S/N(N-1)$.

Column 1: (Rank of X*)*	*Column 2: (Rank of* Y*)*	*Column 3: Ranks lower in table larger than rank of* Y	*Column 4: Ranks lower in table smaller than rank of* Y	*Column 5: (Column 3 – Column 4)*
1				
2				
3				
4				
5				
6				
7				
8				
9				
10				
11				
12				
13				
14				
15				
				S =

τ =

The listing below gives some values of Kendall's τ for 95% confidence ($p = 0.05$) that two ranked variables are correlated. If the calculated value of τ is equal to or greater than the value given for the number of items used to determine the correlation, then the 95% confidence level is met or exceeded. The N (number of items used) for Question 11 is 15.

Is there 95% confidence that the two variables are correlated?

Values for Kendall's τ at $p = 0.050$ (two-sided test) for $N = 11$ to $N = 30$.

N	τ	N	τ
11	0.491	21	0.314
12	0.455	22	0.305
13	0.436	23	0.295
14	0.407	24	0.290
15	0.390	25	0.287
16	0.383	26	0.280
17	0.368	27	0.271
18	0.346	28	0.265
19	0.333	29	0.261
20	0.326	30	0.255

EXERCISE 3 Protista

3.1 INTRODUCTION

The current practice of recognizing five kingdoms of organisms rather than just the kingdoms Plantae and Animalia is due in large part to a more detailed understanding of single-celled organisms than formerly existed. Previously all were forced into either the plant kingdom or animal kingdom depending upon whether they were dominantly **autotrophs** (manufacturing organic compounds from inorganic components) or dominantly **heterotrophs** (feeding on other organisms or at least on organic compounds). At present, the concepts of plant and animal (plus fungi) are reserved for multicellular organisms. Single-celled organisms lacking a cell nucleus (such as bacteria) are placed in the Kingdom Monera, or if a cell nucleus is present in the Kingdom Protista. This exercise focuses on protists that have an extensive fossil record because of their mineralized skeletal parts.

Table 3.1 presents a summary of the classification and characteristics of a few of the subdivisions of the Kingdom Protista. It does not give a balanced biological consideration of all protists, but concentrates on those with mineralized skeletons.

Although foraminifera have been the most extensively studied protist fossils, several other protist groups have important fossil records. This is especially true for radiolaria, diatoms, coccolithophores, and macroscopic calcareous algae. Each of these groups is abundant and virtually ubiquitous today in the oceans, but their local abundance depends on regional oceanographic conditions. Radiolarian, diatom and coccolithophore skeletons are of pre-eminent importance in biostratigraphic and paleoceanographic studies of sediments deposited in open oceans because of their:

1 abundance in oceanic sediments;
2 widespread occurrence;
3 ease of recovery in large numbers; and
4 rapid evolution shown by many lineages.

Calcareous chlorophyte and rhodophyte algae are restricted environmentally, so their presence is important for paleoenvironmental and paleogeographical reconstructions.

Table 3.1 Characteristics of a few of the paleontologically more important phyla of the Kingdom Protista. Adapted from Barnes (1984).

√ Phylum Dinophyta	Mainly unicellular with two dissimilar flagella, one helical or undulating and the other whiplike Cell covering is a complex theca constructed of distinct internal cellulosic plates Chloroplasts brownish or golden when autotrophic Some heterotrophic and lack chloroplasts Food reserves starch and oil Nucleus conspicuous, with condensed chromosomes Most haploid, with equal-sized gametes Freshwater and marine, also as symbionts and parasites
√ Phylum Sarcodina	Single-celled and ameboic, producing transient extensions of the cell surface (pseudopodia) that are used for locomotion or feeding Some with endosymbiotic algae Contractile vacuoles present in freshwater ones Many with organic or mineralized internal skeletons or tests No chloroplasts
√ Phylum Bacillariophyta	Isolated cells or colonies Cell wall an elaborate bivalved silica box (frustule) Single nucleus present and usually several yellow–brown chloroplasts Food reserves are oil and other organic materials Marine and freshwater, planktic and benthic
Phylum Chlorophyta	Single-celled or multicellular of various shapes Flagellate stages normally with two anterior flagella, usually with their bases at right angles Cell covering varied, commonly cellulosic Chloroplasts bright green Food reserve is starch Predominantly freshwater
Phylum Rhodophyta	Single-celled or multicellular of various shapes No flagella Much gelatinous material in cell wall Some calcified Most contain red chloroplasts Food reserve is starch Most marine

The purposes of this exercise are:
1 to introduce the general morphology of radiolaria, diatoms, coccolithophores, and foraminiferans;
2 to analyse some of the conditions that determine the local abundance of planktic diatoms;
3 to compare the features of the extinct receptaculitids with those of living chlorophyte algae; and
4 to determine probable age and paleoenvironment of foraminiferan assemblages.

3.2 RADIOLARIA

The Class Radiolaria is a subdivision of the phylum Sarcodina, as it possesses deformable extensions (**pseudopodia**) of the ectoplasm. The pseudopodia are narrow threads, some of which are stiffened by tiny filaments that radiate from the center of the cell, whereas others are capable of more movement and may anastomose with one another (Fig. 3.1). A thin organic wall, the **central capsule**, divides the **endoplasm** from the surrounding **ectoplasm**. Frothy material termed the **calymma** is located within or on the outer portion of the ectoplasm. Low-density gas vacuoles of the calymma may compensate for the high density skeleton in order to achieve neutral density for flotation at a particular depth. Low-density fats have also been reported and would also counteract sinking caused by high density of the skeleton.

Apparently all radiolarians are planktic heterotrophs, feeding on phytoplankton and bacteria by their movable set of pseudopodia. They range in size from about 100 to 2000 μm, although most are nearer the lower end of the size range. Various species float at characteristic depths or depth ranges, from the photic zone to abyssal depths. Those that live in the **photic zone** (less than 200 m depth) may contain **endosymbiotic** algae, which are algae that live within an organism. Although many different types of **prokaryotic** (lacking a cell nucleus) and **eukaryotic** (containing a cell nucleus) algae may live as endosymbionts, the most common are **zooxanthellae**. Zooxanthellae are dinoflagellate algae that are found within many organisms, including foraminifera, reef corals, and the giant clam *Tridacna*. Radiolarians with endosymbionts do not grow to colossal sizes, however.

Skeletons of radiolarians are composed of opaline silica, strontium sulfate, or silica-enriched organic material, and each major subdivision is characterized by a particular skeletal composition. Near-shore radiolarians are mostly members of the Order Phaeodaria, which have silica-enriched organic skeletons with low preservation potential. Most fossil radiolarians belong to the characteristically open-ocean Polycystina, which have skeletons of silica.

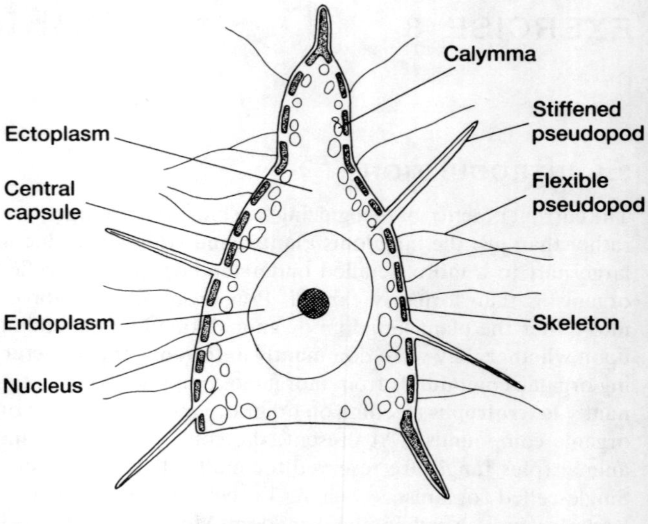

Fig. 3.1 Cellular organization of radiolarians.

Polycystine radiolarians comprise two orders, Spumellaria and Nassellaria (Fig. 3.2). The skeletons may have radial skeletal elements (either discrete spicules which constitute the simplest kind of skeleton, or radiating spines from a more extensive skeleton) and tangential elements that constitute a spherical or conical lattice. The Spumellaria are characterized by spherical, ellipsoidal, or discoidal skeletons and (together with the Acantharia, which have skeletons of strontium sulphate) generally dominate open-ocean radiolarian assemblages within the photic zone. The Nassellaria have conical or pyramidal lattice shells, with radial symmetry about a central axis, and (with the Phaeodaria) generally dominate radiolarian assemblages below 2000 m depth.

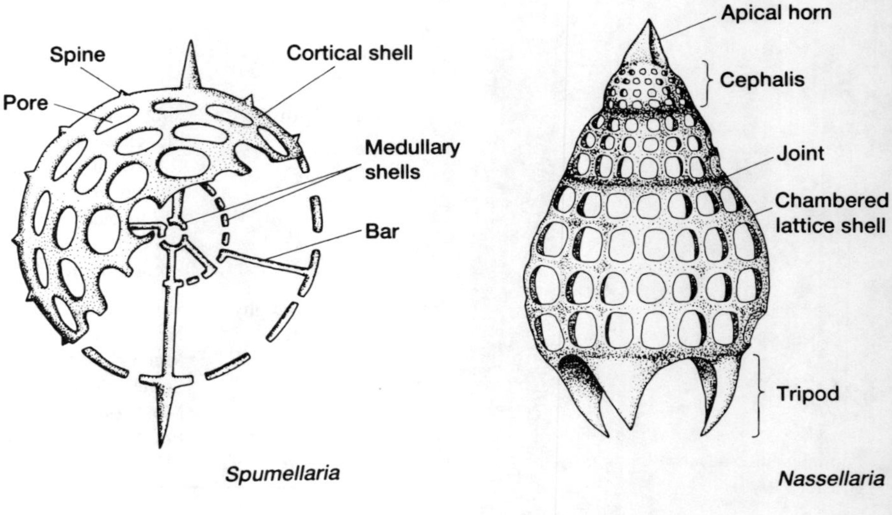

Fig. 3.2 Basic skeletal characteristics of spherical Spumellaria and radially symmetrical Nassellaria. Modified from Brasier (1980) and Casey (1987).

Spumellarid radiolarians have a geologic range from Cambrian to Recent. Nassellarids may have originated as early as the Silurian and range to the Recent. Radiolarians are quite useful for biostratigraphy, particularly for deep oceanic deposits that accumulated below the **calcite compensation depth** (generally 4000–5000 m), which is the level below which calcareous materials such as tests of foraminifera and plates of coccolithophores completely dissolve.
(Questions 1 and 2)

3.3 DIATOMS

Diatoms are a kind of golden-brown algae in which the cell wall constitutes a silica-rich shell (**frustule**) composed of two overlapping **valves** (similar in construction to a pillbox, with the margin of the larger valve lapped over the margin of the smaller one; Fig. 3.3). Tiny perforations constitute up to one-third of each valve, lessening its weight. The frustules range in diameter from less than 1 μm to greater than 1000 μm, although the vast majority fall between 10 and 200 μm. Diatoms are photosynthetic, so are restricted to the photic zone (0 to about 200 m depth in the clearest open ocean waters). Along with coccolithophores, they are the most productive primary producers of the open sea.

Diatoms are considered to constitute the Class Bacillariophyceae of the Phylum Chrysophyta by some specialists, while others consider them to be a distinct phylum, the Bacillariophyta. They are subdivided into two groups (Pennales and Centrales) based primarily on symmetry of the frustule.

The Pennales are bilaterally symmetrical and elongate, either elliptical or rectangular in profile as viewed from the valve surface. Most pennales have a central linear feature that extends down the axial line of each valve, either a slit (**raphe**) or a nonperforate band of silica. They occur in fresh, brackish, and shallow marine waters, and most are benthic.

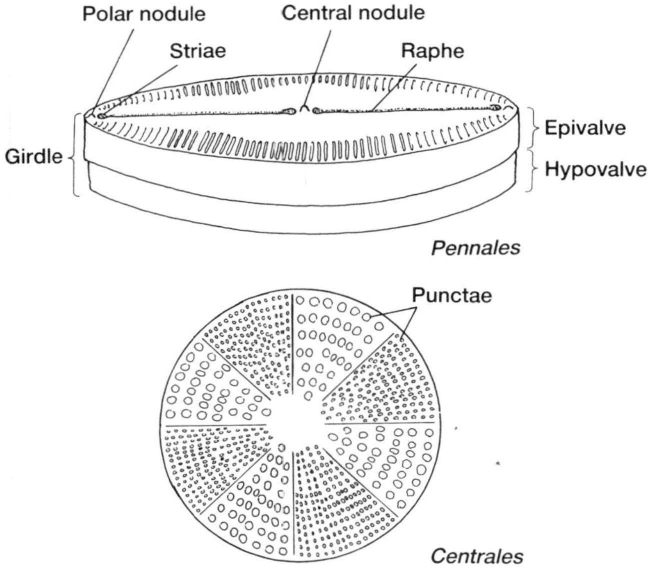

Fig. 3.3 Basic skeletal characteristics of bilaterally symmetrical Pennales and radially symmetrical Centrales diatoms. Modified from Brasier (1980).

Frustules of the Centrales are radially symmetrical, being circular, triangular, or quadrate in profile as viewed from the valve surface. The Centrales are planktic and are predominantly marine. They are especially prolific in regions of nutrient-rich, upwelling waters and in subpolar regions.

The oldest certain diatoms are Early Cretaceous (Barremian–Aptian Ages). However, abundant and diverse diatom assemblages are known only back to the Late Cretaceous (Campanian Stage). Siliceous rocks composed predominantly of diatoms are termed **diatomites**, and most are of Paleogene age.
(Question 3)

3.4 COCCOLITHOPHORES (CALCAREOUS NANNOPLANKTON)

A wide variety of autotrophic (photosynthesizing) protists with sizes of about 5–60 μm are referred to collectively as **nannoplankton**. The most abundant of these are golden algae known as **coccolithophores**. Coccolithophores have two flagella of equal length and a third, the **haptonema**, which is a whip-like modified flagellum. Presence of the haptonema places the coccolithophores into the haptophytes, which some specialists place within the Chrysophyta and others recognize as a distinct phylum, the Haptophyta.

Coccolithophores are predominantly photosynthetic, although they occasionally function as heterotrophs, consuming bacteria and small protists. They range into fresh water, but most are marine and occur in the upper 200 m of the ocean waters (although they have been reported down to 4000 m depth). Along with the diatoms, coccolithophores are the most productive primary producers in the seas.

Coccolithophores are generally 3–25 μm in diameter and are covered by a set of calcareous plates termed **coccoliths**. Individual coccoliths are secreted in vesicles within the cell and then move individually to the outer surface of the cell, where they may replace others that have been shed. The coccolith-covered cell is termed a **coccosphere** (Fig. 3.4), and occasionally these are found entire as fossils also.

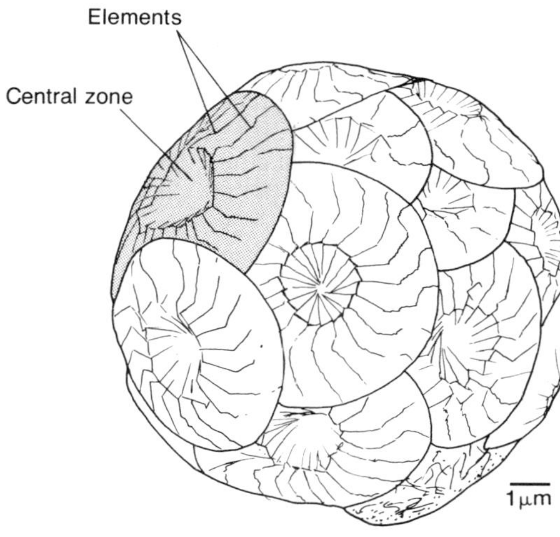

Fig. 3.4 Coccosphere of *Cyclococcolithus leptoporus*. After photo by H. Okada in Haq (1978).

There are two varieties of coccoliths: *holococcoliths*, which are composed of rhombs of calcium carbonate joined along their edges, and *heterococcoliths*, which are built of a variety of elements such as rods and plates that are rigidly cemented together. The more durable holococcoliths constitute most of the recognizable fossil coccoliths, and heterococcoliths are found only rarely. Despite the relatively simple organization, coccoliths display a large variation in form and micromorphology. The central area may be crossed by bars or a lattice, built up as a spine, or may be devoid of any central structures.

Calcareous nannofossils appeared in the Late Triassic and have been widely distributed in marine sediments from Jurassic to Recent. There may be as many as $10^{12}/cm^3$. In oceanic oozes and chalks, and even in fine terrigenous sands deposited on Inner shelves there may be $10^3-10^4/cm^3$. Coccoliths and other calcareous nannofossils are highly useful in biostratigraphy and paleoceanography, but because of their small size and light weight, they are easily reworked and transported. Thus, care must be taken to minimize the impact of reworked and transported specimens.
(Question 4)

3.5 CONDITIONS FAVORABLE FOR PLANKTIC DIATOMS

Living diatoms are heavier than seawater and therefore sink. This is both a curse and a blessing. In oceanic waters where the sea floor is deep, planktic diatoms can — and many do — sink out of the photic zone and die. Obviously, if every member of a diatom species sinks out of the photic zone and dies, the lineage leading to that species terminates. However, since planktic diatoms lack either cilia or flagella and are not capable of swimming, they must sink because marine plants come into contact with and gain nutrients only by differential movement with respect to nutrient-bearing water. The phenomenon by which they gain nutrients (the blessing) is the same by which they sink out of the photic zone (the curse).

The overall density of diatoms is determined by the proportions of their siliceous skeleton (denser than seawater) and fats (less dense than seawater). Diatoms are generally only a little denser than seawater (density = 1.03 g/cm^3. This part of the exercise is based on hypothetical spherical diatoms that are 2% denser than the surrounding water. The sinking rate of such diatoms of different diameters is shown in (Fig. 3.5).

Upwelling and wave-generated turbulence help to maintain diatom populations within the photic zone. Like many other photosynthetic protists, diatoms reproduce by asexually dividing into two new cells. (Each offspring cell takes one half of the two-part frustule, which serves as the larger of the two nested parts of the new frustule, with the smaller part being newly secreted. Therefore, one offspring diatom has the same size as the parent, while the other is necessarily smaller. After several generations some of the descendant diatoms are substantially smaller than the original generation. At a certain minimum size, either sexual reproduction occurs, resulting in full-sized offspring, or the frustule is shed and the naked diatom grows until it reaches the appropriate size, at which time a new, large, two-part frustule is secreted.) The time required for enough divisions to occur within the population

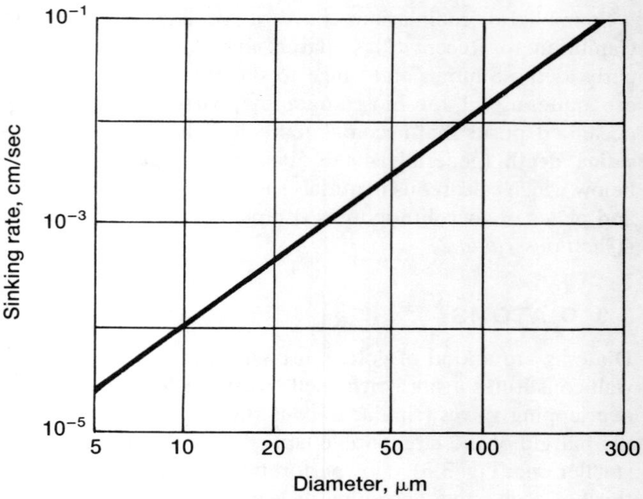

Fig. 3.5 Sinking rate of spherical diatoms of 5–300 μm diameter, based on a density of 2% greater than seawater. After Weyl (1970) from information in Munk & Riley (1952).

so that there are twice the original number is referred to as the **doubling time**. Many animals and heterotrophic protists graze on diatoms and other phytoplankton, so that the effective doubling time results from interaction of the rate of division with the grazing rate. Here we deal with doubling time based only on rate of division.

After one doubling time (t) has elapsed, two similar planktic diatoms that were originally at the same depth will have produced a total of four offspring. In the meantime, with sinking rate of v cm/s, the diatoms would have sunk an amount equal to vt (Fig. 3.6a). However, **diffusion** (the tendency for small particles to disperse due to random motions, in this case the random motion of wave turbulence) will have caused the two original diatoms to move some distance apart. Diffusion is as likely to move particles upward as downward. If mean vertical displacement (z) during t is at least equal to vt (the distance that the diatoms have fallen), the diatom population will be maintained, because on average one of the original diatoms probably diffused upward while the other diffused downward (Fig. 3.6b) by the time that each divided into two new cells.

Sinking rate increases with increased density and size, while doubling rate increases with nutrient concentration in the water and with sinking rate but decreases with increased size. In order for diatoms to be maintained in the photic zone, there must be an appropriate balance of nutrient concentration, diffusion rate, excess density of the diatoms over seawater, and size of the diatoms. Phosphate is usually the critical limiting nutrient. For Fig. 3.7, the diffusion constant is 10 cm^2/s, excess density is 2%, and phosphate concentration is 10^{-7} mol/kg. The two curves plot: (1) the time to double nutrients (n) for spherical planktic diatoms of different sizes; and (2) minimum doubling time (d) for them to be maintained in the photic zone. Where $n<d$, populations will be maintained, but where $n>d$, the population will diminish and disappear. Therefore, for the conditions stipulated, populations of spherical diatoms of 130 μm or less (left of the crossover of n and d) can be maintained in the photic zone.
(Questions 5–8)

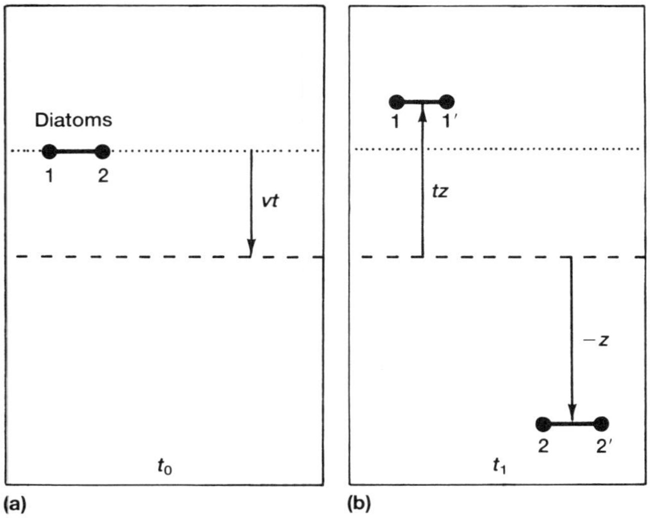

Fig. 3.6 Fate of diatoms after time t with sinking rate v, diffusion rate z, and doubling time equal to t: (a) time 0, (b) time 1. After Weyl (1970).

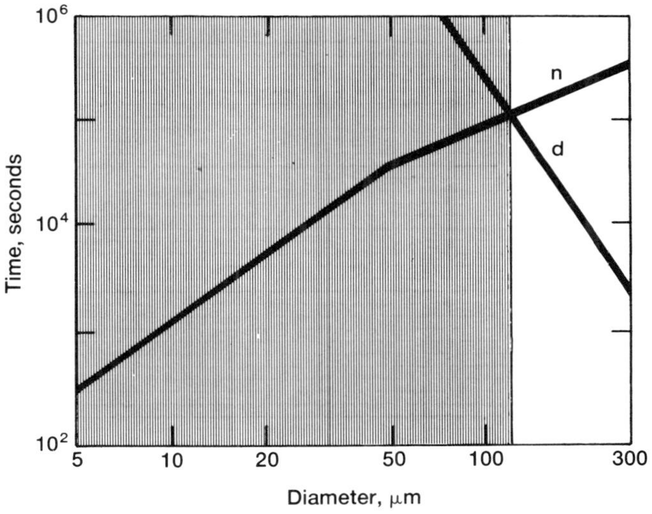

Fig. 3.7 Time required to double nutrient content (n) and duplication time required to maintain spherical diatoms of a given diameter in the photic zone (d), assuming diffusion = 10 cm^2/s, excess density of 2%, and phosphate concentration of 10^{-7} mol/kg. After Weyl (1970) from information in Munk & Riley (1952).

3.6 MACROSCOPIC CALCAREOUS ALGAE

The green algae (Chlorophyta) and the red algae (Rhodophyta) have large calcareous forms and have been important since the Ordovician. They have been volumetrically important in generation of carbonate sediments, and the environmental restrictions on their distribution has been of exceedingly great importance in reconstruction of paleoenvironments and paleogeographies. Calcareous chlorophytes are restricted to tropical marine waters, are by far most prolific within depths of 10–15 m, and generally grow no deeper than 70 m (Wilson, 1975). Calcareous rhodophytes grow from sealevel to depths exceeding 100 m, in some instances growing down to at least 250 m depth.

One group of large calcareous algae, the receptaculitids, was considered to be closely related to sponges until well after the middle of the 20th century. Macroscopic algae are quite abundant in many carbonate rocks, co-occurring with invertebrate fossils, being collected along with and often confused with them.

Receptaculites is now considered to belong to the chlorophyte order Dasycladales. Figure 3.8 shows a reconstruction of the middle Paleozoic *Receptaculites* and the modern dasyclad *Acetabularia*, which is common in shallow carbonate environments in the Florida Keys and the Caribbean. Figure 3.9 compares selected features of some dasyclads and receptaculitids.
(Question 10)

Your instructor may provide a list of calcareous algae to be studied in this exercise. Although specific questions are not asked about most of them, you should be able to recognize them in future encounters in both hand specimen and thin section.

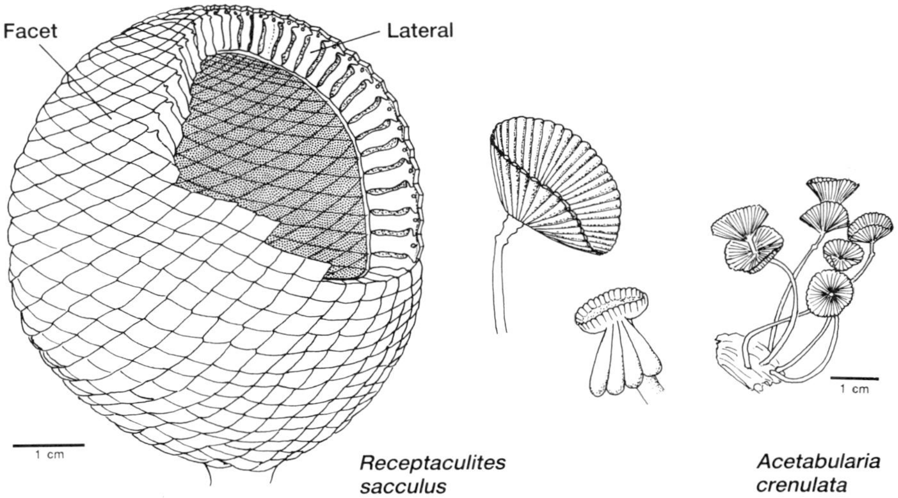

Fig. 3.8 (a) Reconstruction of the Paleozoic green alga *Receptaculites*, and (b) the modern dasyclad green alga *Acetabularia*. From Nitecki (1972) and Dawes (1974).

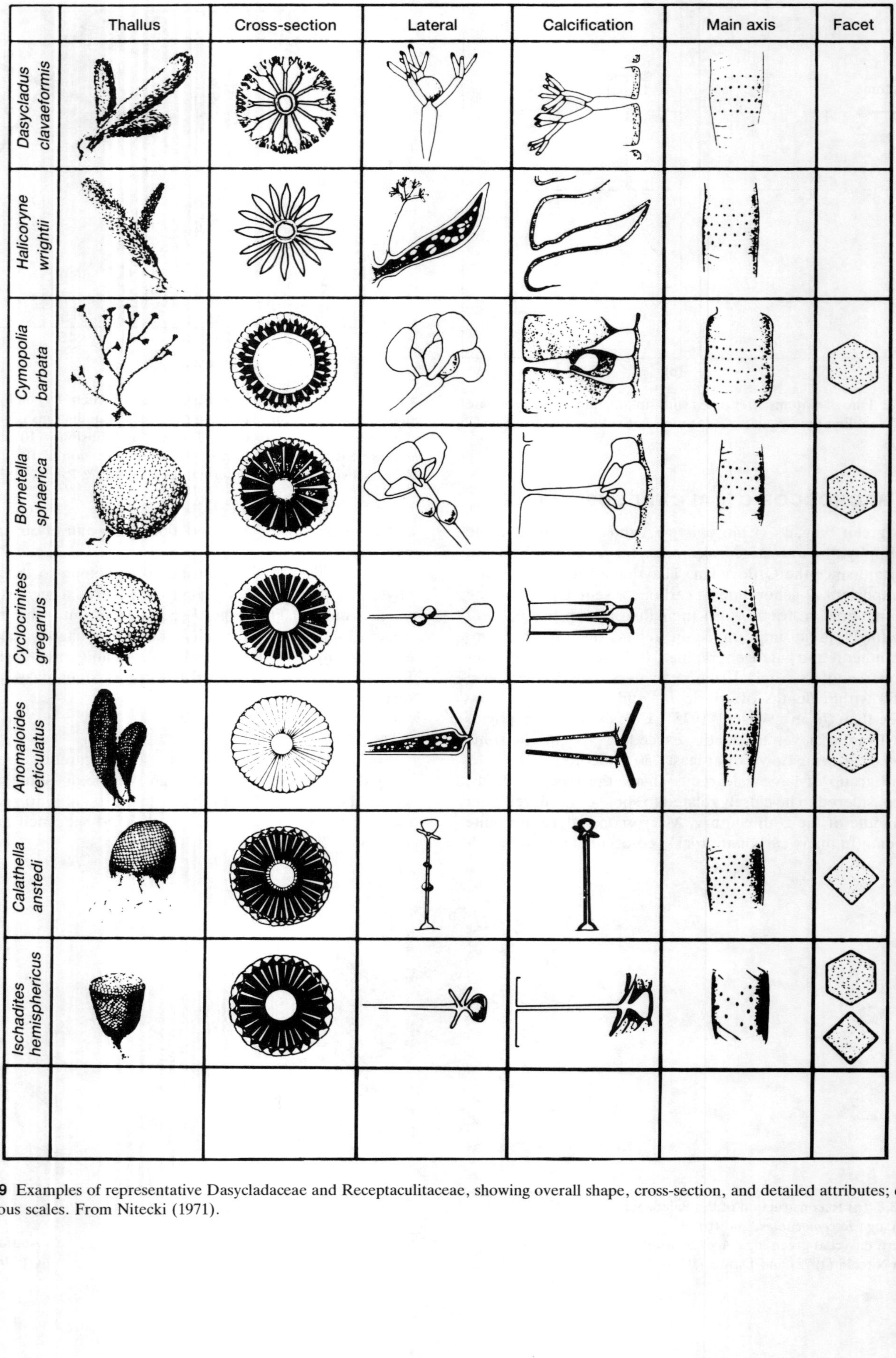

Fig. 3.9 Examples of representative Dasycladaceae and Receptaculitaceae, showing overall shape, cross-section, and detailed attributes; drawn at various scales. From Nitecki (1971).

3.7 FORAMINIFERIDA

The geologically most important subdivision of the Protista is the Order Foraminiferida. They have been extremely important for stratigraphic correlation of Late Paleozoic and Cretaceous through Pleistocene marine deposits, as well as being useful for paleoceanographic studies.

Foraminiferan cells have a deformable cell surface (Fig. 3.10), so are classified along with amoebas as members of the Sarcodina. Their shell (the **test**) is engulfed by and contains cytoplasm; that which is outside the test is the ectoplasm, and that which is inside is the endoplasm. The primitive condition is for the test to consist of a single chamber, but it is more common that several interconnected chambers are successively added as the foraminiferan grows.

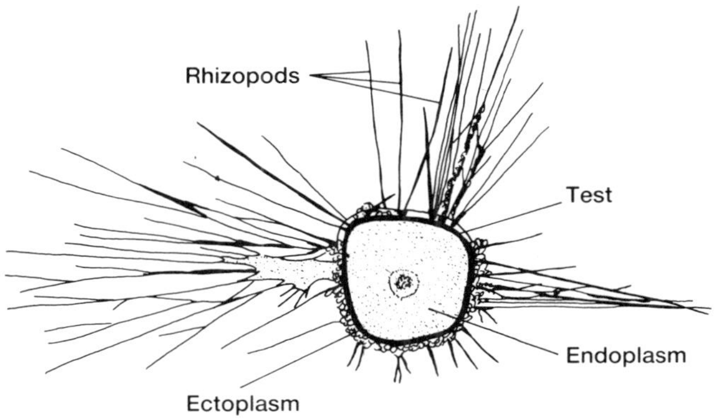

Fig. 3.10 Organization of foraminiferal cells, as illustrated by a single-chambered form, *Myxotheca*. After Loeblich & Tappan (1964).

3.7.1 Classification and morphology

The taxonomic subdivisions of the Foraminiferida are based largely on composition and microstructure of the test (Table 3.2). This is true for the widely used informal subdivision of foraminiferans as well (Table 3.3). In addition, although similar gross test morphology occurs in many groups, detailed characteristic morphology is commonly distinctive for each group. Figure 3.11 illustrates representatives from each of the major groups of foraminiferan fossils.

Four combinations of skeletal composition and structure are commonly fossilized. **Agglutinated** skeletons are built of silt, fine sand grains, or shell fragments held in an organic cement. **Microgranular** calcareous tests consist of minute, equidimensional calcite grains, perhaps set in calcareous cement, that have a characteristic, finely granular surface texture. Other calcareous skeletons are either **perforate** or

Table 3.2 Diagnoses of the order Foraminiferida and its suborders. From Loeblich & Tappan (1988).

Order Foraminiferida	Cytoplasmic body enclosed in test (shell) of one or more interconnected chambers Wall may be homogeneous, or of similar or differing layers or laminae, may be imperforate or finely to coarsely perforate, basically proteinaceous, but may have agglutinated particles, or may deposit calcite, aragonite, or rarely silica, on the organic base Calcareous wall porcelaneous, microgranular, or hyaline and optically or ultrastructurally radiate or granular Canal or stolon systems of varied complexity may be present Commonly test has one or more main apertures through which pseudopodia protrude Commonly sexual and asexual generations alternate, or rarely one generation may be suppressed Gametes biflagellate, triflagellate, or amoeboid Free living or attached to the substratum, rarely parasitic Benthic or pelagic, in marine and brackish water, rare in freshwater Cambrian–Recent
Suborder Allogromiina	Test unilocular or may tend to become multilocular Wall membranous or proteinaceous, may have ferruginous encrustations or small quantity of agglutinated particles Upper Cambrian–Recent
Suborder Textulariina	Test agglutinated, foreign particles held in organic or mineralized groundmass Cambrian–Recent
Suborder Fusulinina	Test of homogeneously microgranular calcite, of tightly packed equidimensional subangular crystals, a few μm in diameter Advanced forms have wall differentiated into two or more layers Ordovician–Triassic
Suborder Involutinina	Proloculus followed by enrolled tubular second chamber Wall calcareous, perforate, radiate, probably originally aragonitic, but may be recrystallized to homogeneous microgranular structure, thickened or with pillar-like structures in umbilical region of one or both sides Triassic–Cretaceous

Table 3.2 *(Continued)*

Suborder Miliolina	Test of porcelaneous calcite, commonly with organic lining and may have added adventitious material Generally imperforate in post-embryonic stage May have flexostyle or spiral passage between proloculus and later chambers Carboniferous–Recent
Suborder Silicoloculinina	Wall imperforate, of secreted opaline silica Middle Miocene–Recent
Suborder Spirillinina	Proloculus followed by enrolled tubular undivided chamber, or by a few chambers per whorl, coiling planispiral to high trochospiral Chambers may be secondarily subdivided Wall of calcite, optically a single crystal or few to a mosaic of crystals a-axis preferred orientation along axis of coiling and c-axis parallel to umbilical surface May have pseudopores or micropores filled with organic matter and closed by sieve plates, wall formed by accretion at edge, not by calcification of an organic template produced by pseudopodia ?Triassic, Jurassic–Recent
Suborder Lagenina	Wall of monolamellar, optically and ultrastructurally radiate calcite, with crystal c-axes perpendicular to surface Crystal units enveloped by organic membranes Primitive taxa without secondary lamination, later ones secondarily lamellar Middle Carboniferous–Recent
Suborder Robertinina	Test planispirally to trochospirally enrolled Chambers commonly with internal partition related to apertural foramen Wall of hyaline, perforate, ultrastructurally and optically radiate aragonite, the hexagonal prisms with c-axis normal to the wall surface Prisms in bundles surrounded by organic sheaths Extremely fine wall perforations may be localized in pore fields Middle Triassic–Recent
Suborder Globigerinina	Planktic habit Test wall of perforate hyaline calcite, optically radiate, preferred crystal orientation with c-axis normal to surface Primary lamination bilamellar, with secondary lamination due to addition of material at formation of new chamber Surface crust may develop at time of gametogenesis Lower Jurassic–Recent
Suborder Rotaliina	Test multilocular, typically enrolled, but may be reduced to biserial or uniserial, or chambers may proliferate with an encrusting habit Chambers simple or subdivided by secondary partitions Wall calcareous, of perforate hyaline, lamellar calcite, formed by calcification at each side of an organic membrane Wall surface smooth, papillate, costate, striate, or cancellate Aperture may be simple or have an internal toothplate Internal canal systems or stolon systems present in some Triassic–Recent
Suborder Carterinina	Trochospiral test of secreted spicules, each of which is a single elongated crystal of calcite, held in a groundmass of smaller spicules and poorly cemented organic matrix that readily disintegrates, hence suborder has meager fossil record Eocene, Recent

imperforate, based on presence or absence of abundant small perforations in the test wall. Perforate calcareous tests are commonly referred to as **hyaline**, and the imperforate tests are commonly termed **porcelaneous**.

The microgranular skeletons characteristic of the Late Paleozoic suborder Fusulinina potentially consisted of several distinctive layers (Fig. 3.12). Two basic variations in the wall structure existed, differing in the degree of development of the layers. **Fusulinellid** wall structure is initially deposited as a dark outer layer (the **tectum**) and a lighter inner layer (the **diaphanotheca**), which may eventually have an additional dark layer (the **tectorium**) added to both the inner and outer

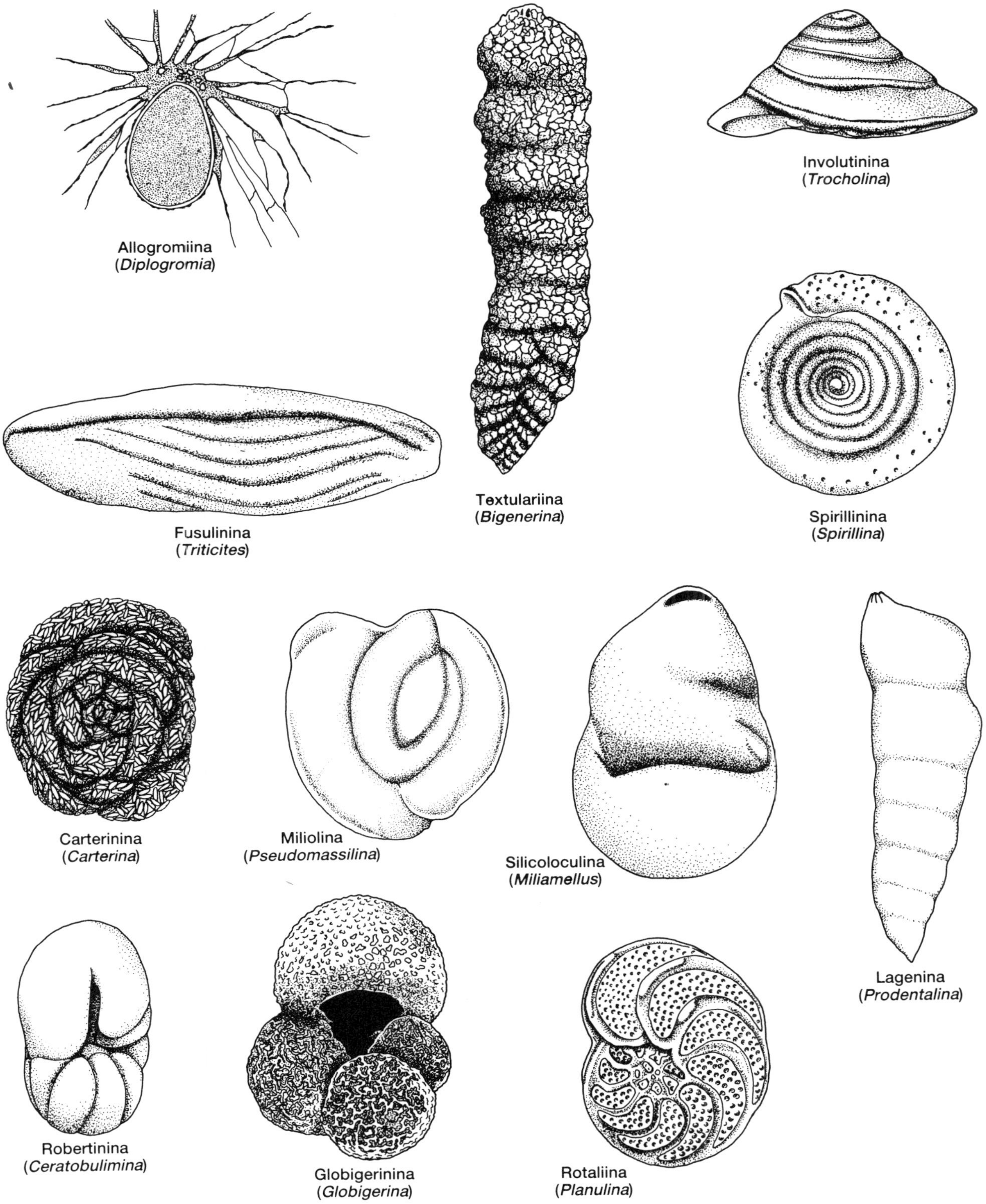

Fig. 3.11 Representatives of the 12 foraminiferan suborders listed in Table 3.2. All except *Triticites* are magnified several tens of times; *Triticites* is magnified less than 8×. Drawn after illustrations in Loeblich & Tappan (1988).

Table 3.3 Informal subdivisions of the Foraminiferida. After Buzas in Boardman *et al.* (1987).

Allogromids	Test organic Upper Cambrian–Recent
Textularids	Test agglutinated, composed of material cemented together by the organism Cambrian–Recent
Fusulinids	Test composed of microgranular calcium carbonate, some with two or more wall layers, usually large Ordovician–Permian
Miliolids	Test calcareous, imperforate, porcelaneous Pennsylvanian–Recent
Rotalids	Test calcareous, perforate Permian–Recent

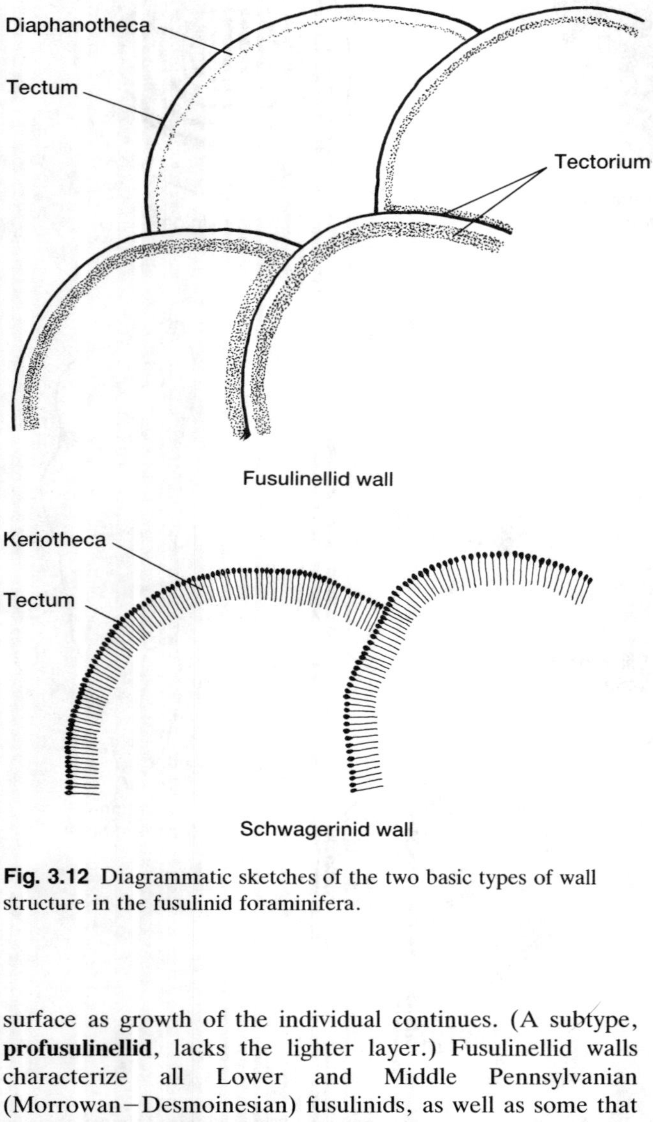

Fig. 3.12 Diagrammatic sketches of the two basic types of wall structure in the fusulinid foraminifera.

surface as growth of the individual continues. (A subtype, **profusulinellid**, lacks the lighter layer.) Fusulinellid walls characterize all Lower and Middle Pennsylvanian (Morrowan–Desmoinesian) fusulinids, as well as some that are younger. **Schwagerinid** wall structure consists of a tectum and a thick inner lining (the **keriotheca**) that has a minute honeycomb structure. Schwagerinid walls developed late in Pennsylvanian time and are characteristic of many Late Pennsylvanian (Missourian and Virgilian) and Permian fusulinids.

Although foraminifera are single-celled, a staggering diversity of morphologies has evolved among the more than 3600 genera that have been described. Some of the more important features that exhibit substantial variation and are important for taxonomy include:

Test symmetry (Fig. 3.13).
General chamber shape (Fig. 3.14).
Aperture position (which may be at the margin of the chamber or nonmarginal).
Aperture shape
circular

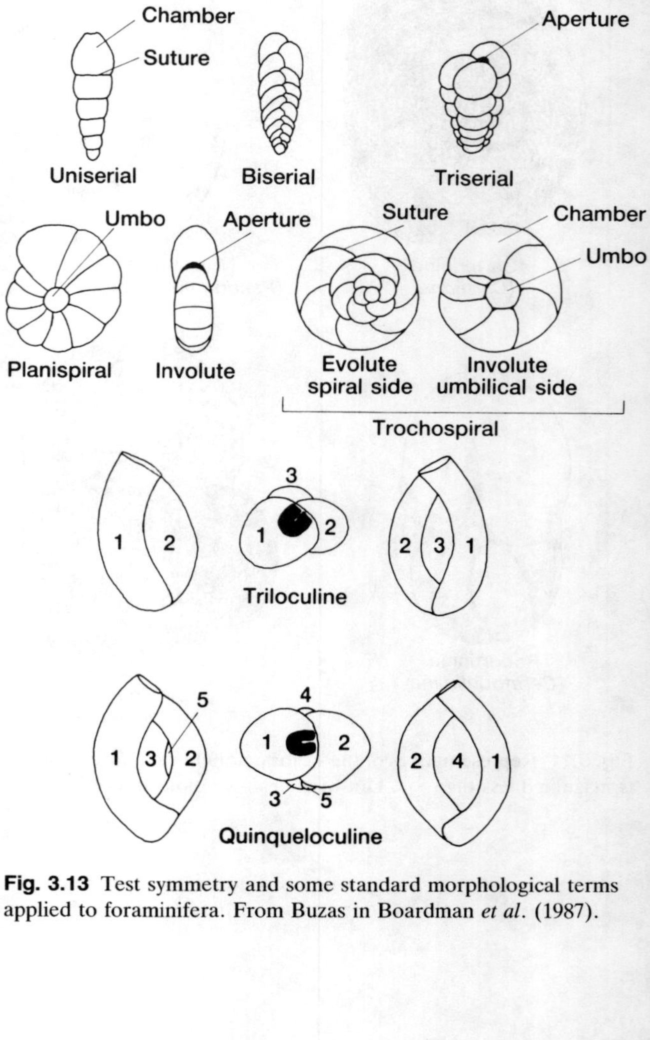

Fig. 3.13 Test symmetry and some standard morphological terms applied to foraminifera. From Buzas in Boardman *et al.* (1987).

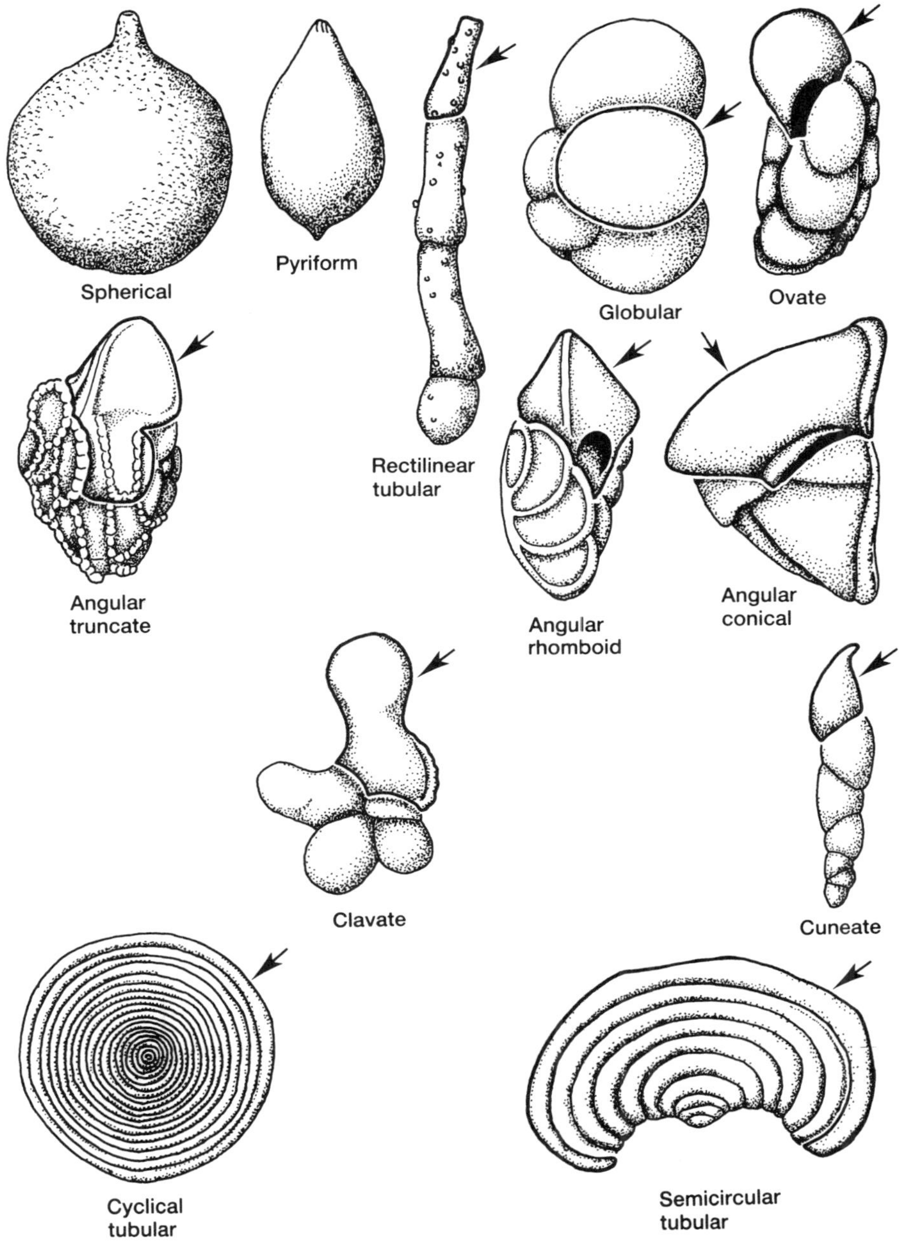

Fig. 3.14 Representative chamber shapes that commonly occur in foraminifera. After Loeblich & Tappan (1964).

semicircular
radiate
slit
multiple
cruciform (cross-shaped)
cribrate (sieve-like)
dendritic

Accessory skeletal features (Figs 3.13 and 3.15)

Test symmetry is one of the more obvious features and is relatively easy to categorize. A single series of chambers is **uniserial**; alternating chambers are **biserial**; spiralled tests that add a new chamber for every 120° rotation are **triserial**; **planispiral** tests add chambers around the periphery and are coiled in a flat plane; **trochospiral** tests also add chambers around the periphery, but each chamber is slightly offset so that a very low spire or cone resuts; **triloculine** tests have only the three most recently formed chambers exposed at the surface, with each new elongate chamber added at 120° offsets and alternately oriented towards opposite ends of the test; **quinqueloculine** tests have the five most recently formed chambers exposed at the surface, with each new chamber added at 144° offsets and alternately oriented toward opposite ends of the test. A few other basic morphologies also occur, and some foraminifera are characterized by ontogenetic change from one type of symmetry to another, so that the resulting test has compound morphology.

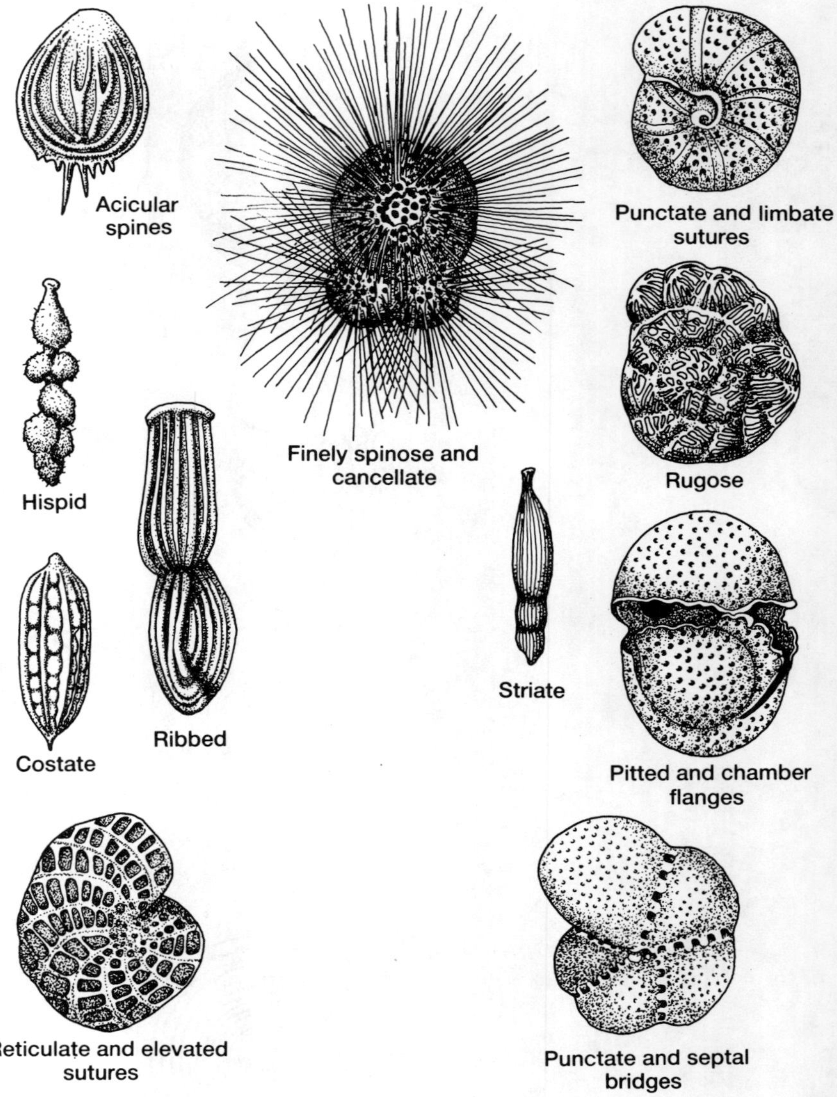

Fig. 3.15 Examples of surface textures and accessory structures found in foraminifera. After Loeblich & Tappan (1964).

Large foraminifera, those with maximum dimensions well over 1 mm, are known from all the major mineralized skeletal types. The larger Foraminiferida do not have as broad range of variations in their overall test morphologies as do the smaller forms. For example, the surface sculpture is generally smooth or finely textured, and the apertures typically consist of a series of small pores. The following list of characters and character states gives a general idea of the range of basic morphology of the larger Foraminiferida:

- Test symmetry
  - **rectilinear** (tubular or uniserial)
  - planispiral discoidal
  - planispiral **fusiform** (axis of coiling is longest dimension)
  - annular discoidal
- Chamber overlap in spiral tests
  - **involute** (chambers completely overlapping previously formed whorls)
  - **evolute** (chambers not overlapping previously formed whorls)
- Wall structure or composition
  - agglutinate
  - calcareous perforate
  - calcareous imperforate
  - fusulinellid
  - schwagerinid
- Chambers (in larger foraminifera, such as fusulinids)
  - without chamberlets
  - chamberlets irregular
  - chamberlets regular

(Question 10)

3.7.2 Planktic and benthic rotalids

Planktic (floating) foraminifera that have mineralized tests belong to the rotalid group. Most planktic rotalids have one of the three basic morphologies (Fig. 3.16). The geologically most persistent of these is the **globigerine** form, which is a trochospiral test consisting of globular, rapidly enlarging chambers. Derived from this are **spherical** forms, in which successive chambers nearly or completely engulf the entire

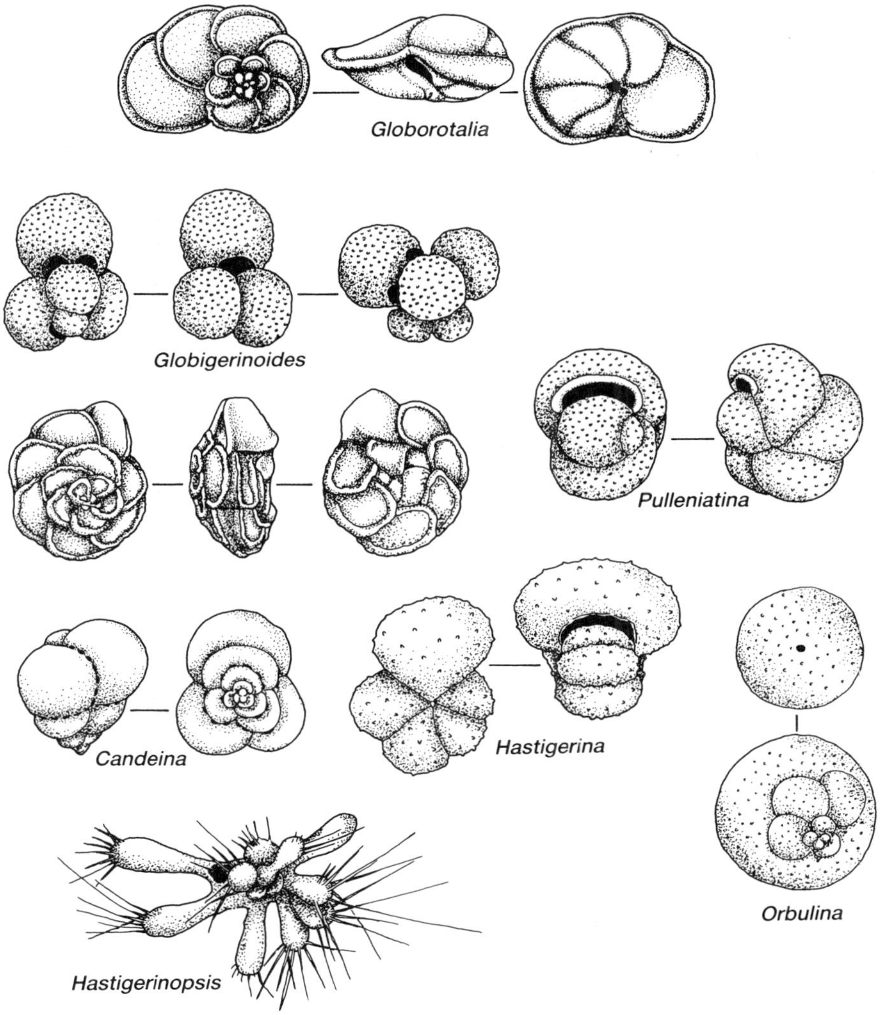

Fig. 3.16 Common morphologies of planktic foraminifera. Redrawn from Galloway (1933).

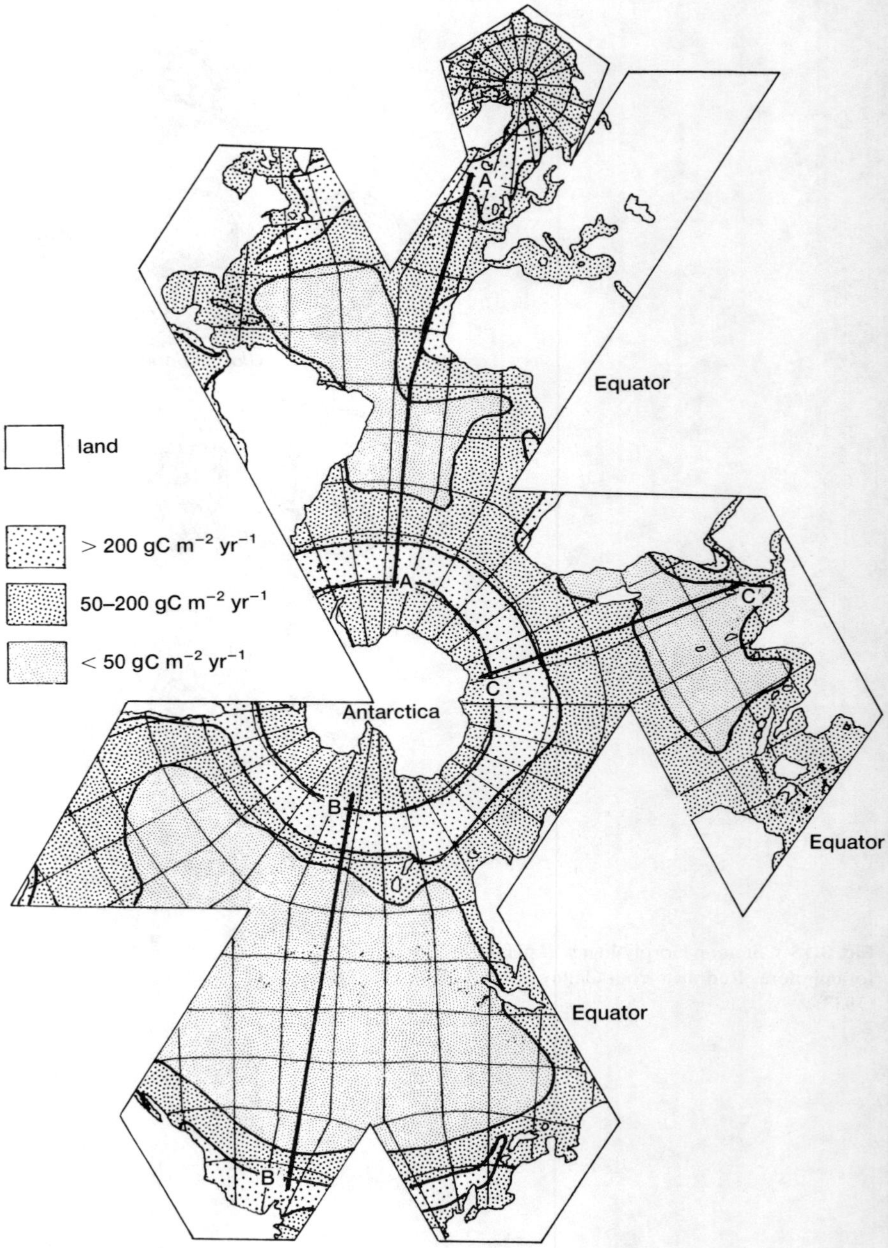

Fig. 3.17 Distribution of primary productivity (in g C/m^2 per year) in the world's oceans. Cross-sections showing phosphate concentrations in Fig. 3.18 are located approximately along lines A–A′, B–B′, and C–C′. Modified from Weyl (1970).

Table 3.4 Generalized distribution of foraminifera in modern oceans, listing only a few of the most diagnostic genera

	Environment	Characteristic foraminifera	
Marine marginal	marshes	Few species, most are *Trochammina* and *Ammobaculites* (both are agglutinated); virtually no planktic tests	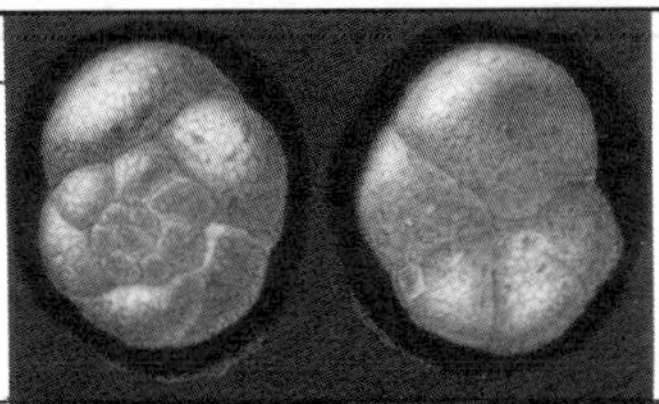 *Trochammina* 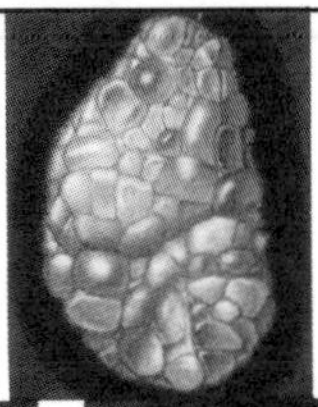 *Ammobaculites*
	estuaries and lagoons	Species diversity low, but greater than marshes, perforate calcareous species generally dominant, e.g. *Ammonia*, *Elphidium*, and *Rosalina*; virtually no planktic tests	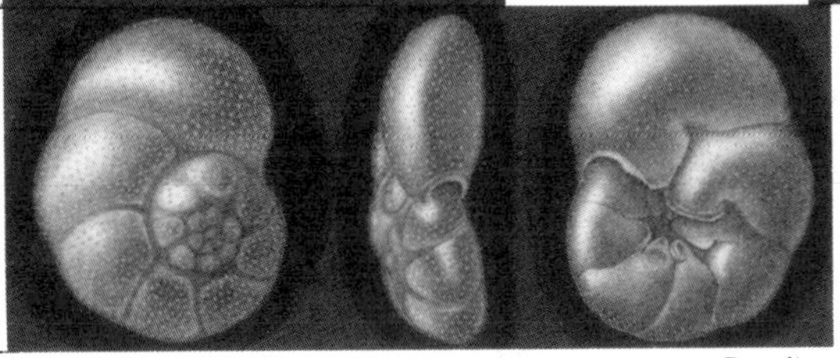 *Rosalina* 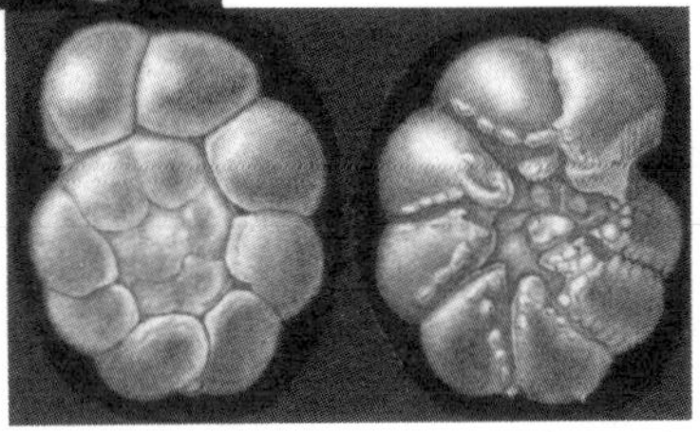 *Ammonia* 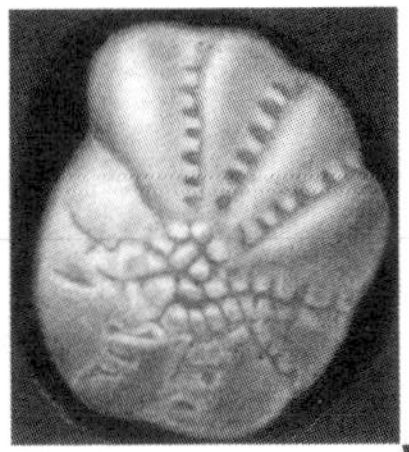 *Elphidium*
Neritic	near-shore (low tide – 50 m)	Resembles estuarine fauna plus *Quinqueloculina*; relatively few planktic tests	 *Quinqueloculina*
	outer shelf (50–200 m)	Commonly dominated by perforate forms such as *Cassidulina*, *Nonion*, and *Nonionella*; planktic tests common	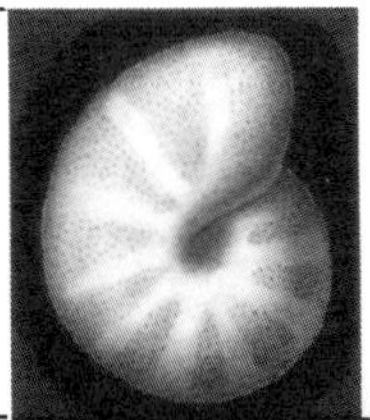 *Nonion* 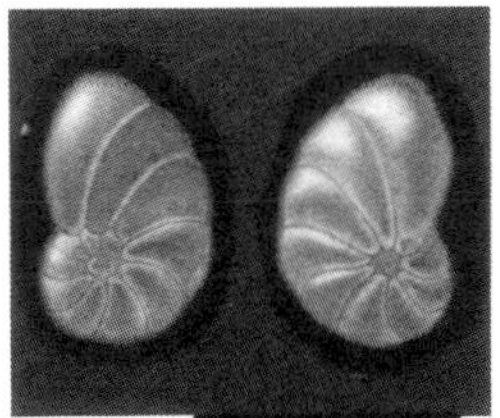 *Nonionella* 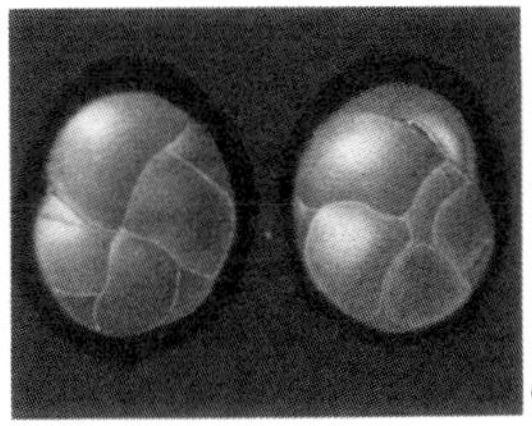 *Cassidulina*
Oceanic	bathyal zone (200–~2500 m)	Benthic forms dominated by perforate genera such as *Bolivina*, and *Cassidulina* and imperforate *Pyrgo*; planktic tests usually at least as abundant as benthic	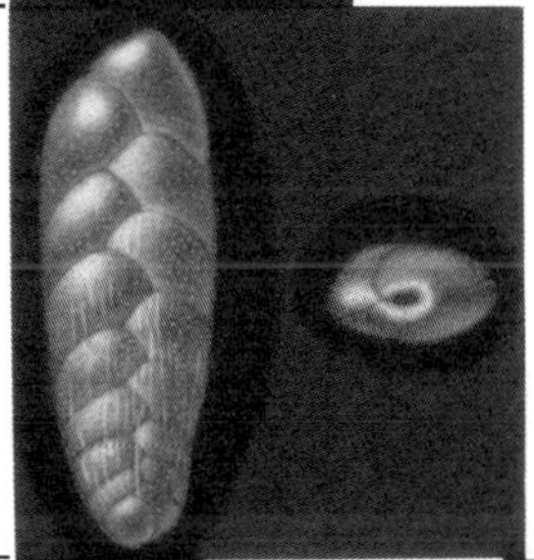 *Bolivina*  *Pyrgo*
	abyssal zone (~2500–>7000 m)	Common benthic forms in shallower parts are *Cassidulina* and *Lagena*, but agglutinated *Reophax* dominant below carbonate compensation depth (CCD); above CCD, planktic tests may be more than 10 times more abundant than benthic tests	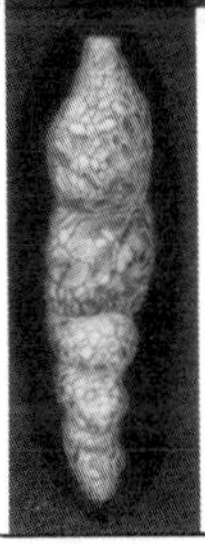 *Reophax* 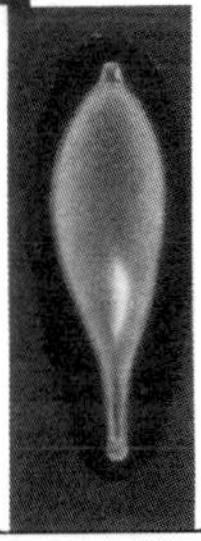 *Lagena*

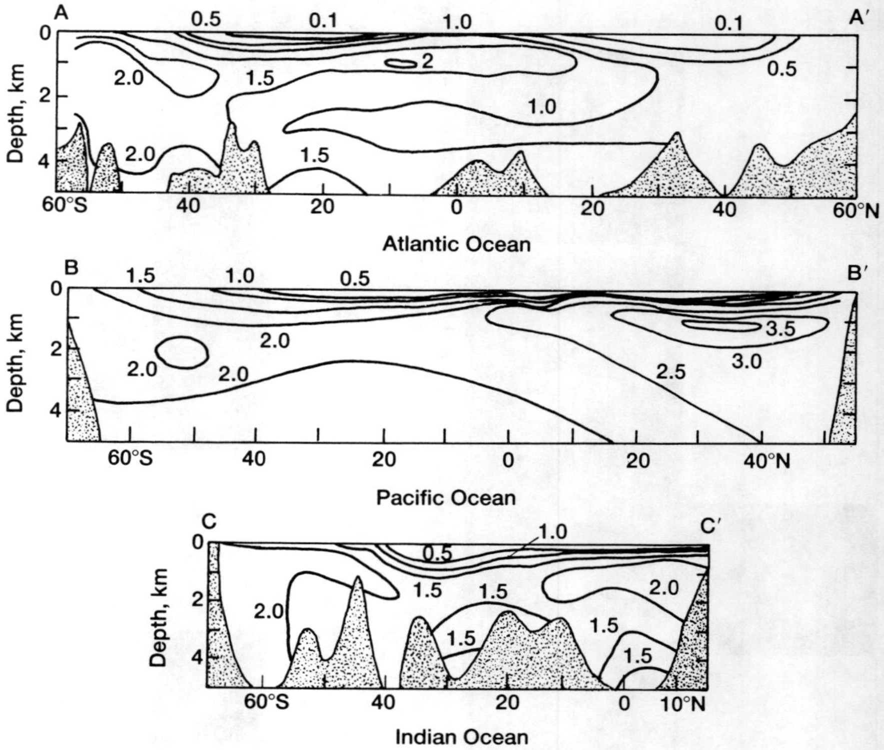

Fig. 3.18 Distribution of phosphate (in μg atoms/l) in oceanic waters. After Sverdrup *et al.* (1942) with permission from Prentice Hall, Inc.

earlier-formed test, and **globorotalid** forms, which are trochospiral and have moderately inflated chambers with angular margins that may be extended by keels. The thin shells and commonly spherical chambers of planktic foraminifera cause only minimal increase in overall density, but any increase over density of the surrounding water (approximately 1.03 for normal seawater) must be offset by low-density fat or gas vacuoles to maintain neutral buoyancy for the depth at which the organism floats.
(Question 11)

Planktic and benthic foraminiferal distribution, ranging from their occurrence over or within 1 cm^2 of the ocean floor to areal distribution on a global scale, is determined by a host of biological, physical, and chemical factors. As is true for any organism, such factors include conditions such as substrate type, oxygen concentration, nutrient availability and type, predators, competitors, turbulence, and temperature. As a result, distributional patterns are complex and shifting. Nonetheless, some readily usable general patterns exist (Table 3.4), both for living foraminifera and their occurrence in ancient sediments. The patterns have apparently held throughout much of the Cenozoic and are therefore quite useful in biostratigraphy, paleoceanography, paleoclimatology, paleoecology and paleogeography, for this part of the geologic column.
(Question 12)

FURTHER READING

Brasier, M. D., 1980. *Microfossils*. George Allen & Unwin: London

Buzas, M. A. & Sen Gupta, B. K. (eds), 1982. *Foraminifera*. University of Tennessee Department of Geological Sciences Studies in Geology 6. 219 pp.

Culver, S. J., 1987. *Foraminifera*. University of Tennessee Department of Geological Sciences Studies in Geology 18. pp. 169–212.

Haynes, J., 1981. *Foraminifera*. John Wiley & Sons: New York.

Haq, B. U. & Boersma, A., 1978. *Introduction to marine micropaleontology*. Elsevier: New York.

Lipps, J. H. (ed.), 1987. *Fossil prokaryotes and protists. Notes for a short course*. University of Tennessee Department of Geological Sciences Studies in Geology 18.

Loeblich, A. R. Jr & Tappan, H., 1964. Sarcodina chiefly "Thecamoebians" and Foraminiferida, in R. C. Moore (ed.), *Treatise on invertebrate paleontology*, Part C, Protista 2. Geological Society of America and University of Kansas Press: Lawrence, KS.

Loeblich, A. R. Jr & Tappan, H., 1988. *Foraminiferal genera and their classification*, 2 vols. Van Nostrand Reinhold: New York.

Moore, R. C. (ed.), 1954. *Treatise on invertebrate paleontology*, Part D, Protista 3. Protozoa (chiefly Radiolaria and Tintinnina). Geological Society of America and University of Kansas Press: Lawrence, KS.

EXERCISE 3 Worksheet

Name

QUESTION 1

Slides on which Recent radiolarians are strewn are available for one or two oceanic localities. Carefully examine each slide through a binocular microscope at about ×50–×100 magnification. Take note of the apparent diversity of forms, the symmetry of each (spherical versus radial), and the most abundant form. In the space provided below, carefully draw a representative of the most abundant radiolarian form from one of the slides. Immediately below the drawing, write in the magnification. If you drew the specimen at the same size that it appears through the microscope, the magnification is equal to the power of the objective lens just above the stage times the power of the ocular lens through which you look. If you drew it at some other size, you must multiply the two lens powers together and then multiply the result by the proportional increase in linear dimension of your drawing over the apparent size as seen in the microscope.

To which polycystine order does the specimen belong?

QUESTION 2

Use a handlens or binocular microscope at low power to examine a handspecimen of radiolarian chert. Then examine a thin section of the rock at ×50–×100 through a petrographic microscope. Slowly scan most or all the thin section, developing familiarity with the various appearances of the radiolaria in section. Sketch the best radiolarian seen in the slide and a few of those that you would not have recognized if you had not seen the better specimens and did not know that the rock is composed largely of radiolarians.

QUESTION 3

Preparations of a Recent open-ocean diatom deposit and a Cenozoic diatomite are provided for study. Carefully examine one of the slides of Recent diatoms through a binocular or petrographic microscope at about ×100–×200 magnification. Carefully take note of the apparent diversity of forms, the symmetry of each (centric versus pennate), and the most abundant form. Which is dominant, centric or pennate symmetry?

____

What does this suggest about environment of deposition?

____

Examine the slide prepared from a Cenozoic diatomite. Which is the dominant symmetry, centric or pennate?

Does this indicate anything about environment of deposition? If so, what?

QUESTION 4

Skeletons of calcareous nannoplankton are a dominant component of chalks and Recent calcareous oozes. A broad variety of calcareous nannoplankton constitute this component, but coccoliths are predominant. To get an idea of the abundance of the calcareous nannoplankton in chalks and calcareous oozes, place a small fragment (about 1 mm^3) of any Cretaceous or Cenozoic chalk or Recent calcareous ooze on a glass microscope slide, then put a drop of water on the fragment. After it has soaked for 2–3 min, gently crush it so that it is partially dispersed in the water. (You may be provided with previously prepared slides.) Let the water evaporate fully, then examine at ×300 or higher power through a petrographic microscope, in both plane and cross polarized light.
What shapes occur commonly among the grains?

Sketch those that exhibit a general shape consistent with the morphology of coccoliths.

It should be obvious at this point why detailed study of modern coccolithophores and biostratigraphic usefulness of coccoliths depend on scanning electron microscopy at magnifications of a few thousand times.

QUESTION 5

Which direction would curve *n* be displaced in Fig. 3.7 if the phosphate concentration is increased? Towards the upper left, or towards the lower right?

What will be the effect on maximum possible size of diatoms sustained in the photic zone?

QUESTION 6

Which direction would curve *d* be displaced in Fig. 3.7 if excess density is increased? Towards the upper right, or towards the lower left?

What will be the effect on maximum possible size of diatoms sustained in the photic zone?

____

QUESTION 7

Which direction would curve *d* be displaced in Fig. 3.7 if the diffusion constant is increased? Towards the upper right, or towards the lower left?

____

What will be the effect on maximum possible size of diatoms sustained in the photic zone?

____

QUESTION 8

Figures 3.17 and 3.18 show distribution of productivity and nutrient concentration in the world oceans. The largest average wave heights and lengths are typically found in the "Roaring 50s" in the uninterrupted oceanic stretch surrounding Antarctica at about 50° latitude, and large waves are also characteristic of the high latitudes of the Atlantic and Pacific. On Fig. 3.17, mark with red the condition most favorable to diatoms and in blue the condition least favorable to diatoms. Write a paragraph summarizing oceanographic conditions and latitudes favorable for diatom populations and where in the present oceans such conditions are best attained. What paleoceanographic conditions would a diatomite of planktic diatoms indicate?

____

____

____

____

____

____

QUESTION 9

A row has been provided across the bottom of Fig. 3.9 for *Receptaculites*, which should be filled in from the characteristics visible in the specimen(s) available in the laboratory, supplemented from Fig. 3.8 if necessary.

QUESTION 10

Some of the most common genera from each of the four commonly fossilized major types of foraminifera are provided so that you can become familiar with them and develop an idea of the range of morphologies developed in the order. For each genus of smaller foraminifera designated by the laboratory instructor, describe in the table on p. 46 all the visible skeletal features for the morphologic characters listed. Use Figs 3.13–3.15, plus the brief list in the text, to determine the states for each of the characters. Pay particular attention to the transparency, color, and surface texture, which are determined by skeletal composition and structure (agglutinated, micro-granular, hyaline, porcelaneous). If geologic range is not given on the specimen

Genus	Suborder	Geologic range of genus	Test symmetry	Chamber overlap if spiral	Chamber shape or organization	Aperture: Shape	Aperture: Position	Accessory features

label, it may be determined from the *Treatise on invertebrate paleontology*, Part C (1984) or Loeblich & Tappan (1988), *Foraminiferal genera and their classification*. Be aware that the list of characters for this question and similar questions throughout the laboratory manual do not fully describe and differentiate the genera, but are intended to cover much of the essential morphology of the group.

QUESTION 11

Find out from the instructor which of the rotalid genera studied for Question 10 are planktic. List their names here:

________________ ________________ ________________

All the other rotalids examined for Question 10 are benthic. Contrast the typical differences in morphology between benthic and planktic rotalid foraminifera as exhibited in the suite of rotalids examined in this exercise. Note especially chamber shape, overall test shape, test wall thickness, and any other consistent differences.

	Benthic	*Planktic*
Chamber shape	________	________
Overall test shape	________	________
Wall thickness	________	________
Others	________	________

QUESTION 12

Population samples of at least 100 specimens each from two or three Recent or Neogene foraminiferan assemblages are available. For each sample, record the number of specimens of each taxon or category listed below. Then, for each assemblage, determine the probable depth range and, for shallow-water assemblages, record the general latitude (low, mid, high). Specify the bases for your answer.

	Sample 1	*Sample 2*	*Sample 3*
Trochammina	____	____	____
Ammobaculites	____	____	____
Other textularids	____	____	____
Total textularids	____	____	____
Quinqueloculina	____	____	____
Other miliolids	____	____	____
Total miliolids	____	____	____
Ammonia	____	____	____

Elphidium			
Rosalina			
Cassidulina			
Nonion			
Nonionella			
Bolivina			
Pyrgo			
Lagena			
Other benthic rotalids			
Total benthic rotalids			
globigerines			
Orbulina			
Globorotalia			
Other planktic			
Total planktic			

Probable depth range and approximate latitude

Sample 1 ______

Reasons ______

Sample 2 ______

Reasons ______

Sample 3 ______

Reasons ______

EXERCISE 4 Porifera and Archaeocyatha

4.1 INTRODUCTION

The two groups included in this exercise are treated within one unit because of similar skeletal structure or symmetry of at least some members of each group, and because they are thought to have roughly similar levels of biological complexity and function. The Paleozoic receptaculitids have superficial resemblance to Porifera and Archaeocyatha but are now considered to be algae so were included in Exercise 3. This exercise consists of four parts:

1 examination of a variety of sponge spicules;
2 drawing and labeling examples of one or more grades of sponge organization;
3 detailed examination of skeletal morphology in representatives of each group; and
4 determination of some general biomechanical aspects of simple spongelike structure.

4.2 PORIFERA

The sponges (phylum Porifera) have cells that show indications of independence from one another and the ability to change functions during their existence. Sponges have the least cellular integration among the multicellular animals. Independence of cells within sponges is graphically demonstrated by their ability to regenerate new, independently

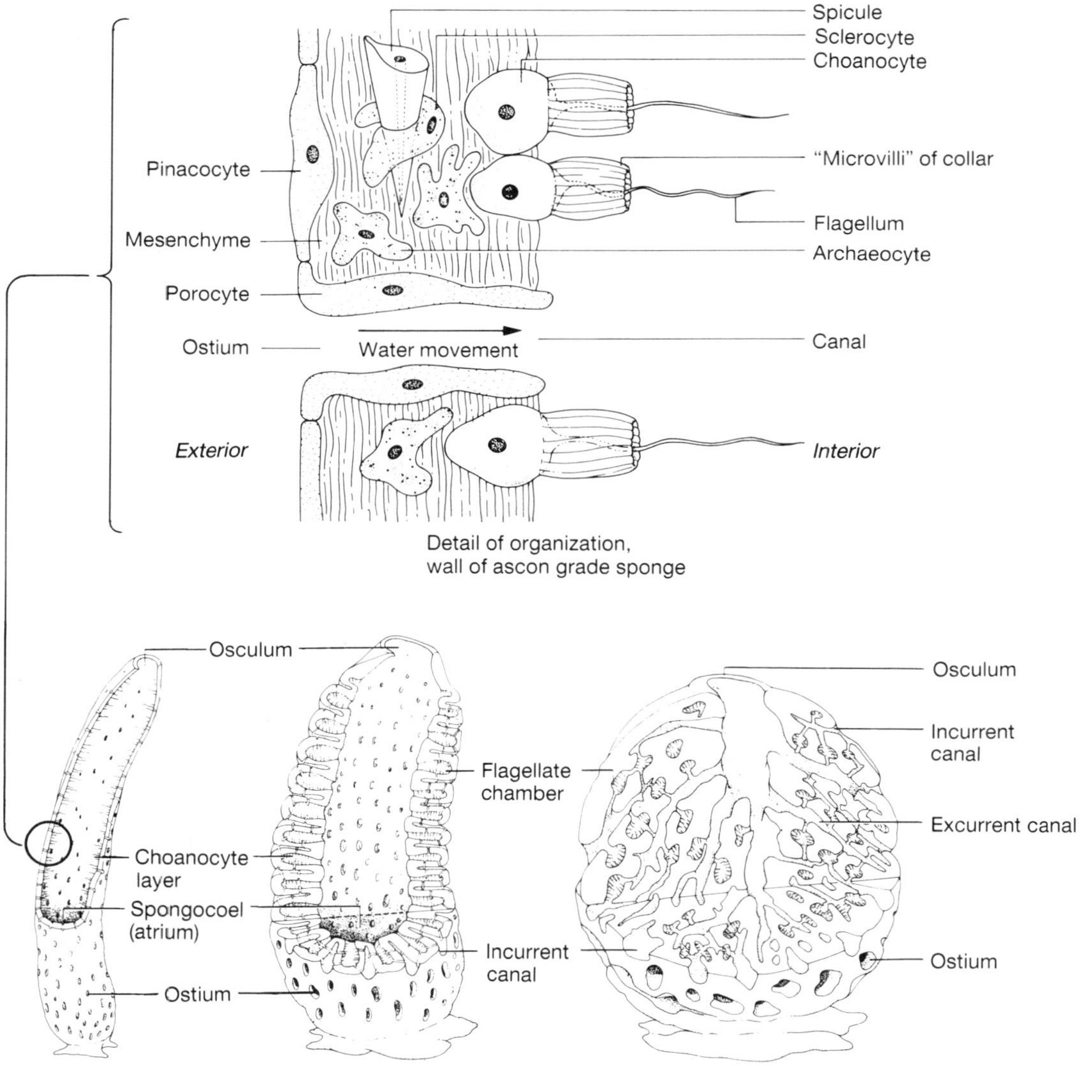

Fig. 4.1 The three grades of organization of sponges, with some morphology labeled, and a detail of a portion of a sponge with ascon structure to show cell types and distribution. From Rigby in Boardman *et al.* (1987).

growing sponges from a single living sponge even after it has been forced through a fine sieve.

Three cell types (Fig. 4.1) are particularly important in the paleontology and functional interpretation of sponges.

1 **Sclerocytes** secrete the mineralized portions of skeletons (those secreting the organic skeleton are termed **spongocytes**).

2 **Choanocytes** are cells that bear flagellae that generate the feeding currents.

3 The most versatile are the **archaeocytes**, which are amoeboid cells that carry out several functions but may also change into any other type cell.

Sponges may have an organic skeleton composed of a material called **spongin**, or they may have calcareous or siliceous skeletons comprised of **spicules**, which are small, basically rod-like, branched, or spherical skeletal elements. Spicules have four symmetries (Fig. 4.2):

Monaxon — single rod, with or without ornamentation on the ends.

Triaxon — each spicule with rods corresponding to three axes; the spicule has three to six rays, depending on where the rods meet.

Tetraxon — basically, four axes 120° apart, resembling the axes of a pyramid; in many, one of the axes is reduced

Polyaxon — many rays radiating from a central point.

In addition, they fall into two general categories, relatively large spicules termed **megascleres** and relatively small spicules termed **microscleres** (Fig. 4.2).

The paleontological record of the sponges is dominated by spicules, which can commonly be seen in thin sections of marine rocks where no sponges are otherwise visible. Other groups of organisms, such as gorgonian corals and holothurian echinoderms, also produce calcareous spicules. Examples may be provided for comparison, but the term spicule used in the remainder of this exercise refers only to sponge spicules. Several spicules are provided for your examination; the com- position of each is given, but their morphology and possible class assignment are to be determined according to your observations and the information on spicules given in Fig. 4.2 and information on class characters given in Table 4.1. *(Question 1)*

Fossil and Recent sponges have three kinds or grades of organization around the path through which water is driven through the body. These grades are generally readily dis- tinguishable (Fig. 4.1). The simplest grade (**ascon**) has thin porous walls, and the choanocytes line the atrium. The intermediate grade (**sycon**) has a somewhat thicker wall that is perforated by dead-end canals from the exterior and from

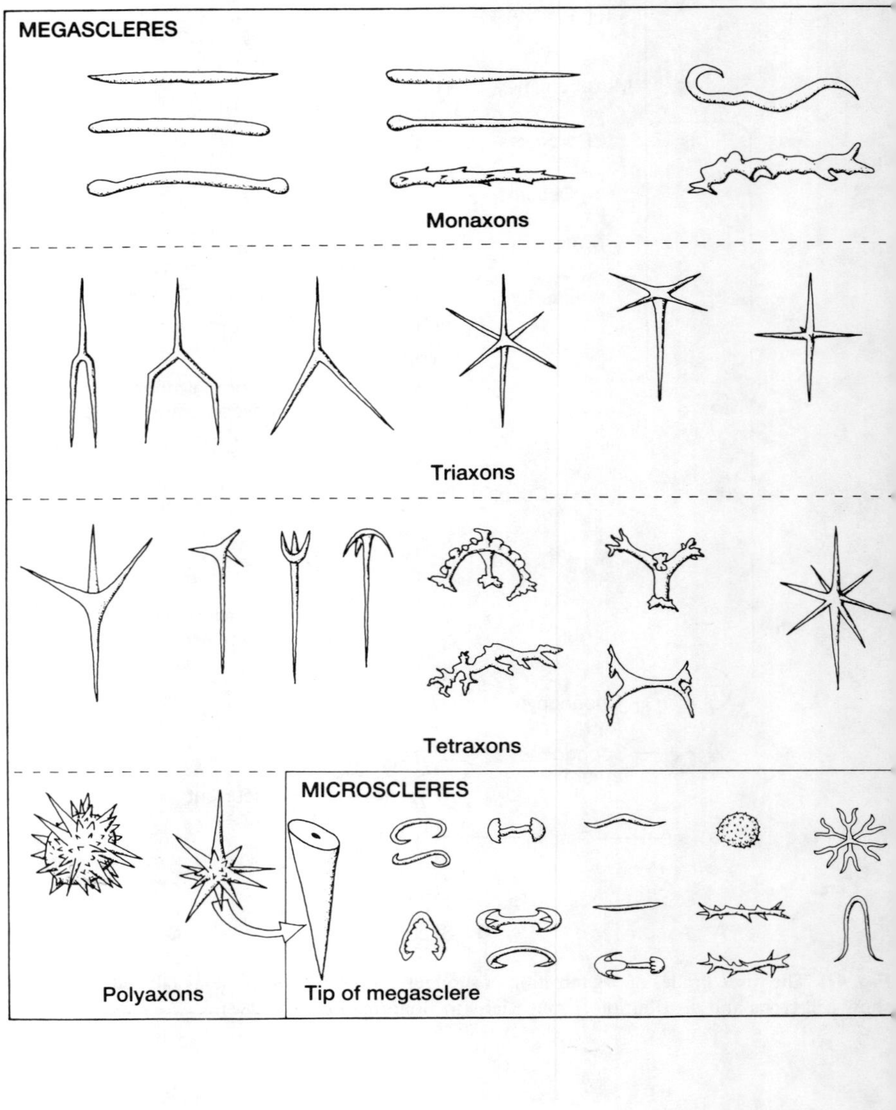

Fig. 4.2 Types of sponge spicules, with several examples of each type. Magnification is variable, but microscleres are much smaller than megascleres. From Rigby in Boardman *et al.* (1987).

the inner atrium; pores link the two sets of canals, and choanocytes line those extending from the atrium. The most complex grade (**leucon**) is characterized by choanocyte-lined chambers located along a dendritic canal system that leads finally to one or more openings (**osculum**) for outflow from the sponge. Preserved examples of modern sponges exhibiting one or more of the different grades are provided. Study them carefully to see how each matches its assigned grade.
(Question 2)

Overall shapes of sponges are so variable that a clearly defined set of descriptive terms has not been agreed. Most are **encrusting sheets, globular, conical, basket-shaped, branched**, or **irregular**. Members of the genus *Cliona* infest calcareous shells by excavating interconnected globular chambers.

Representatives from each class of sponges are provided so that you can become familiar with their characteristics. As you examine representatives of each class, first read its diagnosis in Table 4.1 and examine the line drawings in Figs 4.3 and 4.4.
(Questions 3 and 4)

4.3 ARCHAEOCYATHA

The Archaeocyatha are a predominantly Lower Cambrian phylum with a very few representatives that lived into the early Middle Cambrian and perhaps into the Late Cambrian. Archaeocyathids generally have skeletons in the form of a calcareous cup composed of porous granular walls that are most commonly double (outer and inner), infrequently single, surrounding a central cavity (Fig. 4.5). The space between the outer and inner walls is termed the **intervallum** and contains variously porous **septa** (longitudinal plates), **tabulae** (horizontal plates), and nonporous rods, bars, and **dissepiments** (curved, cyst-like plates).

Two classes of Archaeocyatha are recognized. Above the seldom-seen proximal tip of a specimen, the class **Regulares** is characterized by septa, tabulae, or both in the intervallum (Fig. 4.5). Pores in the septa are in longitudinal rows that diverge upwardly toward both bounding walls. Other features, especially dissepiments, also may occur within the intervallum. Tabulae are bent such that they are higher in the middle of the intervallum than along either edge.

Above the proximal tip, the class **Irregulares** is characterized by invariable presence of dissepiments in the intervallum, and commonly by tabulae, rods, bars, or septa (Fig. 4.5). Longitudinal rows of septal pores, where septa are present, curve upward from inner wall to outer wall. Inner edges of tabulae are higher than outer edges.

Archaeocyathids are poorly represented and not identified to genus in many college and university collections. Therefore, directions for their study in laboratory are very generalized

Table 4.1 Characteristics of the classes of the Porifera. Largely after Rigby in Boardman *et al.* (1987).

Class Demospongea	Sponges with skeletons of spongin, spongin and siliceous spicules, or fused opalline silica Spicules commonly monaxon, tetraxon, or polyaxon, but never triaxons Two spicule sizes generally present Cambrian–Recent
Class Hexactinellida	Sponges with skeletons of siliceous spicules that are almost always triaxons and are frequently fused Two spicule sizes (**megascleres** and **microscleres**) generally present Commonly make well-preserved sponge fossils because (1) skeleton is siliceous and (2) it may be fused Late Cambrian–Recent
Class Calcarea	Sponges that have calcareous (calcitic) spicular skeletons or porous nonspicular calcareous walls Spicules are characteristically monaxons and/or tetraxons Spicules are usually separate, not fused to one another at tips Typically only one size category of spicules is present Cambrian–Recent
Class Sclerospongea	Sponges with basal calcareous (aragonitic or calcitic) sheet-like skeleton overlain by living tissue that also secretes microscleres of silica along with spongin fibers Canal systems in many join dendritically to empty from an elevated outlet Ordovician–Permian?, Triassic–Recent
Class Stromatoporata	Apparent sponges (prior to 1970 assigned most commonly to Coelenterata) with basal calcareous sheet-like skeleton Lacking siliceous microscleres Early representatives with laminae parallel with substratum and pillars perpendicular to laminae Later with large cysts or with closely appressed vertical polygonal tubes containing tabulae or dissepiments Ordovician–Cretaceous

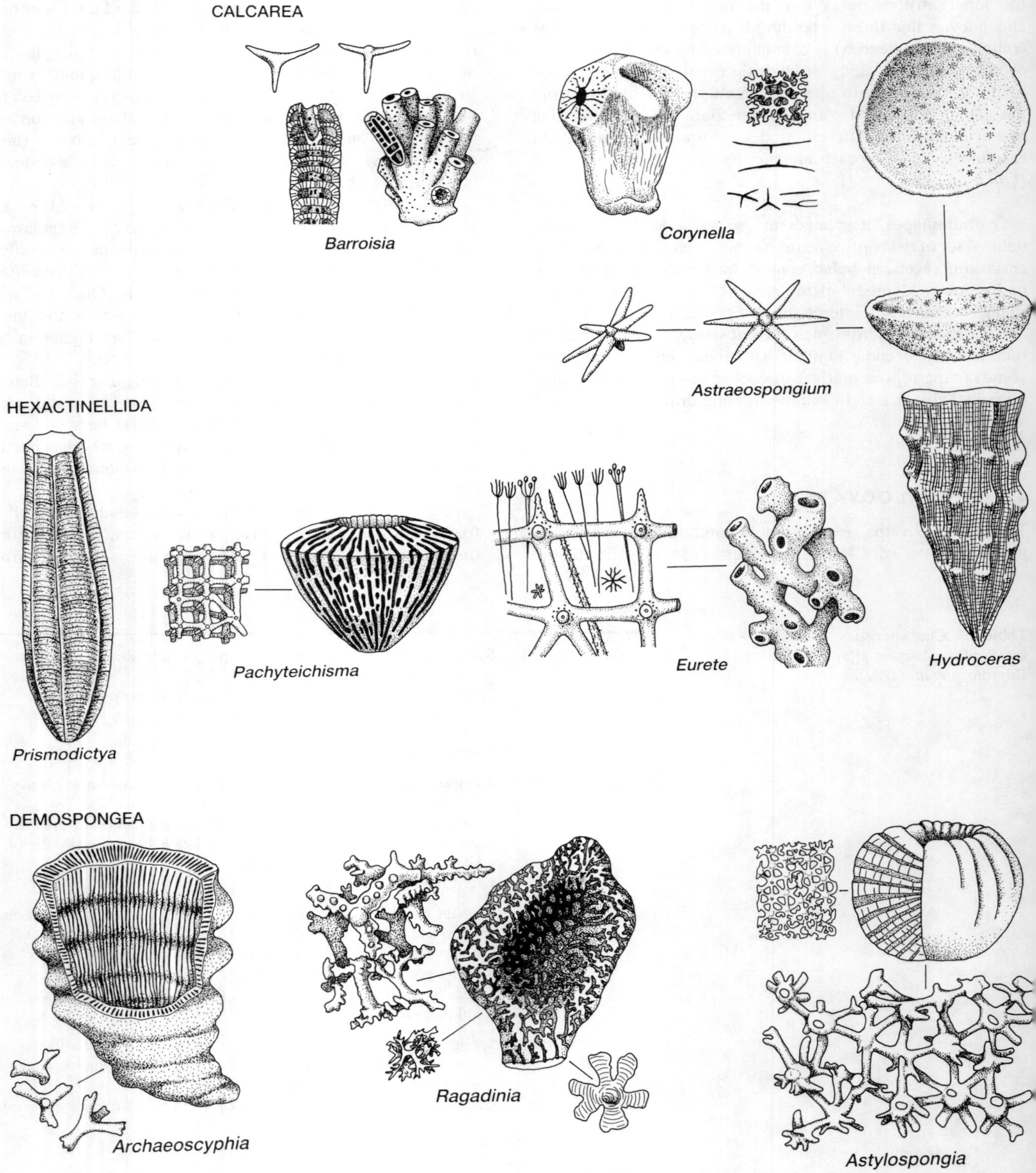

Fig. 4.3 Examples of the sponge classes Calcarea, Demospongea, and Hexactinellida. Whole specimens are slightly reduced from actual size; spicules are variably magnified. Largely after de Laubenfels (1955) with permission from the Geological Society of America.

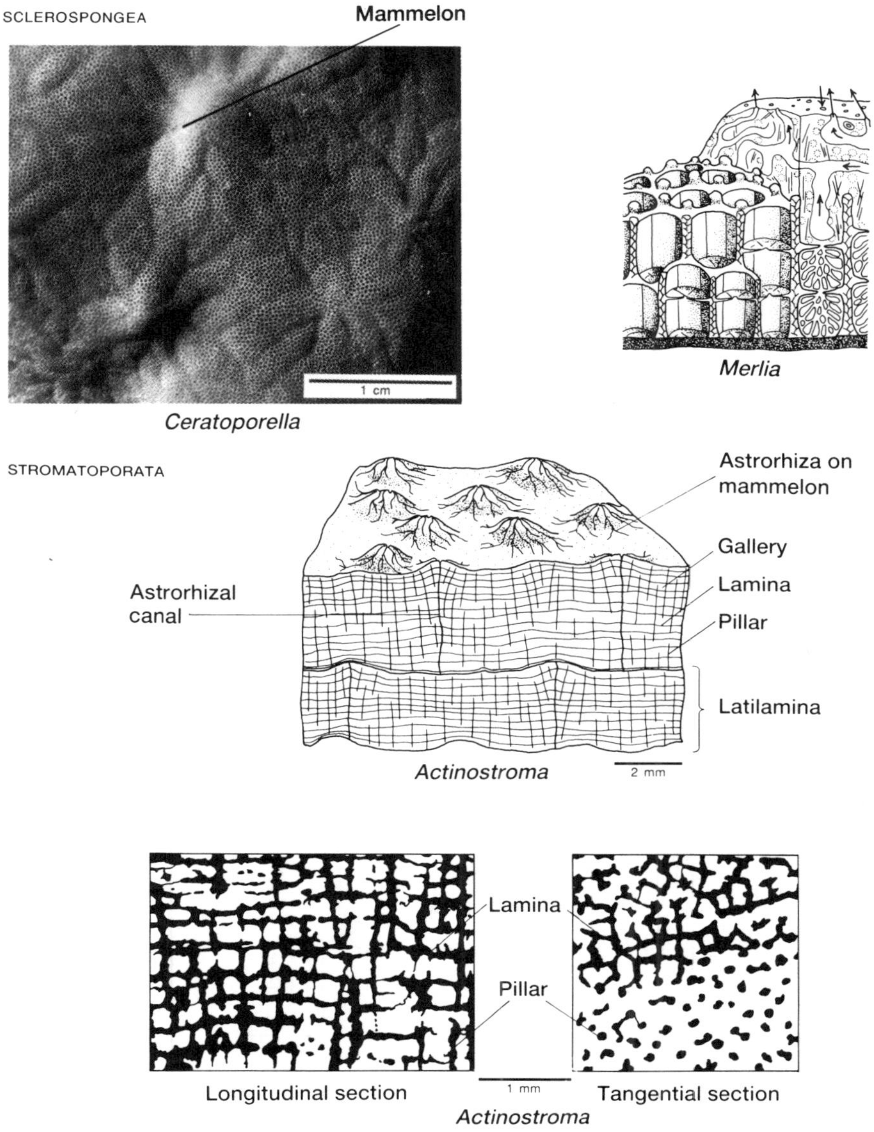

Fig. 4.4 Examples of the sponge classes Sclerospongea and Stromatoporata. After Kirkpatrick (1911), Stearn (1972, 1983), and Rigby in Boardman *et al.* (1987). *Ceratoporella* reprinted from Stearn (1972) with permission from Norwegian University Press (Universtetsforlaget AS), Oslo.

here. If a good assortment of archaeocyathids is available, your instructor may give an additional assignment. *(Question 5)*

4.4 SOME BIOMECHANICS OF SPONGE-LIKE ORGANISMS

Sponges interact with their fluid environment, as do all organisms in various ways. The porous structure of sponges allows the choanocytes to drive a flow of water through the body, which enters through the ostia and leaves through the osculum. However, erect sponges are constructed in such a way that they may use environmental conditions to do part of the work of driving their currents.

Keep in mind while working on biomechanics of spongelike organisms that sponges are active suspension feeders. The flagellae of their choanocytes pump a flow of water into the ostia and out of the osculum regardless of whether the sponge is in still water or in flowing water. Flow through sponges that is generated by their interaction with the movement of surrounding water is in addition to the choanocyte-driven flow.

Where a body projects into the path of flow of a fluid within a bounded area and thereby causes a reduction in cross-sectional area (Fig. 4.6), velocity past the obstructing object is increased. This is because

$$Q = VA \qquad (1)$$

where Q = volume/unit time, V = velocity and A = cross-sectional area. Where Q is constant as in the case given, a decrease in A must result in an increase in V. **Bernoulli's principle** relates increase in velocity to decrease in pressure. In other words, pressure is greater upstream of the obstruction in Figure 4.6 than it is over the obstruction. A simplified form of the Bernoulli equation for two points along a single stream is

$$1/2\rho\ (V_2^2 - V_1^2) + (p_2 - p_1) = 0, \qquad (2)$$

where ρ is the density, V_1 is the velocity at the upstream

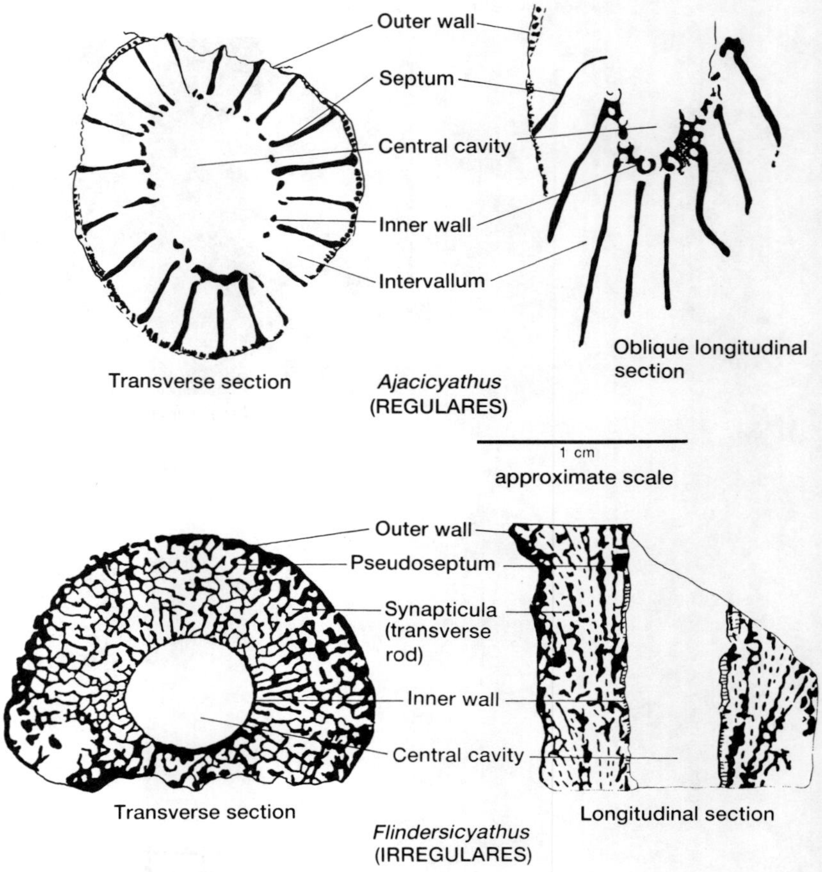

Fig. 4.5 Examples of regular and irregular archaeocyathids. Source: Energy Mines and Resources, Geological Survey of Canada. Reproduced from Handfield (1971) with permission of the Minister of Supply and Services, Canada, 1989.

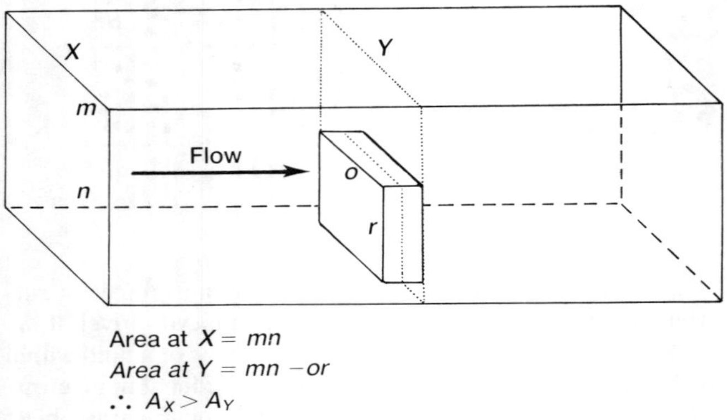

Fig. 4.6 Idealized flow through rectangular column with locally constricted cross-section.

point, V_2 is the velocity at the downstream point, p_1 is the pressure at the upstream point, and p_2 is the pressure at the downstream point. As V_2 increases, p_2 must decrease at a rate equal to one-half the square of V_2, where V_1 remains constant. The decrease in pressure over an obstructing sponge tends to draw water upward out of the osculum, which is replaced by water that is at higher pressure on the upstream and downstream sides.

The Bernoulli effect is calculated above as though the fluid lacks viscosity and therefore also lacks frictional resistance to flow. However, seawater and other fluids do have viscosity, which affects the efficiency of the Bernoulli effect but does not negate it. The viscosity and resulting frictional resistance to flow, however, generate a second phenomenon that must be taken into account. This is the reduction in velocity towards the stationary surfaces (in this case the sea floor) that bound a flowing fluid, which results in a zone of zero to very low flow velocity near the bounding surface. This zone of reduced flow is referred to as the **boundary layer**. Therefore an erect sponge, attached to the seafloor and growing out into flowing water, has inherently higher velocity past the osculum than past its lower portions. The higher velocity past the osculum is accompanied by lower pressure than exists nearer the base of the sponge, which causes a flow upward out of the osculum just as results from the Bernoulli effect. Both these flows operate in the same direction as that generated by the choanocytes, and they operate whether the ambient flow is unidirectional, caused by a current, or reverses due to wave motion.

(Questions 6 and 7)

Any erect sponge growing into flowing water will experience a lateral bending force. The intensity of the bending force may be measured by the bending moment (M_b), which is determined in such conditions by

$$M_b = F \cdot d \tag{3}$$

where F is the lateral force and d is distance along the structure from the point of its application. The lateral bending force due to flowing water is known as drag (D). For a cylindrical object

$$D = 1/2C_d \rho A V^2 \quad (4)$$

where C_d = the coefficient of drag and A is the cross-sectional area of the profile of the cylinder (height times diameter). This is a simplification and involves a term (C_d) that is difficult to determine accurately. However, since C_d will remain essentially unchanged in the comparisons here, it will not need to have values assigned. Substituting D for F in equation 3,

$$M_b = D \cdot d \quad (5)$$

or

$$M_b = (1/2C_d \rho A V^2)(d).$$

For simplicity, we will consider sponges to be simple cylinders growing upward into the water column. That way, deforming stress at the base of sponges of the same diameter is directly proportional to M_b, which can stand in the following comparisons as a surrogate for stress if the sponge is rigid and has not absorbed stress by bending. However, sponges are flexible to varying degrees, and to consider them as rigid bodies that translate all stress to the base does not fully match nature but does give a starting point for understanding mechanical forces that they experience.
(Questions 8 and 9)

FURTHER READING

Hartman, W. D., Wendt, J. W. & Wiedenmayer, F. 1980. *Living and fossil sponges*. University of Miami Rosenstiel School of Marine and Atmospheric Science, Sedimenta VIII, Miami, FL.

Hill, D., 1972. *Archaeocyatha*, in C. Teichert (ed.), *Treatise on invertebrate paleontology*, Part E, Vol. 1, 2nd edition. Geological Society of America and University of Kansas Press: Lawrence, KS.

Rigby, J. K. & Stearn, C. W. (eds), 1983. *Sponges and spongiomorphs*. University of Tennessee Department of Geological Sciences Studies in Geology 7.

Vogel, S., 1981. *Life in moving fluids*. Willard Grant Press and Princeton University Press: Princeton, NJ.

Vogel, S., 1988. *Life's devices*. Princeton University Press: Princeton, NJ.

EXERCISE 4 **Worksheet** Name

QUESTION 1

For each spicule sample provided, list the composition, number of axes, name of type, and possible class(es) to which the spicule-producing sponge belongs. Refer to Fig. 4.2 for appropriate terminology and to Table 4.1 to relate characteristics of the sponge classes to characteristics of spicules.

Specimen	*Composition*	*Number of axes*	*Type*	*Class(es)*

QUESTION 2

After megascopic and microscopic or handlens examination of the whole specimens, and if available, dissected specimens or sections, accurately draw and label a portion of a specimen of each of the available grades of sponge organization to demonstrate the characteristics of the grade.

Ascon *Sycon* *Leucon*

Genus:________ Genus:________ Genus:________

QUESTION 3

For each demosponge, hexactinellid, and calcarean, give: the class, overall shape of the sponge; grade; and composition, number of axes and type spicules, in the table on p. 58. Some genera may be represented by both whole specimens and thin sections, both of which must be studied.

				Spicules		
Genus	*Class*	*Shape*	*Grade*	*Composition*	*Number of Axes*	*Spicular type*

QUESTION 4

Examine each specimen of sclerosponge and stromatoporate both megascopically as a whole specimen, and microscopically in thin section or polished surface. Draw an area of each in longitudinal and tangential section that shows its characteristics and label them. See labelled examples in Fig. 4.4.

Genus:____________________ Genus:____________________

Genus:____________________ Genus:____________________

QUESTION 5

Specimens of the classes Regulares and Irregulares are provided, either as etched whole specimens, thin sections, or both. They may have generic names on the labels, but their class is not given. Determine the appropriate class for a specimen from each class (see Fig. 4.5 for examples), list the features labeled on Fig. 4.5 that are visible on each specimen, and place an asterisk to the left of each feature that is diagnostic for the class to which you assigned it. What is the composition and grain size of the sediment that surrounds each specimen? Is this consistent with the observation that archaeocyathids are most abundant in carbonate rocks with 10–20% clastic debris but tolerate up to 33% clastic debris, and that they are limited to rocks with little or no dolomite?

Specimen:

Class:

Characters exhibited

1 ____________________

2 ____________________

3 ____________________

4 ____________________

5 ____________________

Sediment:____________________

Consistent?____________________

Specimen:

Class:

Characters exhibited

1 ____________________

2 ____________________

3 ____________________

4 ____________________

5 ____________________

Sediment:____________________

Consistent?____________________

QUESTION 6

Place into a ring stand a vertical glass tube with the lower end of the glass tube 2 mm from the level surface on which the ring stand rests (Fig. 4.7). In the presence of a lateral flow of air, place a point source of smoke at the upper end of the glass tube, and then immediately beside the lower end of the glass tube.

Does air flow through the tube?____________________

If so, in which direction?____________________

What is the cause?____________________

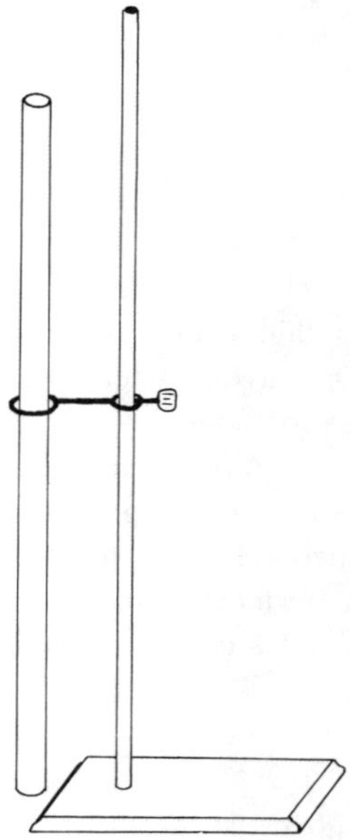

Fig. 4.7 Apparatus consisting of vertical glass tube held by clamp attached to ring stand, to demonstrate effect of lateral flow of fluid (air, in this case) on the similar fluid inside the tube.

QUESTION 7

Take a glass tube that has been bent in right angles so that one end stands up 20–30 cm above the surface on which it rests and the other end is parallel with the surface but no more than 1 cm above it (Fig. 4.8). With the lower opening upstream in a lateral flow of air, place a point source of smoke at the upper opening and then at the lower opening.

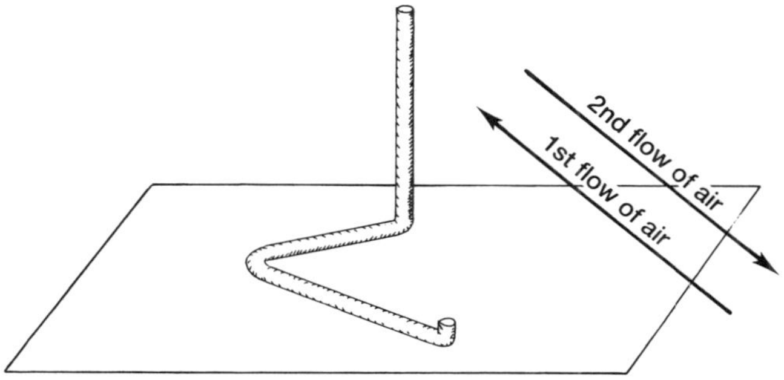

Fig. 4.8 Apparatus consisting of bent glass tube to demonstrate effect of lateral flow of fluid (air, in this case) on the similar fluid inside the tube when upper and lower openings are reversed in the upstream and downstream positions.

Does air flow through the tube?____________________

If so, in which direction?____________________

Now turn the apparatus so that the lower opening is downstream, repeat the operation and answer the same questions.

Does air flow through the tube?____________________

If so, in which direction?____________________

Does it matter whether the lower end is upstream or downstream? Why or why not?

____

Prairie dogs build a mound at one end of their burrows but not at the other (Fig. 4.9). What happens when the wind blows from the elevated end toward the level end?

____

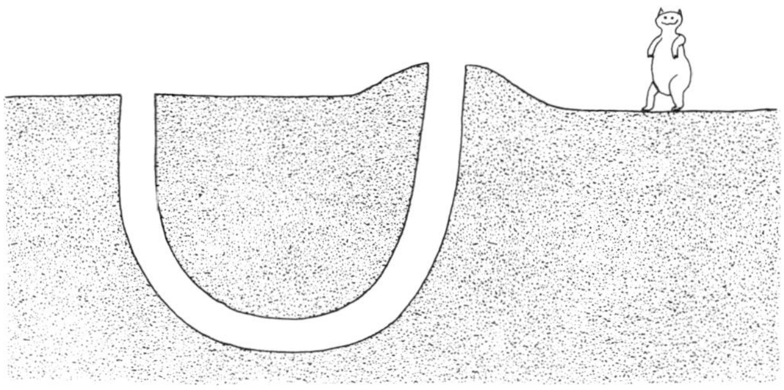

Fig. 4.9 Satisfied prairie dog beside its burrow.

What happens when it blows the opposite direction?

____

QUESTION 8

The table below gives the sizes of some cylindrical sponges and some velocities to which they may be exposed. Calculate the M_b at the base for each sponge at each velocity. Density of seawater = 1.03 g/cm^3, and for each consider d = height/2.

Sponge dimensions		*Flow velocity*			
Width	*Height*	*1 cm/s*	*2 cm/s*	*5 cm/s*	*10 cm/s*
1 cm	1 cm	______	______	______	______
1 cm	10 cm	______	103 g cm^2/s^2C_d	______	______
1 cm	100 cm	______	______	______	______
10 cm	0.1 cm	______	______	______	______
10 cm	1 cm	______	______	______	______
10 cm	10 cm	______	______	______	______

QUESTION 9

It has long been observed that tall, narrow sponges are more typically found in relatively gently moving water, whereas lower, broader and globular sponges may occur in more vigorously moving water. What would be the benefits and problems for a sponge in these two different conditions? What physical reasons might cause the growth forms to sort out roughly as they do? Keep the various points of Questions 6–8 in mind as you consider how to best answer this question.

____

____

____

____

____

____

____

____

____

____

EXERCISE 5 Cnidaria

5.1 INTRODUCTION

The phylum Cnidaria includes both solitary and colonial species and, together with Bryozoa and graptolites, is one of the important groups of invertebrate fossils that form colonies. A **colony** of invertebrate animals is one in which modular functional units are budded asexually (i.e. grown mitotically so are genetically uniform throughout) and continue to be connected by tissues, skeleton, or both. If each of the newly budded modular units becomes totally disconnected from its origin, then multiple genetically identical units result, and most zoologists do not consider that they constitute a colony. Colonies vary in their **integration**, that is, degree of interconnectedness and interdependence of modules within a single colony. At one extreme they are connected in the most minimal possible way, and at the other extreme — such as in *Physalia*, the Portuguese man-of-war (Fig. 5.1) — the modular units perform as organs with only one function within a colony that looks and behaves overall like an individual organism.

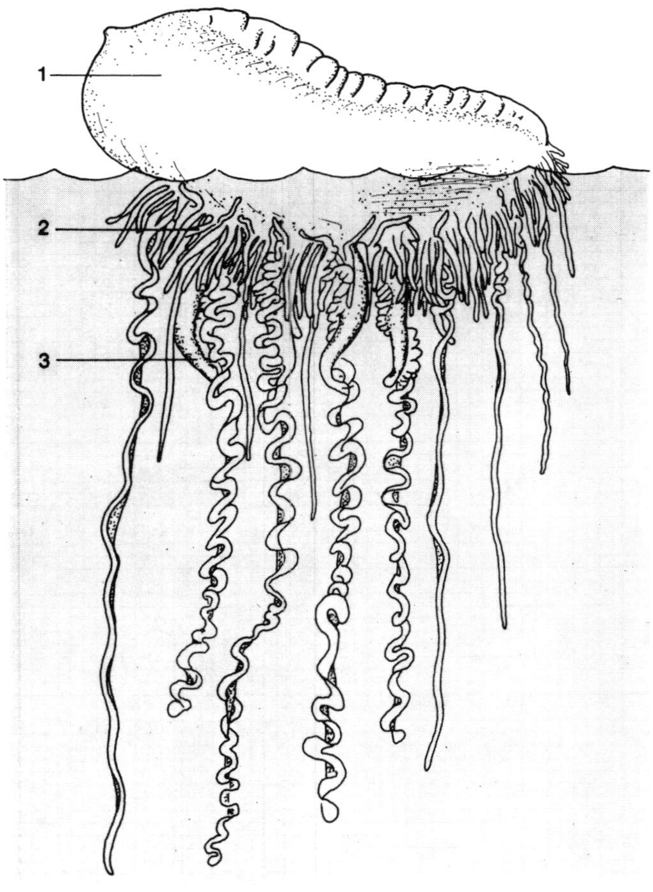

Fig. 5.1 The Portuguese man-of-war (*Physalia*), a colonial cnidarian with highly specialized polymorphs so that the colony resembles an individual organism. The polymorphs include a medusoid module, the **pneumatophore**, by which the colony floats; polypoid modules, some of which are specialized for feeding; elongate modules specialized for capturing prey; and other modules that are specialized for swimming and for reproduction. Approximately 50% reduction. After Hyman (1940).

The purposes of this exercise are:
1 to provide an introduction to the basic structure of cnidarians;
2 to examine skeletal morphology of various cnidarians; and
3 to investigate degrees of coloniality and their general evolutionary history within the zoantharian corals.

5.2 CLASSIFICATION AND MORPHOLOGY

The Cnidaria exhibit radial or modified radial symmetry and are characterized primitively by alternation of generations, with attached **polyps** alternating with free-swimming **medusae** (jellyfish). Polyps and medusae (Fig. 5.2) are organized on the same basis: a body wall consisting of two layers separated by a jelly-like material called the **mesoglea.** The body wall is penetrated by a mouth, which is the only opening into a central **enteron**, the cavity for digestion. Tentacles in a single or multiple row are extensions of the body wall and surround the mouth. The outer layer of the body wall contains several types of cells, including cells with **nematocysts**, which are stinging structures that are typically confined to the tentacles.

Aside from polyps being attached and medusae being free-swimming, the two cnidarian forms are characterized by different positions of the mouth: it is upward-facing in polyps and downward-facing in medusae. Among living cnidarians, only polyps produce mineralized ($CaCO_3$) or strongly sclerotized skeletons. The soft tissues of polyps and medusae are preserved only in exceptional circumstances.

Presence of nematocysts and the specialized cells (cnidocytes) that contain them in cnidarians are one of the features that distinguish this group from the generally similar phylum Ctenophora. (For many years Cnidaria and Ctenophora were considered to be subdivisions of a single phylum called the Coelenterata). Ctenophores exist only as medusae which, in contrast to cnidarians, have connective tissue and muscles within the mesoglea. The fossil record of ctenophores extends back to the Cambrian, but it is very sparse because they lack a durable skeleton and generally decompose even more rapidly than cnidarian medusae.
(Questions 1 and 2)

5.2.1 Hydrozoa

The class Hydrozoa is characterized in all instances by a shelf on the inside of the medusae at their elongated oral ends, typically by **polymorphism** (i.e. two or more distinct types of polyps), and commonly by tetrameral symmetry. Although the medusa may be the only form for many hydrozoans, many others alternate between polypoid and medusoid generations. Among the several orders of hydrozoans, two (Milleporina and Stylasterina; Fig. 5.3) have calcareous skeletons and

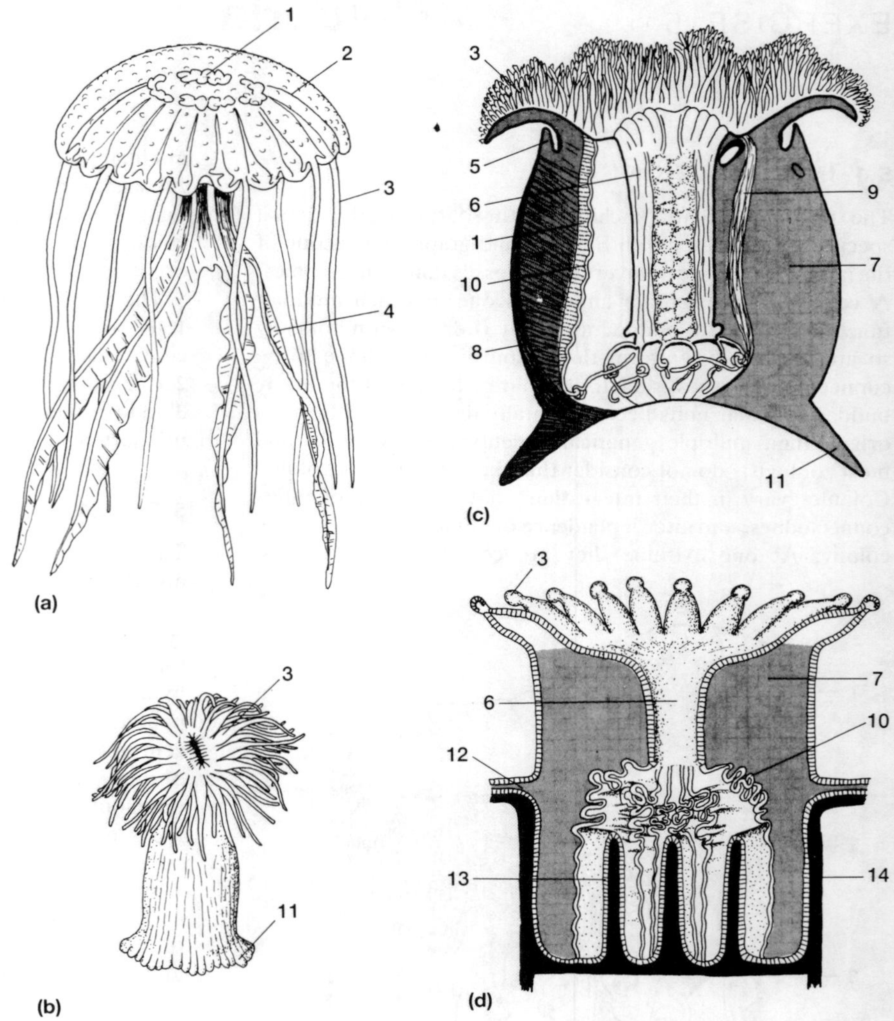

Fig. 5.2 Forms of Cnidaria. (a) Medusa, (b) and (c) **anemones**, which are a nonskeletal polyp, and (d) a coral, which consists of polyps and their skeletal supports. Diagrams (a)–(c) are approximately natural size; (d) is magnified approximately ×4. (1) Gonad, (2) radial canal, (3) tentacle, (4) oral arm, (5) collar, (6) pharynx, (7) complete septum, (8) parietal, (9) retractor muscle, (10) septal filament, (11) pedal disc, (12) connecting sheet, (13) skeletal septum, (14) skeletal theca. After Hyman (1940).

commonly are represented in Cenozoic tropical reef deposits. [Some paleontologists consider the Spongiomorphida (Triasic–Jurassic) to be hydrozoans, but the skeletal structure of pillars and connecting bars plus presence of astrorhizae suggest that they are more likely related to sclerosponges]. The other hydrozoans are preserved only under unusual circumstances.
(Question 3)

5.2.2 Scyphozoa

The class Scyphozoa has greatly reduced polyps in the sedentary generation but have conspicuous, generally tetrameral medusae. Scyphozoans are all marine and may be divided into two subclasses: the true jellyfish (Cambrian–Recent, Scyphomedusae; Fig. 5.2a) and extinct, cone-shaped organisms (Middle Cambrian–Lower Triassic, Conulata; Fig. 5.4). Few jellyfish are preserved. There are rather more Conulata, which are preserved as chitinophosphatic, tetrameral cones to elongate pyramids or subcylinders, occasionally with attachment disks at the apex. They commonly bear transverse or longitudinal markings. Placement of the Conulata among the Cnidaria is based essentially on the tetrameral symmetry of the generally thin skeleton and on apparent tentacles at the open end in some.
(Question 4)

5.2.3 Anthozoa

By far the most abundant fossil cnidarians belong to the class Anthozoa, all of which are polypoid and may be either solitary or colonial. Two of the three subclasses secrete $CaCO_3$ skeletons: the subclass Octocorallia generate minute calcified spicules, although some species also produce solid calcified branch axes; three orders (among eight) of the subclass Zoantharia secrete rigid calcareous skeletons; only the subclass Ceriantipatharia lack species that secrete $CaCO_3$.

The three zoantharian orders of calcified corals are Tabulata, Rugosa, and Scleractinia. Tabulate and rugosan corals characterize many Paleozoic marine deposits, but they went extinct at about the end of the Permian Period. Scleractinians first appear in Middle Triassic strata and extend to the present, forming the framework of reefs during parts of

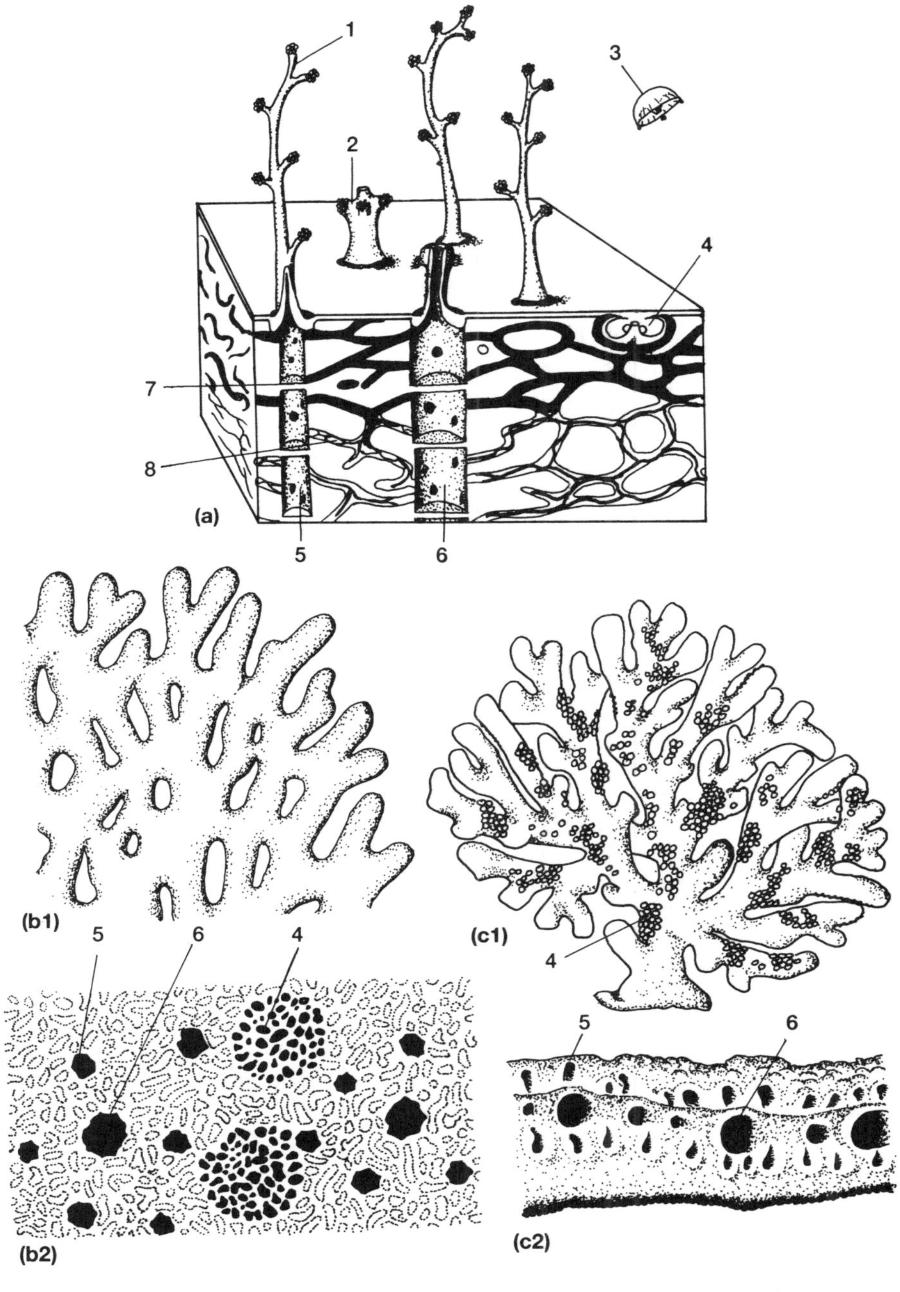

Fig. 5.3 Hydrozoans with calcified skeletons: (a) and (b) the milleporine *Millepora*, and (c) the stylasterine *Distichopora*. Diagrams (b2) and (c2) are surface views, magnified about ×20; (c2) is a lateral view of the edge of a flattened branch. (1) Dactylozooid, (2) gastrozooid, (3) medusa, (4) ampulla, (5) dactylopore, (6) gastropore, (7) tabula, (8) canal system. Modified from Bayer *et al.* (1956) with permission from the Geological Society of America.

Mesozoic and during Cenozoic time. Characteristics of the zoantharian orders are listed in Table 5.1.

Common structures and shapes of solitary rugose and scleractinian coral skeletons are shown in Fig. 5.5. The skeleton of a single coral polyp, whether solitary or within a colony, is termed a **corallite**. Many solitary corallites are high and conical, and depending on length and sinuosity may be:

Conical — continuously expanding, with a straight axis.

Horn-shaped (ceratoid) — continuously expanding, with a curved axis.

Cylindrical — parallel-sided after initial expansion, with a straight axis.

Scolecoid — parallel-sided, with a sinuous axis.

Patellate — horn-shaped with width much greater than height.

Discoid — broad, with flat bottom and gently domal upper surface.

Other conical to pyramidal shapes also exist, some with variously flattened sides bounded by pronounced corners.

Usually a bowl-shaped depression, the **calyx**, develops at the expanded top of a corallite, above the upper edges of radiating partitions termed septa. There are six **primary septa** in rugose and scleractinian corals, and in rugosans they are named in order of their appearance in young developing corallites: **cardinal**, **counter**, two **alar**, and two **counter-lateral** septa. In curved corallites, the cardinal septum is typically on

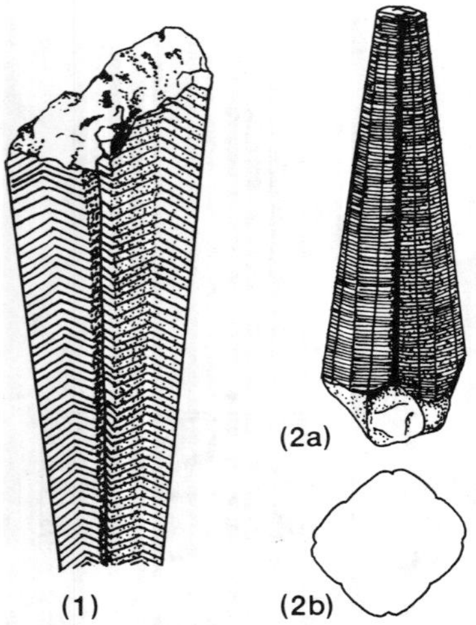

Fig. 5.4 Representative Conulata: (1) *Paraconularia*, (2) *Archaeoconularia*; (2a), exterior, (2b) cross-section. Approximately natural size. After Bayer *et al.* (1956) with permission from the Geological Society of America.

the concave side, with the counter septum opposite, on the convex side. Additional septa are added between the primary septa with continued growth of the corallite, equally in all six intervening spaces in scleractinians. However, additional septa are added only in four of the intervening spaces in rugosans, never in the two spaces between counter and counter-lateral septa. If one of the primary septa (usually the counter septum) stops growth while the other septa continue to extend upwardly, the resulting gap in septal spacing is termed a **fossula**. If septa extend all the way to the axis of the corallite, they may form a central elevated structure, the **axial boss**. Septa in diverse solitary corals exhibit a large number of morphologies: they may be **perforated**, bear minute ridges (**carinate**) or spines along their inner margin (**acanthine**), be **dilated** towards their outer edges or swollen (**rhopaloid**) along their inner margins, etc. Additional structures within solitary corals include horizontal plates that extend across the corallites (**tabulae**) and curved, overlapping plates (**dissepiments**) that are commonly concentrated just inside the bounding surface of the corallite.

The compound skeleton of colonial corals is referred to as a **corallum** (Fig. 5.6). The coralla of colonial corals may be broadly grouped into **encrusting**, **massive**, and erect **branched** forms. Corallites within a corallum exhibit diverse structures generally similar to those found in solitary corals. In addition corallites within a corallum may have solid or variously perforated walls, and they may be polygonal because they are compressed against one another or circular in cross-section with from little to a great deal of **colonial skeleton** that cannot be assigned to any particular corallite between them. Several types of coralla result (Fig. 5.6):

Phaceloid — corallites subparallel, separated except for points of origin and occasional lateral skeletal projections.

Cateniform — single series of parallel corallites, adjoining along a line of contact, single series commonly dividing into two similar series, resulting in open boxwork with continuous walls formed of the parallel corallites.

Cerioid — corallites polygonal, walls of adjacent corallites closely appressed and united; no intervening colonial skeleton between corallites.

Astreoid — corallites polygonal but with no lateral walls, septa terminating where walls would be; no intervening colonial skeleton between corallites.

Thamnasterioid — corallites polygonal, lacking lateral walls, septa continuous between adjacent corallites, no intervening colonial skeleton between corallites.

Aphroid — like asteroid but adjacent corallites united by a dissepimental zone; no intervening colonial skeleton between corallites.

Meandroid — corallites meandering, bounded by walls; intervening colonial skeleton present or absent.

Hydnophoroid — corallite centers arranged around protuberances; lateral walls incomplete or lacking; no intervening colonial skeleton between corallites.

Table 5.1 Characterization of the calcified orders of the subclass Zoantharia.

Tabulata	Typically colonial Calcitic skeletons of lateral wall (epitheca) and cross-partitions (tabulae) within the corallites Radially disposed longitudinal plates (septa, which are formed in association with infolded sheets of body wall termed mesentaries) absent or diminutive Lower Ordovician–Permian
Rugosa	Solitary or colonial Calcitic skeletons of epitheca, conspicuous septa, and usually tabulae and dissepiments, which are cyst-like structures formed inside the periphery of corallites Six primary septa, with subsequent septa added in four of the resulting spaces but not in the two dorsolateral spaces Lower Ordovician–Permian
Scleractinia	Solitary or colonial Aragonitic skeleton of epitheca, septa, tabulae, and dissepiments Six primary septa, with subsequent septa added in all six of the resulting spaces Middle Triassic–Recent

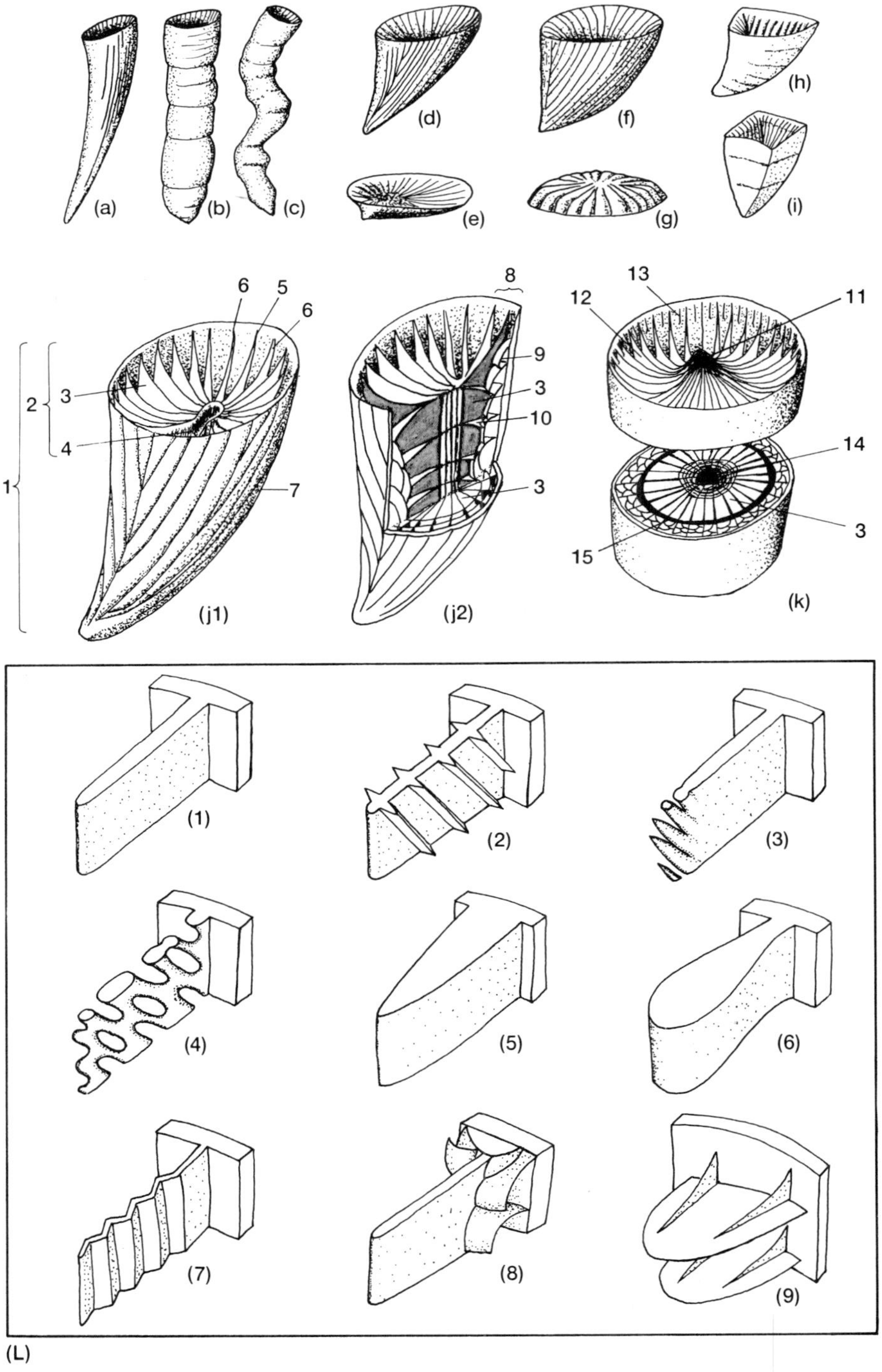

Fig. 5.5 Some morphology of solitary corals, particularly Rugosa. (a)–(i) *Types of corallites*: (a) ceratoid, (b) cylindrical, (c) scolecoid, (d) trochoid, (e) patellate, (f) turbinate, (g) discoid, (h) calceoloid, (i) pyramidal. (j) and (k) *General morphology of corallites*: (1) corallite, (2) calyx, (3) septum, (4) cardinal fossula, (5) counter septum, (6) counter-lateral septum, (7) epitheca, (8) dissepimentarium, (9) dissepiment, (10) tabula, (11) axial boss, (12) major septum, (13) minor septum, (14) columella, (15) stereozone. (l) *Types of septa*: (1) continuous, (2) carinate, (3) acanthine, (4) perforate, (5) dilated, (6) rhopaloid, (7) zigzag, (8) lonsdaleoid, (9) amplexoid. Modified from Moore, Lalicker & Fischer (1952).

Coenostoid — possessing colonial skeleton between corallites.
(Question 5)

5.3 COLONIAL INTEGRATION

The colonial state is inferred to be derived from the solitary state through increased connectedness of asexually budded individuals. This involves weakening of the "individuality" of the asexually budded units (**modules**), and increase in "individuality" of the colony. The interconnectedness and degree of interdependence of the modules within a colony is referred to as colonial integration. Commonly, development of functional groups of modules, termed **subcolonies** or **cormidia**, occurs within the colony. Zoantharian corals lack cormidia, but other cnidarians and other colonial invertebrates commonly exhibit them.

Table 5.2 is a listing of measures of colonial integration

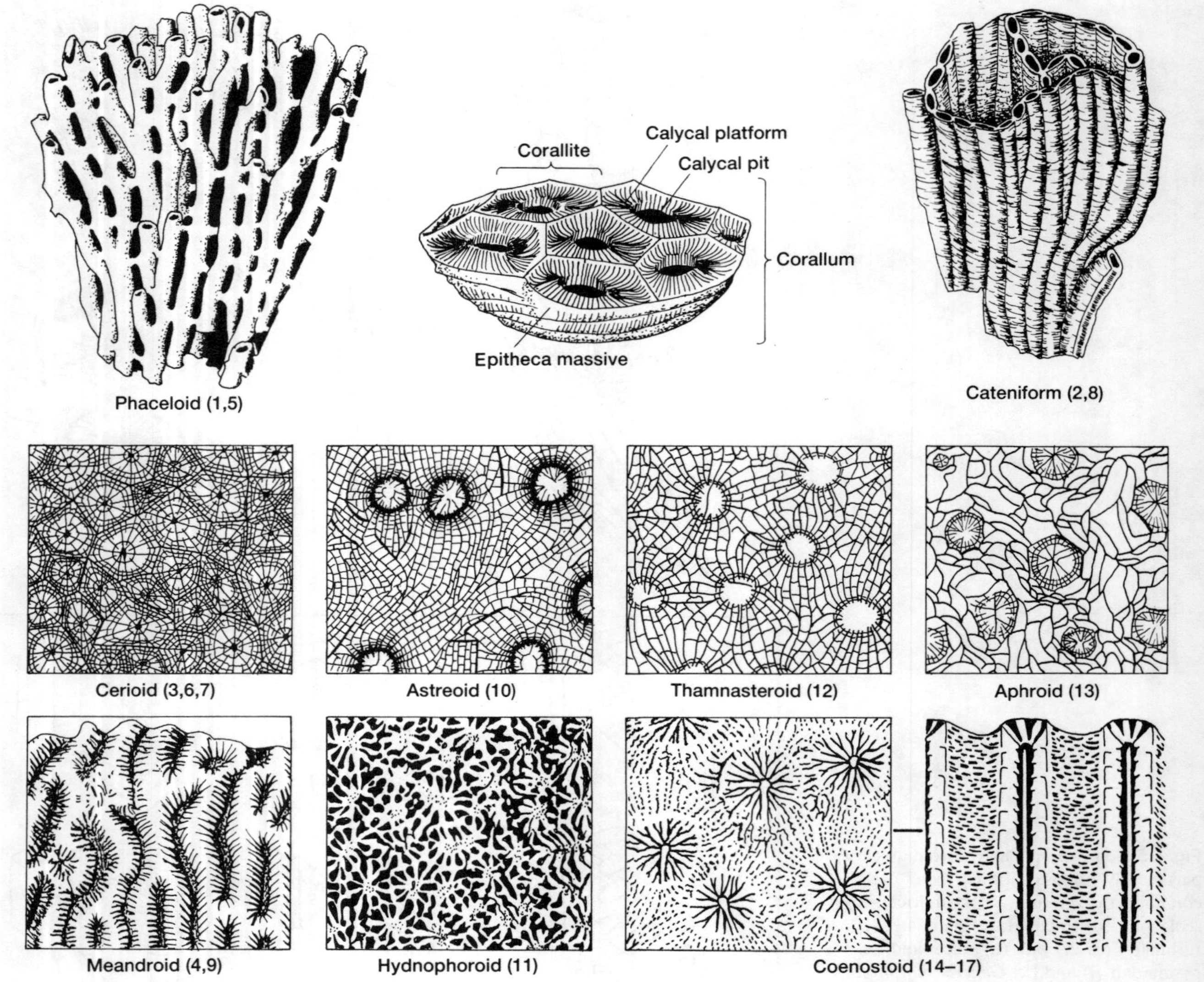

Fig. 5.6 Morphology and types of colonial corals. The lower two lines of illustrations show morphology of the various massive compound coral types as seen on the surface or in sections cut parallel with the surface, except the lower right figure, which is a longitudinal cut through a coenostoid coral. The cateniform morphology is represented by Halycites, which has recently been suggested to be a sponge (Kazmierczak, 1989); its assignment to the Anthozoa had not previously been questioned, and further investigation is needed to settle the matter. After Moore *et al.* (1952), Bayer *et al.* (1956), Coates & Oliver (1973), and Clarkson (1986).

in zoantharian corals. There are three different ways of measuring colonial integration given in Table 5.2: (1) skeleton, (2) tissues, and (3) polymorphism. The best indication of integration is to use all three types of measures in combination. For fossils, however, the degree of skeletal integration may be determined, but integration of tissues can only be inferred. Polymorphism is the existence within a colony of two or more distinctive types of modules, and each module belonging to that type is referred to as a **polymorph**. If only one type of module is present, the modules are said to be **monomorphic**. Polymorphism is rare in zoantharian corals and may be inferred only if the colony origin is preserved with different-appearing founding individuals, or if the colony is **dimorphic**, consisting of two distinctive types of modules. Presence of monomorphic corallites does not necessarily mean lack of polymorphism, because living colonies exhibit a functional difference for dimorphic polyps in monomorphic corallites. Therefore, polymorphism is demonstrated if dimorphic corallites are seen, but the converse is not necessarily true: monomorphic corallites do not demonstrate absence of polymorphism.

(Questions 6–10)

FURTHER READING

Bayer, F. M. *et al.*, 1956. Coelenterata, in R. C. Moore (ed.), *Treatise on invertebrate paleontology*, Part F. Geological Society of America and University of Kansas Press: Lawrence, KS.

Boardman, R. S., Cheetham, A. H. & Oliver, W. A. Jr,

Table 5.2 Scales of colonial integration in zoantharian corals. Morphologies of corals represented by the terms in scale A (skeleton) are illustrated in Figure 5.6. Modified from Coates and Oliver (1973).

	Inferred common position on the other scales (B and C) given below	
	B	C
A Skeletal scale		
Corallites bounded by imperforate wall with or without epitheca		
1 Phaceloid	1	1
2 Cateniform	2	1
(a) Dimorphic	2	3
3 Cerioid	2	1
4 Meandroid	2–4	1
Wall perforate		
5 Phaceloid	1–2, 3?	1
6 Cerioid	2–3	1
(a) Dimorphic	4	3–4
Separating wall incomplete or lacking		
7 Cerioid	3–4	1
8 Cateniform	2–3, 4?	1(3?)
9 Meandroid	4	1
10 Astreoid	2–3	1
11 Hydnophoroid	3–4	1
12 Thamnasterioid	4	1
13 Aphroid	4?	1
Coenosteoid (skeleton between polyps)	4	1
14 Corallite diameter > intercorallite distance; plocoid		
15 Corallite diameter < intercorallite distance		
16 Corallites vague or hardly distinguishable		
17 Meandroid coenosteoid		
B Tissue-level scale		
1 Polyps separated in space		
2 Polyp walls in apposition		
3 Interpolyp connections by body wall cells or tissue		
4 Confluent body activity		
C Polymorphic scale		
1 Monomorphic		
2 First-formed module of colony (protopolyp or protocorallite) different, all others uniform		
3 Dimorphic		
4 Dimorphic with functional difference		

1973. *Animal colonies – development and function through time*. Dowden, Hutchinson & Ross: Stroudsburg, PA.
Hill, D., 1981. Coelenterata, Supplement 1, Rugosa and Tabulata, in C. Teichert (ed.), *Treatise on invertebrate paleontology*, Part F. Geological Society of America and University of Kansas Press: Lawrence, KS.
Hyman, L. H., 1940. *The invertebrates: Protozoa through Ctenophora*. McGraw-Hill: New York.

EXERCISE 5

Worksheet

Name

QUESTION 1

Examine the preserved specimens of a medusa, an anemone, and a colonial coral with polyps preserved. Carefully draw the medusa and label all visible features. List the similarities and differences between the medusa and the preserved polyps of the anemone and coral.

Labeled drawing of medusa:

Generic name of medusa drawn: ______________________________

Comparison of medusa and polyps:

Similarities	*Differences*
______________________________	______________________________
______________________________	______________________________
______________________________	______________________________

QUESTION 2

Figure 5.7 illustrates Silurian favositid corals (Class Anthozoa, Subclass Zoantharia, Order Tabulata). Figure 5.7b is an exceptional specimen that must be compared with the more typically preserved specimen in Fig. 5.7a, with the favositid coral provided for study in the laboratory, and with the preserved coral that was examined for Question 1. What is unusual about the specimen in Fig. 5.7b? Describe the unusual structures and compare them in detail with their apparent analogs in the modern preserved coral.

Unusual structures in Fig. 5.7b: ______________________________

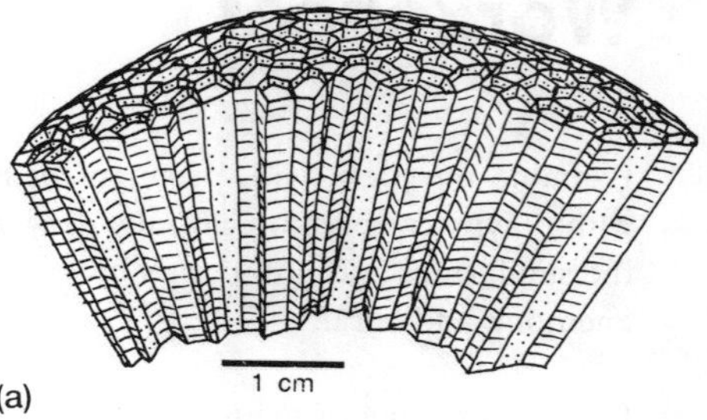

Fig. 5.7 *Favosites*. (a) Line drawing of typically preserved colony, (b) exceptional preservation. (a) After Bayer *et al*. (1956) with permission from the Geological Society of America, (b) courtesy of Paul Copper, specimen originally illustrated in Copper (1985).

Description and comparison: ______________________________

____

____

____

QUESTION 3

Study the example of a milleporine or stylasterine hydrozoan skeleton that is provided and answer the following questions.
Can cavities or pores that represent positions of polyps be identified?

____

Are these monomorphic (all of a kind), or are they polymorphic?

____

Examine the skeleton between the polyp positions, at about ×50 magnification. Is it (1) solid, (2) continuous but porous, or (3) spicular?

____

QUESTION 4

In the space below, carefully draw the conulatid that is available in the laboratory, and label the midline that longitudinally divides one of the flat surfaces (faces) and the interfacial angle where two faces meet at a 90° angle.

QUESTION 5

Examine the coral specimens provided for this question and give the information required for each on the table on p. 74–75. Range should be determined from the *Treatise on invertebrate paleontology*, Part F, and any supplements provided by the instructor.

Solitary corals

Specimen number	*Genus*	*Order (T,R,S)*	*Geologic range*	*Shape*	*Fosulae (present, absent)*	*Approx. number of septa*	*Special septal structures*	*Dissepiments (present, absent)*	*Tabulae (present, absent)*

Colonial corals

Specimen Number	*Specimen Genus*	*Order (T,R,S)*	*Geologic range*	*Colony form*	*Type corallum*	*Structures in corallites*	*Score from Table 5.2, scale A (see Question 6)*

QUESTION 6

For each of the colonial corals examined in this laboratory, score the level of skeletal integration in the spaces provided on the table for Question 5; enter the appropriate number (1–17) from list A of Table 5.2 that represents the skeletal organization that the specimen displays.

QUESTION 7

Figure 5.8 plots the distribution of most colonial coral types through the Phanerozoic. In the space below the figure, carefully plot the temporal range of each of the colonial corals examined in this laboratory, write the generic name for each, and indicate its state of skeletal integration in parenthesis. (*Lithostrotionella* has been plotted as an example.) Do the ranges that you plot fit within the ranges of the various states as given in Fig. 5.8?

___

___

___

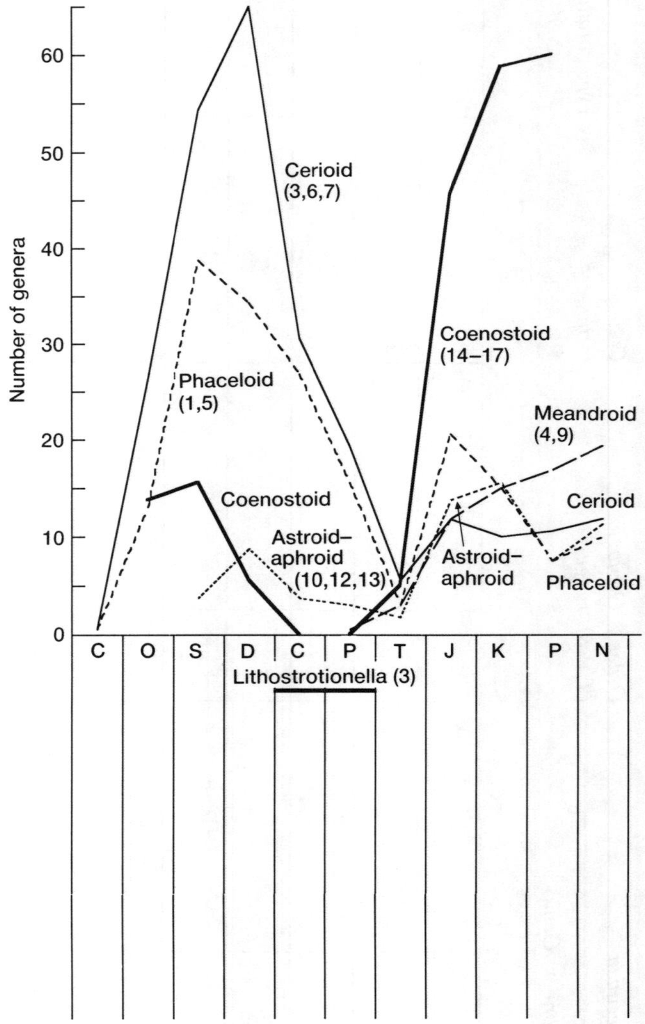

Fig. 5.8 Range and generic diversity of most colonial coral types. Skeletal integration states from Table 6.2 are indicated for each type. States 2, 8 and 11 are not included. From Coates & Oliver (1973).

QUESTION 8

What is the most probable age of a coral fauna consisting of Tabulata and Rugosa and dominated by approximately equal proportions of coenostoid and phaceloid but twice as many cerioid species?

____

QUESTION 9

On average, do the scleractinian corals seem to show an increase, a decrease, or no change in colonial integration through time?

____

QUESTION 10

Taking into account the skeletal features of colonial zoantharian corals seen in this exercise, the ranges of the three orders studied, Figure 5.8, and your answer to Question 9, write a description of the history of evolution of coloniality in zoantharian corals through the Phanerozoic.

____

____

____

____

____

____

____

____

____

____

____

____

EXERCISE 6

Arthropoda I: Chelicerata and Crustacea

6.1 INTRODUCTION

Arthropods are the most familiar invertebrates to most people, because of their success on land. Some of the features that make arthropods generally so easy to recognize include:

1 their segmented exoskeleton of **sclerotized** (i.e. hardened) chitin and protein;

2 the presence of paired, jointed legs that give the group its name (Greek for "*jointed leg*");

3 their bilateral symmetry; and

4 the organization of the body into an anterior region that in most is clearly distinct from the remainder of the body, marked by a conspicuous joint (Fig. 6.1).

In addition to such conspicuous visual clues, arthropods have other unifying characters such as:

1 similar patterns of early embryonic development;

2 reduction of the coelom to a sack surrounding the reproductive organs so that the fluid-filled body cavity is a **pseudocoel** (i.e. a body cavity that is not a coelom); and

3 the near-ubiquitous necessity to **moult** (shed) their rigid exoskeleton so that they can grow rapidly during a brief period before their newly developed exoskeleton hardens.

Important arthropod groups are much too numerous to be covered adequately in one or two laboratory exercises. However, many of the groups have left a poor or spotty fossil record. Some of these, such as insects, are exceedingly important in terrestrial ecosystems, and others, such as copepod crustaceans, are equally important in marine ecosystems. This exercise and the one that follows focus almost exclusively on the arthropods that have left a good fossil record, so omit many of the most important arthropods because they are not commonly encountered by most geologists and paleontologists.

The purposes of this exercise are:

1 to introduce the concepts of "arthropod";

2 to examine basic body regions and appendage specialization in modern chelicerates and crustaceans; and

3 to study examples of selected important fossil arthropod groups.

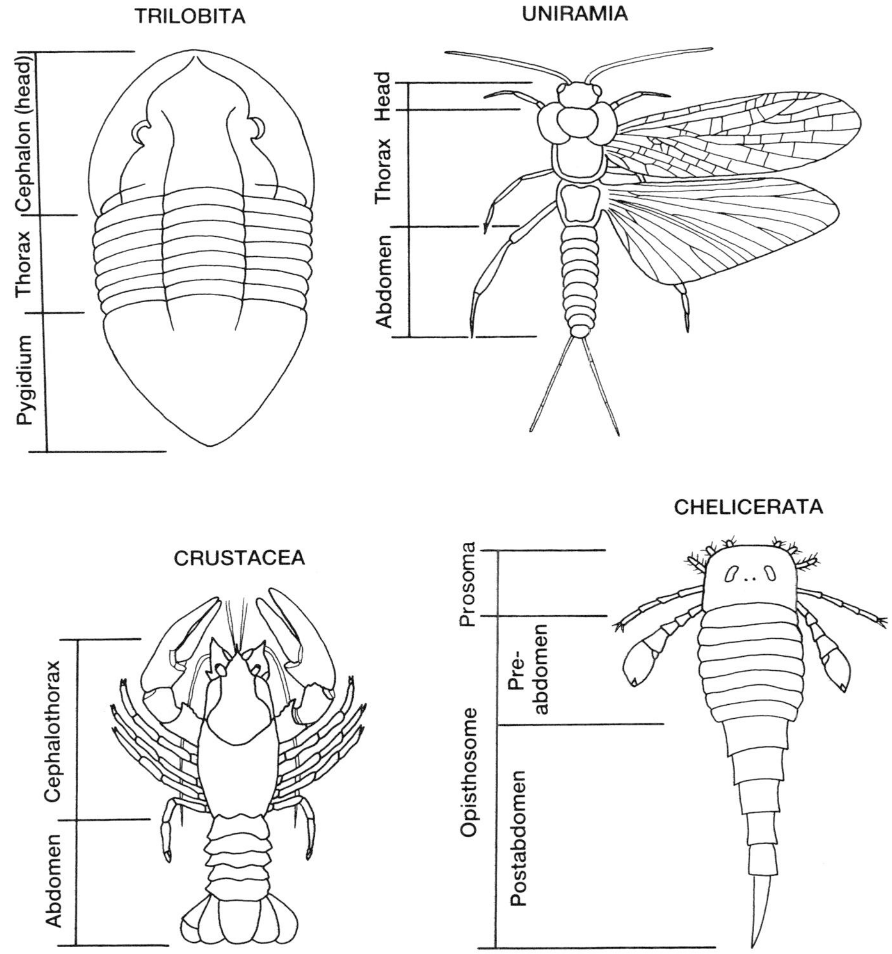

Fig. 6.1 Examples of the four major subdivisions of the arthropods, with nomenclature of the body regions. Modified from Størmer in Harrington *et al.* (1959) with permission from the Geological Society of America.

Table 6.1 Characteristics of the major subdivisions of the arthropods. Abbreviated and modified from Barnes *et al.* (1988) and Clarkson (1986).

Uniramia	Bilaterally symmetrical body divided into anterior "head" of acron plus three or four appendage-bearing segments and "trunk" bearing three to several hundred pairs of walking legs All appendages uniramous, those of head comprising one pair each antennae, mandibles, maxillae, and some a second pair of maxillae, but chelicerae never present One or two pairs of wings in one subgroup Exoskeleton may be calcareous Through, straight gut lacking digestive diverticula Primarily terrestrial Silurian–Recent
Chelicerata	Bilaterally symmetrical body divided into anterior prosoma of acron plus six appendage-bearing segments and posterior opisthosoma with highly modified appendages (not including legs) or opisthosomal appendages absent All appendages uniramous, those of head comprising one pair of chelicerae, one pair of pedipalps, and four pairs of walking legs Only chelicerae function as mouthparts Exoskeleton non-calcareous, commonly thin Opisthosoma may be segmented and divided into distinct anterior and posterior portions Respiration by external book-gills, internal lung-books, or tracheae (tubes perforating body through which gases diffuse) Through, straight gut, with many diverticula ?Cambrian, Ordovician–Recent
Crustacea	Bilaterally symmetrical body divided into head of acron plus five appendage-bearing segments and trunk of two to over 65 segments plus terminal telson One to seven trunk segments may be fused with head to form a cephalothorax Trunk normally divided into anterior thorax and posterior abdomen Cephalothorax only or up to entire body enclosed within dorsal carapace (continuously fused shield) Appendages biramous, those of head comprising two pairs antennae, one pair mandibles and two pairs maxillae and those of trunk highly variable Some limbs may be chelate, but no true chelicerae Exoskeleton commonly calcareous Trunk but not head normally shows external segmentation Respiration through gills on trunk or across body wall of carapace Through, straight gut with two diverticula in midgut ?Late Proterozoic, Early Cambrian–Recent
Trilobita	Bilaterally symmetrical body divided into anterior **cephalon**, medial thorax, and posterior **pygidium**, all organized into three longitudinal lobes (**axial** and two lateral **pleural** lobes) Appendages of one anterior pair uniramous antennae, followed by one pair biramous walking legs per segment Cephalon with central **glabella** (primitively furrowed by traces of segmentation), facial sutures, compound eyes, and doubled-under anterior and lateral margins Thorax articulated, divided into two to approximately 40 segments Pygidium of one to several fused segments Calcareous exoskeleton Respiration through gills on biramous appendages Through, straight gut with paired diverticula in cephalon Exclusively marine Early Cambrian–Late Permian

6.2 HOW MANY PHYLA?

Interpretation of the characteristics that unite arthropods differs among specialists. Some think that arthropods arose from **metameric** ancestors, with a body that consisted of a series of segments each of which repeated at least some internal organs. According to this interpretation, lack of repetition of many internal organs in arthropods results from loss of unnecessarily duplicated organs during phylogeny. In contrast, other specialists think that segmentation in arthropods arose during their evolutionary origin and never necessarily involved more than the superficial expression of metamerism of skeleton, muscles, and multiple paired appendages by which it is currently expressed. This is an interesting debate that will produce much fruitful insight before it is more-or-less settled. And it is related to an equally interesting disagreement: do the arthropods constitute a single phylum, or multiple phyla?

The unifying characteristics of arthropods that are listed in Section 6.1, plus some others not given here, caused them to be placed in a single phylum prior to the mid-20th century. Many specialists continue to adhere to that view, which means that they interpret the phylum as **monophyletic**, i.e. a single evolutionarily related group with a single ancestral origin. Others, however, are convinced that "arthropod" is a grade of evolution that is **polyphyletic**, which means that it arose more than once, perhaps many times (apparently during the Late Proterozoic). In order to have what they infer to be monophyletic phyla, therefore, these specialists divide arthropods into at least three independently originated living phyla (Crustacea, Chelicerata, Uniramia) and consider the extinct Trilobita to probably constitute yet another independently developed phylum. An abbreviated listing of characteristics of these major groups is given in Table 6.1.

The differing classifications appear superficially similar, mainly with shifts in hierarchical level for various groups. See the contrasting classifications given in Table 6.2. However, there is a profound conceptual difference implied in the shift of a name from subphylum to phylum. To place a group in the category subphylum indicates that the group is judged to be evolutionarily related to other subphyla within the phylum.

Table 6.2 Comparison of two classifications representative of the single phylum concept and of the multiple phylum concept of arthropods.

Classification of Robison and Kaesler in Boardman *et al.* (1987)	Classification of Barnes *et al.* (1988) supplemented by Clarkson (1986)
Phylum Arthropoda	
Superclass Trilobiotmorpha	Phylum Trilobita
Class Trilobita	
Class(es) uncertain	
Superclass Crustacea	Phylum Crustacea
Class Branchiopoda	Class Branchiopoda
Class Malacostraca	Class Malacostraca
Class Ostracoda	Class Ostracoda
Class Cirripedia	Class Cirripedia
Class Euthycaricinoidea	Class Remipedia
Class Copepoda	Class Copepoda
Class Cephalocarida	Class Cephalocarida
Class Mystacocarida	Class Mystacocarida
Class Branchiura	Class Branchiura
Superclass Chelicerata	Phylum Chelicerata
Class Merostomata	Class Merostomata
Class Arachnida	Class Arachnida
Superclass Pycnogonida	Class Pycnogona
Class Pantopoda	
Superclass Onychophora	Phylum Onychophora
Class(es) uncertain	Phylum Uniramia
Superclass Myriapoda	Subphylum Myriapoda
Class Archipolypoda	
Class Arthropleurida	
Class Diplopoda	Class Diplopoda
Class Chilopoda	Class Chilopoda
Class Symphyla	Class Symphyla
Class Pauropoda	Class Pauropoda
Superclass Hexapoda	Subphylum Hexapoda
Class Collembola	Class Oligoentomata
Class Insecta	Class Pterygota
Class Protura	Class Myrientomata
Class Diplura	Class Diplurata
Superclass Pentastomida	Phylum Pentastoma
Class Linguatulida	
Superclass Tardigrada	Phylum Tardigrada
Class Eutardigrada	
Class Heterotardigrada	

That is, they are "sister group" subdivisions of a larger evolutionary whole that necessarily are tied together through a single common origin. In contrast, recognition of groups as independent phyla leaves the possibility of interpreting them as evolutionarily related or — as in this case — evolutionarily independent in their origins.

As you work through this exercise and the one that follows, continually assess the similarities and differences among the chelicerates, crustacea, and trilobites, so that you can make an evaluation concerning whether or not arthropods are one or several phyla. The exercises do not include the uniramians, the majority of which are insects that are preserved only in exceptional circumstances.

6.3 BODY PLAN AND APPENDAGES: A COMPARISON BETWEEN CHELICERATES AND CRUSTACEANS

The body plan of all arthropods consists essentially of three divisions. There is an anterior region termed the **acron**, a segmented middle region that bears legs, and a posterior portion (**telson**) without legs. Among the most fundamental differences among the arthropod groups is the number of segments from the middle region that are fused with the acron to make a head-like portion of the body. The organization of appendages of the fused anterior portion and of those remaining in the middle portion behind the head-like structure is also important. The types of differences are illustrated here by comparison of a marine chelicerate and a marine crustacean (Figs 6.2 & 6.3).

6.3.1 Chelicerates

The anterior region of a chelicerate is termed the **prosoma**. It consists of the acron plus six segments from the middle region fused onto it, and it is covered by a continuous skeletal unit, the **carapace**, on the dorsal side. Appendages are located on the ventral side of the prosoma and are **uniramous** meaning that each consists of a single linear series of segments (Fig. 6.4). There are six pairs of appendages on the prosoma which is the reason for interpreting that six segments from the middle region, each bearing a single pair of appendages became fused with the acron to make the prosoma. From the anterior-most set, the appendages are:

one pair of **chelicerae**
one pair of **pedipalps**
four pairs of walking legs

The chelicerae vary from claw forms (**chelae**) to elongate shafts (**stylets**) and are the *only* appendages that form parts of the mouth in chelicerates. The pedipalps may be chelate feeler-like, or leg-like. There are no antennae or jaws in the chelicerates.

The posterior portion of a chelicerate is termed the **opisthosoma**, which has no legs. Any appendages that occur on the opisthosoma are highly modified for particular functions, none of which involve walking.

6.3.2 Crustaceans

The head region of a crustacean is formed of the acron plus five appendage-bearing segments that are fused with it. Appendages are located on the ventral side of the head and are **biramous**, which means that each consists of two linear series of segments beyond the single basal segment (Fig. 6.4). The five pairs of appendages of the head are:

one pair of small antennae (**antennules**)
one pair of larger antennae
one pair of **mandibles**
two pairs of **maxillae**

At least three pairs of appendages are parts of the mouth. They are the mandibles, which are basically crushing units and the two pairs of maxillae, which are chewing/shredding mouthparts. There may be accessory mouthparts consisting of up to three pairs of **maxillipeds**, derived from the appendages

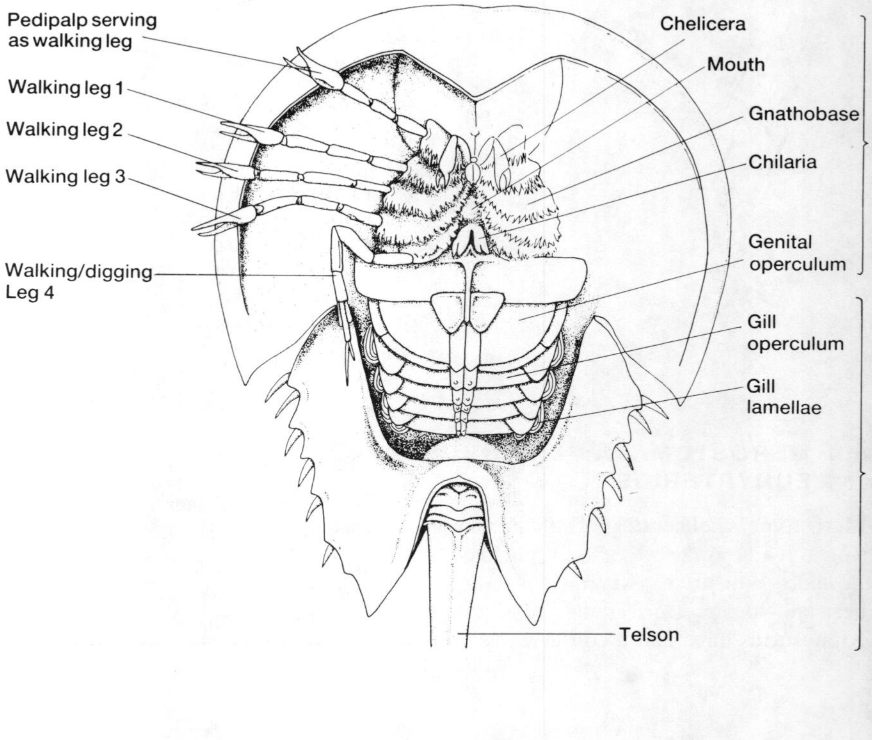

Fig. 6.2 Ventral surface of the merostome *Limulus*. After Barnes (1963), reprinted by permission of Saunders College Publishing, a division of Holt, Rinehart and Winston, Inc.

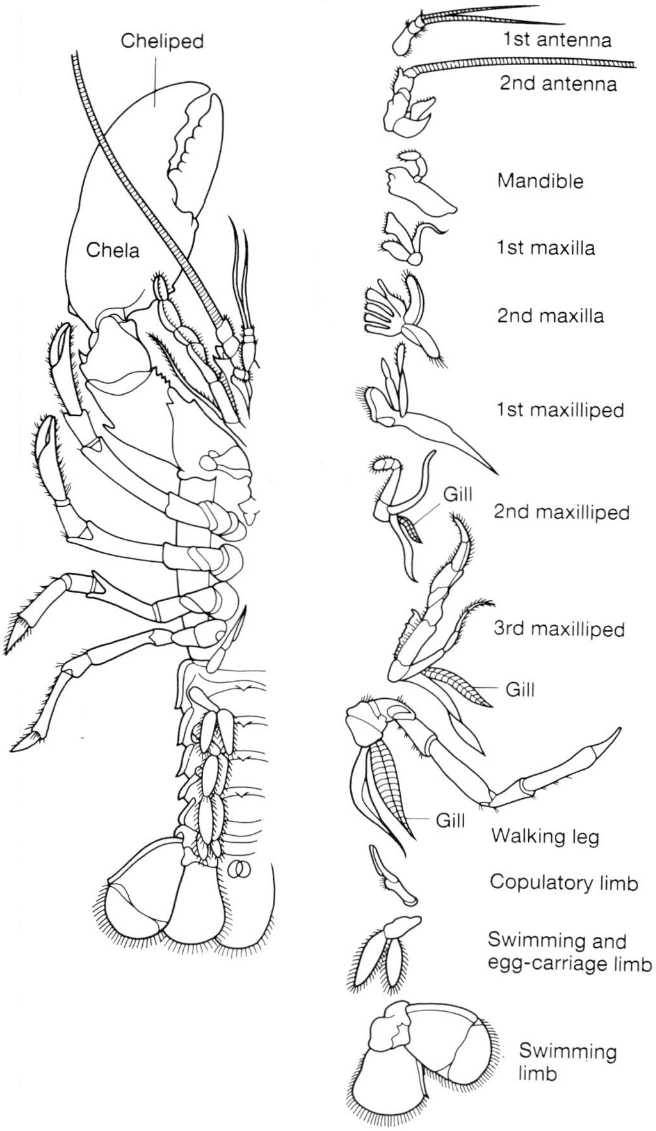

Fig. 6.3 Ventral surface and dissected sequence of appendages from a lobster. The sequence of crustacean appendages from 1st antennae to 2nd maxillae is always the same, but an additional few appendages from fused thoracic segments may also be incorporated into the cephalothorax. Redrawn from *Living Invertebrates* by Pearse & Buchsbaum (1987) Blackwell/Boxwood.

of the forward thoracic segments in the crustaceans that have the head and up to seven thoracic segments fused together into a **cephalothorax**.
(Questions 1–4)

6.4 MEROSTOMATA: XIPHOSURIDS AND EURYPTERIDS

Merostome chelicerates (Table 6.3) are represented by a single living order, the Xiphosurida, or horseshoe crabs. *Limulus*, which was examined in Questions 3 and 4, is the best-known xiphosurid genus and is quite typical of the group. Xiphosurids have had a conservative form and low diversity since their origin and apparently have been confined to shallow coastal waters throughout their existence.

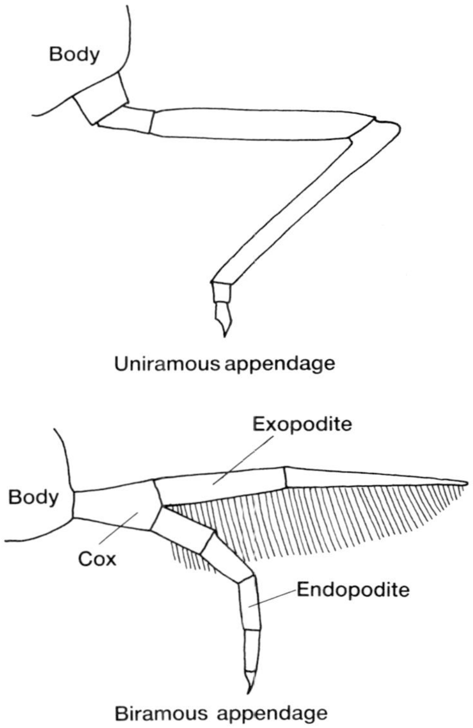

Fig. 6.4 Diagrammatic drawings of uniramous and biramous appendages of arthropods.

The extinct Eurypterida (Fig. 6.5), in contrast, were fairly successful for a substantial part of the Paleozoic Era, being moderately diverse and locally abundant especially during the Silurian Period. In addition, various species lived in hypersaline, normal marine, brackish, or fresh water. However, eurypterids became extinct by the end of the Permian Period. *(Question 5)*

6.5 CRUSTACEA: OSTRACODES AND BARNACLES

Crustaceans exhibit a much wider diversity of body forms than do the other major arthropod groups (Fig. 6.6). They comprise a large number of classes, including the Copepoda, which are highly important as low-order consumers within the plankton and are considered to be the "insects of the sea." The familiar crabs (decapod Malacostraca), which locally have left a substantial fossil record, also are a group of crustaceans. However, this part of the exercise focuses only on two crustacean classes that are common as fossils: Ostracoda and Cirripedia.

6.5.1 Ostracoda: morphology and taxonomy

Ostracodes are characterized by their bivalved carapace and small size, which is usually less than 1 mm and rarely greater than 10 mm. (This characterization does not include parasitic ostracodes, which may have strangely aberrant form as seen in Fig. 6.6b). The carapace usually is heavily calcified, hinged

Table 6.3 Classification of the Class Merostomata. Adapted from Robison & Kaesler in Boardman *et al*. (1987).

Class Merostomata	Aquatic Respiration through gills Prosoma with six pairs of locomotary and feeding appendages, plus one additional small appendage-like single or paired plate along the posterior margin Opisthosoma with up to 12 pairs of covered appendages Most opisthosomal appendages respiratory Cambrian–Recent
Order Xiphosurida	Prosoma approximately equal to opisthosoma in length Opisthosoma of up to 10 segments that may be all articulated, some fused, or entirely fused, plus terminal telson Cambrian–Recent
Order Eurypterida	Prosoma generally less than one-fourth the length of opisthosoma Opisthosoma composed of 12 articulated segments and telson Opisthosoma may be divided into relatively wide **preabdomen** of seven segments and relatively narrow **postabdomen** of five segments plus telson Ordovician–Permian

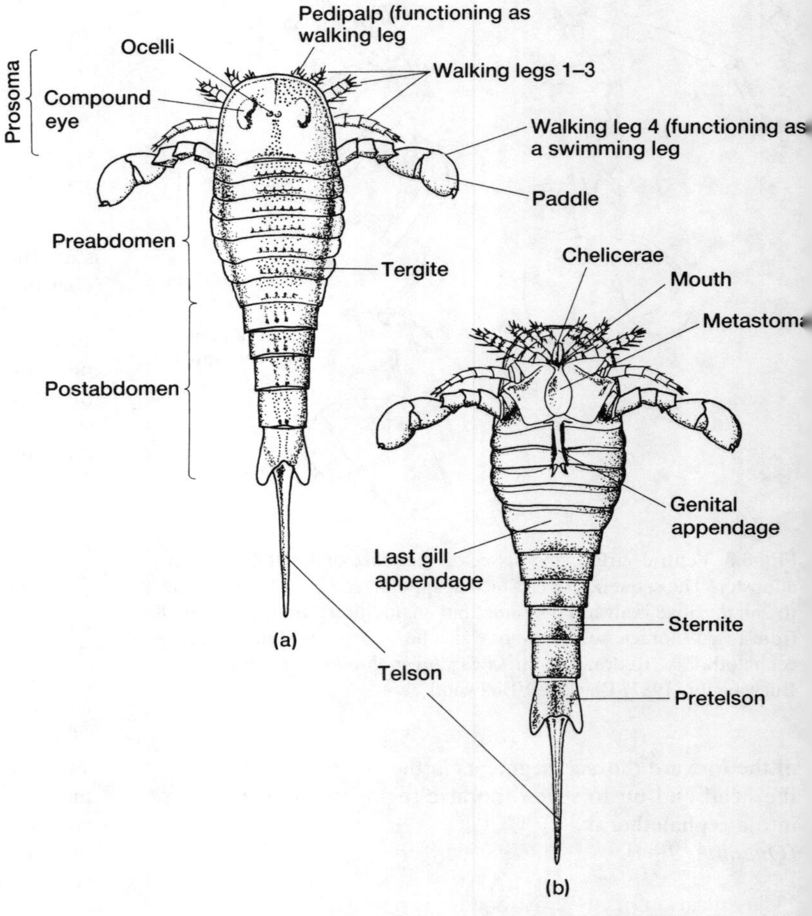

Fig. 6.5 Surface features and terminology of eurypterids. (a) Dorsal surface and (b) ventral surface. After Størmer in Størmer *et al*. (1955) with permission from the Geological Society of America.

on the dorsal side, and overlapped along the other sides. The valves vary from smooth and virtually featureless to highly textured by ridges, pits, bumps, and other structures.

The two valves of the carapace are organized and operated very much like those of bivalve molluscs. The valves are closed by contraction of adductor muscles that connect them. The adductor muscles attach to a small median area and generate scars on the shell that are important in taxonomy There is a ligament along the dorsal side with a resting stat that opens the valves when the adductor muscles relax. Th appendages protrude through the space between the ope valves for locomotion, feeding, and reproduction (Fig. 6.6g) The animal moults the bivalved shell and reforms larger bu similar valves as it grows.

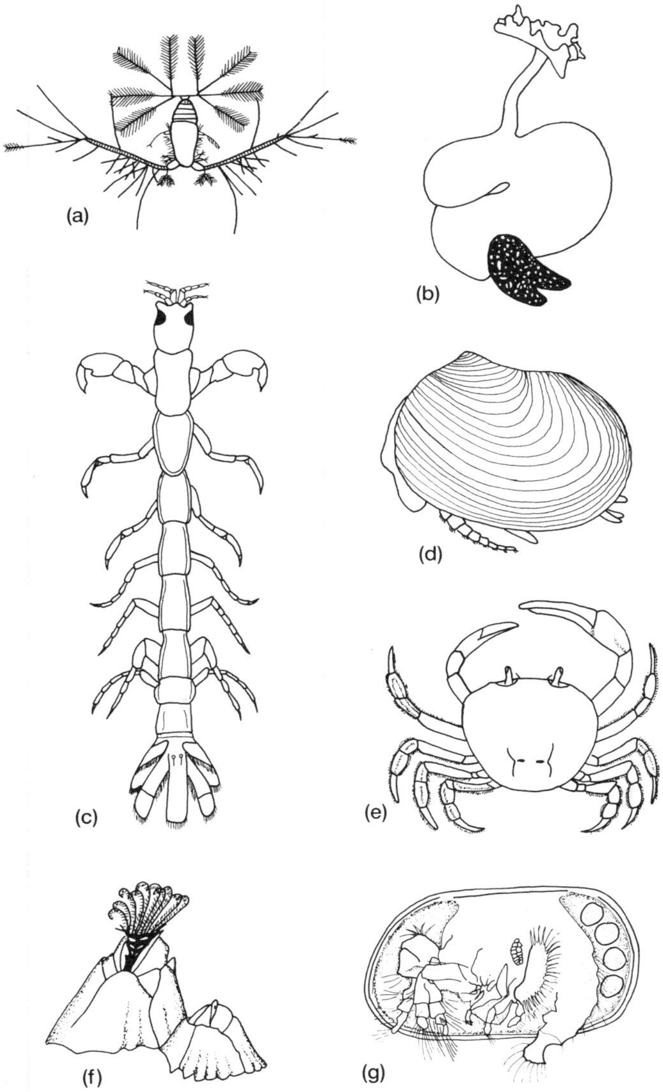

Fig. 6.6 Diverse crustaceans. (a) Planktic copepod about 1 mm across, (b) parasitic copepod about 0.5 mm, (c) an isopod about 5 mm long, (d) a bivalved conchostracan about 5 mm long, (e) a crab about 50 mm long, (f) a barnacle about 10 mm across, (g) an ostracod about 1 mm long. After Miner (1950) and Barnes *et al.* (1988).

The reproductive organs occupy a large proportion of the volume of ostracodes, and sexual dimorphism is common. This includes impression of the organs on the shell interior, shell-inflating brood chambers in females of some species, and differences in length : height ratio. The females commonly are higher relative to length than are males of the same species.

Classification of fossil ostracodes is based largely on shape of the carapace, muscle-scar patterns, hinge structures, eye tubercles, **pore canals** (perforations through the carapace that are usually perpendicular to the surface), and surface sculpture. Characteristics of the ostracode orders are given in Table 6.4, and representatives of the orders are shown in Fig. 6.7.
(Question 6)

6.5.2 Ostracoda: hardy organisms

Ostracode taxa are long-lived, which limits their usefulness for biostratigraphy although they are abundant and easily reclaimed from sedimentary rocks where megafossils have been entirely removed by over-zealous collectors or in certain facies even where megafossils are originally lacking. The paleoenvironmental significance of ostracodes, however, is quite high.

Various members of the group may be found in hypersaline waters, normal marine waters, brackish waters, or fresh waters. (A little mud from a lake shore and some dried vegetation placed together in a liter of water for a week or so will almost invariably generate a thriving community of ostracodes for study.) Ostracode assemblages from each of these environments can be recognized after moderate study of the group. Because of their abundance and almost universal occurrence, ostracodes can be used to recognize non-standard marine environments even by those with minimal acquaintance with them.

Basically, one expects to find representatives of less physiologically tolerant groups such as brachiopods, echinoderms, bryozoans, etc., along with ostracodes in normal marine environments (Fig. 6.8). Sediments deposited from brackish waters may contain pelecypods, gastropods, foraminiferans, stromatolites, and occasionally a few other less commonly fossilized organisms in addition to ostracodes. Hypersaline assemblages are characterized by ostracodes, stromatolites, fecal pellets generated by a limited number of gastropod species that grazed on muddy algal mats, and perhaps occasional gastropod body fossils. This is one of the most distinctive indicators of hypersaline waters, although it can also occur where water is brackish but environmental stresses (e.g. extreme temperature and salinity fluctuations) excluded other normally brackish-water inhabitants. However, if the assemblage occurs in association with even small amounts of evaporite minerals such as gypsum or anhydrite, inference of the assemblage as hypersaline is almost assured.
(Question 7)

6.5.3 Cirripedia

Most cirripedia are barnacles, including:
1 the acorn barnacles such as *Balanus* and its relatives, which grow attached to virtually any solid or semi-solid surface in the sea (Fig. 6.6f);
2 the goose-neck barnacles that grow most commonly on floating pieces of wood; and
3 rock-boring barnacles.

The group also includes some pretty nasty parasites, including one that infests crabs and essentially ramifies in fungus-like threads throughout the crab until it pretty well fills what was the crab's tissues. The focus here is on acorn barnacles, which range back to Upper Cretaceous rocks and are very common in shallow-water Cenozoic deposits.

The larval acorn barnacle attaches to the substratum by its antennules, which expand as a thin attachment disk as the sessile animal grows. The animal has a virtually non-existent head but feeds from suspended matter by its six pairs of throacic appendages that are modified into sweeping sieves of feather-like cirri. The body is encased in a continuously

Table 6.4 Characteristics of the orders of the Ostracoda. From Kaesler in Boardman *et al*. (1987).

Order Archaeocopida	Carapace weakly calcified or phosphatized, flexible Hinge long and straight or lacking Ventral margin convex Eye tubercle usually prominent Surface smooth, punctate, wrinkled, or ribbed, sometimes spinose Inner lamella not calcified Sexual dimorphism and muscle-scar pattern unknown Assignment to Ostracoda is insecure Cambrian, ?Ordovician
Order Leperditicopida	Carapace large, thick, well calcified, and usually smooth with long, straight, hinge and commonly prominent eye tubercle Inner lamella not calcified Large, complex muscle-scar pattern up to one-third height of valve with as many as 200 small, secondary scars Sexual dimorphism unknown ?Cambrian, Ordovician–Devonian
Order Palaeocopida	Carapace well calcified, surface usually smooth, typically with long, straight hinge, broadly convex ventral margin, inner lamella not calcified, and either a deep pit occupied by the adductor muscle scar or prominent lobes, sulci, and ridges Usually with pronounced sexual dimorphism including development of brood pouch by female No valve overlap Muscle scars poorly known Ordovician–Triassic, possibly extending into the Cenozoic
Order Podocopida	Carapace well calcified, surface smooth to highly and variously textured Valves of unequal size, one commonly overlapping the other Dorsal margin curved or, if straight, shorter than total length of carapace Ventral margin commonly concave Inner lamella partly calcified, radial pore canals present, commonly numerous Hinge without teeth or with various tooth arrangements Muscle-scar pattern usually with few secondary scars homologous within superfamilies Males usually more elongate than females Ordovician–Recent
Order Platycopida	Nearly smooth or ribbed, well calcified carapace with oval outline, strongly unequal valves with larger, usually right valve overlapping smaller valve around entire carapace in uninterrupted marginal furrow No calcified inner lamella or lobes and sulci Sexual dimorphism by inflation of posterior part of chamber of female No eyes Appendages modified for burrowing Muscle-scar pattern with numerous secondary scars among Paleozoic forms, fewer among Mesozoic and Cenozoic forms Ordovician–Recent
Order Myodocopida	Carapace commonly large, weakly calcified, with equal to subequal valves and no valve overlap, inner lamella partly calcified, and convex ventral margin Second antenna commonly extended through notch in anterior margin of carapace Muscle-scar pattern consisting of numerous elongate scars Ordovician–Recent

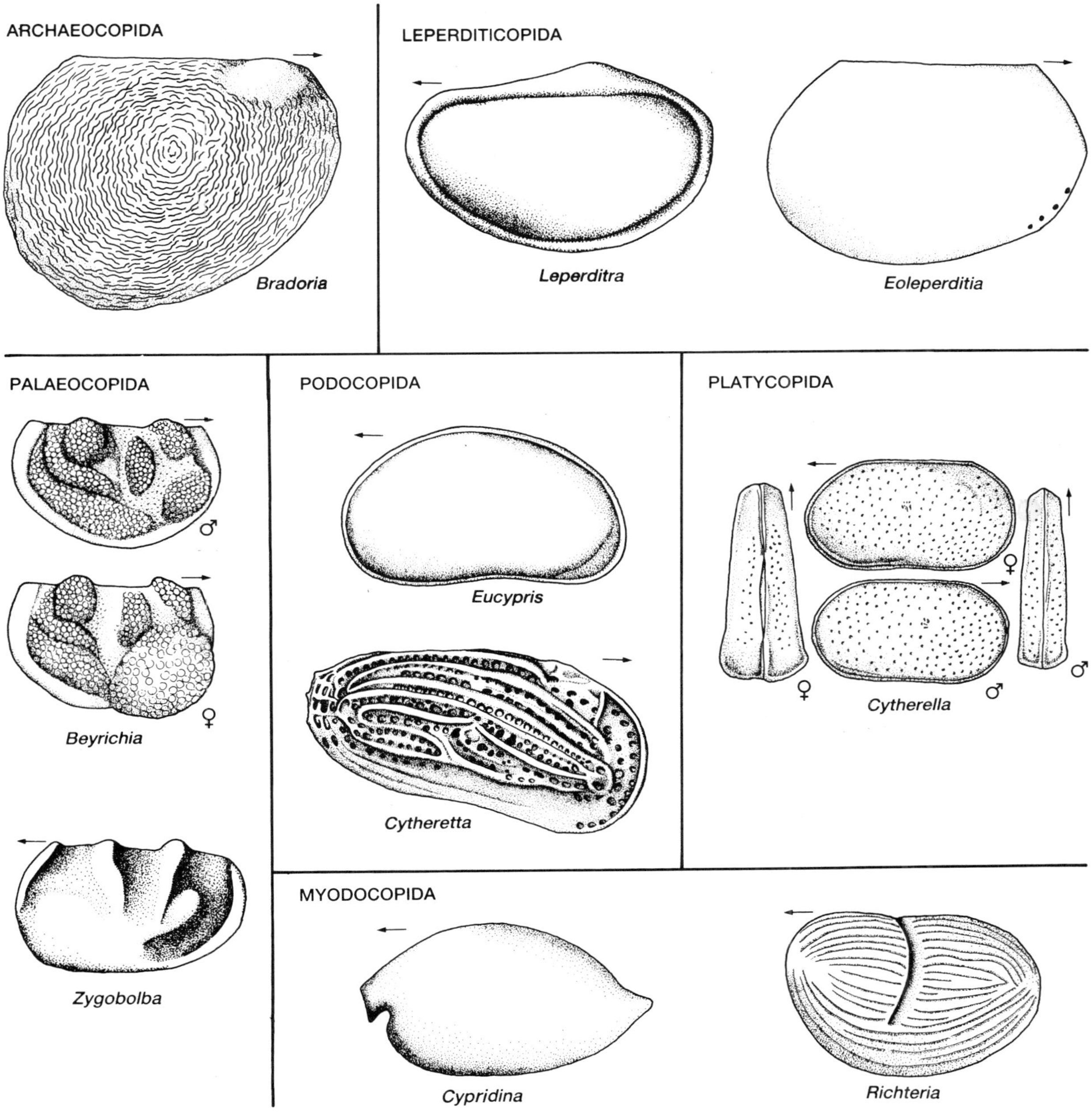

Fig. 6.7 Representative genera of the suborders of the Ostracoda. Most are shown in lateral view so that only one valve may by seen. Male and female are shown for two genera, including dorsal view of the paired valves for one of the genera. The arrow for each valve points toward the anterior direction. Most magnified about ×50; leperditicopids magnified about ×10. Redrawn from Benson *et al.* (1961) with permission from the Geological Society of America.

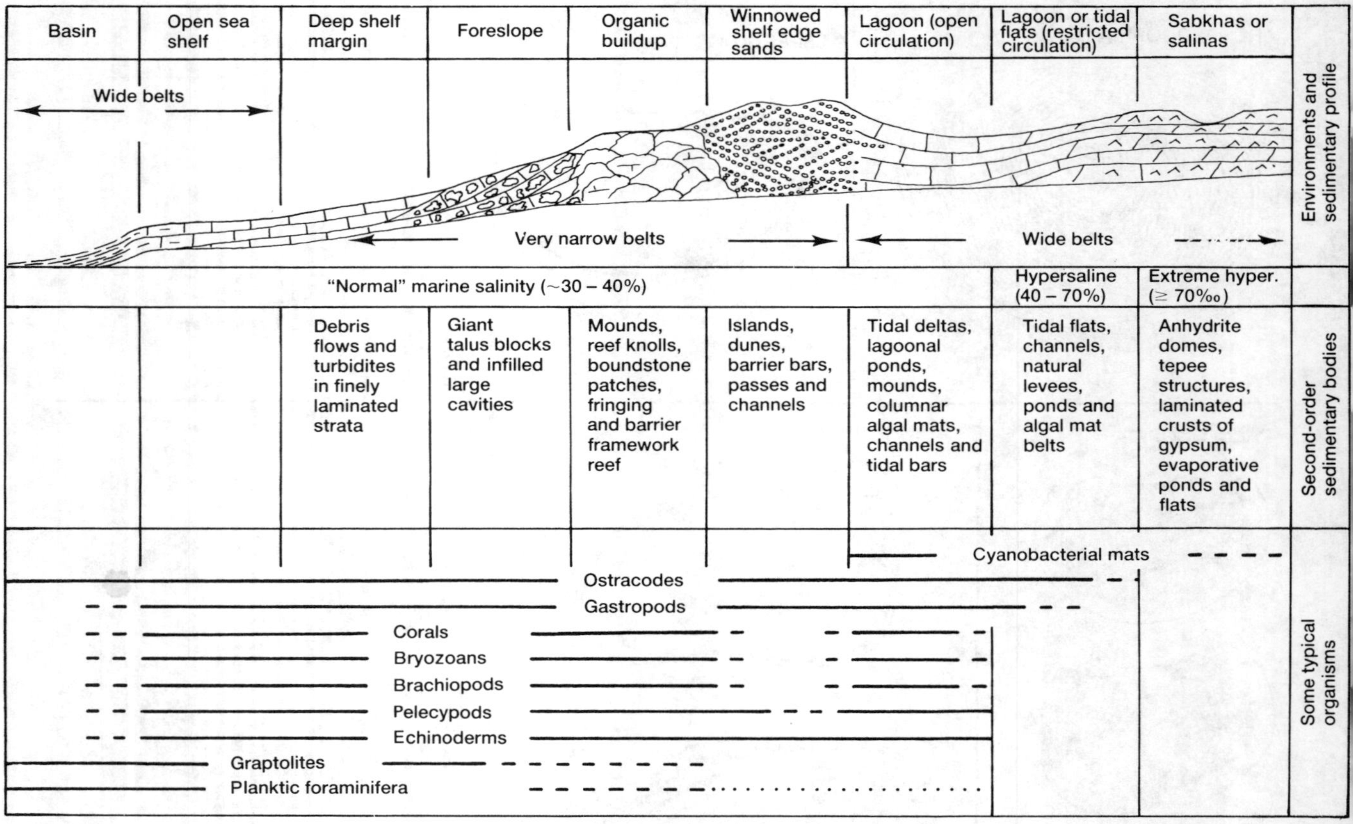

Fig. 6.8 Idealized sequence of carbonate facies belts and their characteristics as recognized by Wilson (1975). Approximate ranges of commonly fossilized marine invertebrates across the facies belts are indicated across the bottom of the figure.

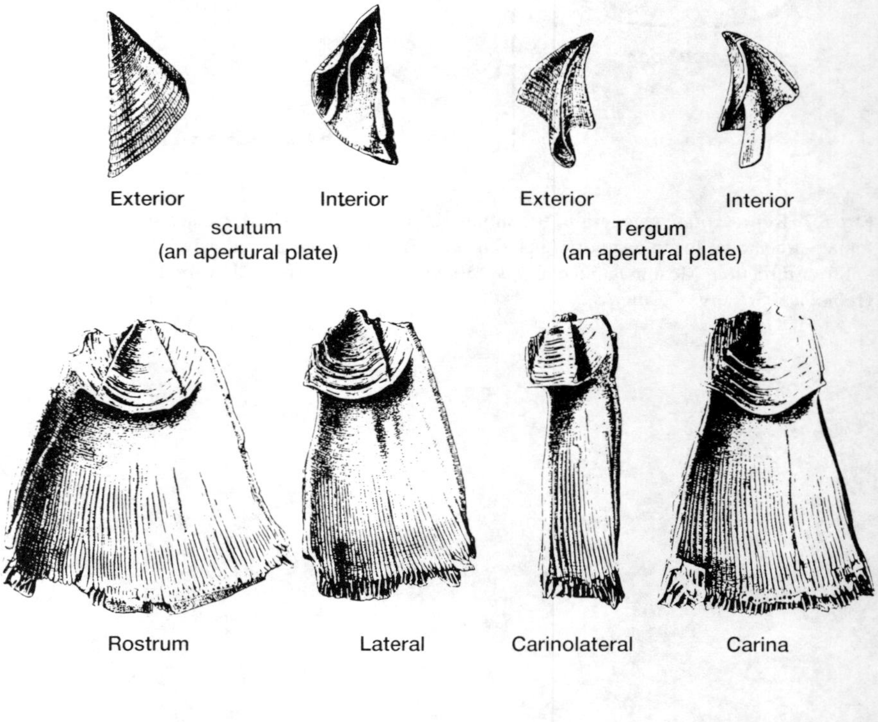

Fig. 6.9 Plates of balanomorph barnacles. Approximately natural size. From Clark (1904).

growing conical shell of calcium carbonate that consists of a single **basal plate**, up to eight tapered **compartmental plates** that make the open-topped cone, and four movable **covering plates** that seal the opening against predators and dessication when the barnacle is not feeding (Fig. 6.9). The covering plates usually are lost from the conical shell upon death of the barnacle, but they are well calcified and are common individually as fossils.
(Question 8)

FURTHER READING

Benson, R. H. *et al.*, 1961. Arthropoda 3. Crustacea, Ostracoda, in R. C. Moore (ed.), *Treatise on invertebrate paleontology*, Part Q. Geological Society of America and University of Kansas Press: Lawrence, KS.

Brooks, H. K. *et al.*, 1969. Arthropoda 4 (Crustacea exclusive of Ostracoda, Myriapoda, Hexapoda), in R. C. Moore (ed.), *Treatise on invertebrate paleontology*, Part R, Vols 1 and 2. Geological Society of America and University of Kansas Press: Lawrence, KS.

Clarke, K. U., 1973. *The biology of the Arthropoda*. Elsevier: New York.

Stormer, L., Petrunkevitch, A., & Hedgpeth, J. W., 1955. Arthropoda 2. Chelicerata with sections on Pycnognida and Palaeoisopus, in R. C. Moore (ed.), *Treatise on Invertebrate paleontology*, Part P. Geological Society of America and University of Kansas Press: Lawrence, KS.

EXERCISE 6

Worksheet

Name

QUESTION 1

Get one of the Crustacea (probably crayfish) provided for dissection, along with a dissection pan or board, a sharp scalpel, and forceps. First, place the crustacean with the dorsal side up and note its division into the carapace-covered cephalothorax and the segmented abdomen. Next, place it on the dissection surface with its dorsal side down and its ventral, appendage-bearing surface up. Carefully spread the appendages out towards the sides as much as possible and examine each in turn from the anterior end to the posterior end of the specimen. With reference to Fig. 6.3, list the appendages, give the number of pairs of each type, and give at least one apparent function.

	Appendage	*Number of pairs*	*Location (cephalothorax or trunk)*	*Apparent function*
Anterior				
Posterior				

QUESTION 2

With the appendages still spread to the sides, take the scalpel and make a *saggital section*, completely bisecting the specimen exactly along its plane of bilateral symmetry, from the anterior end to the tail. Hold one side of the bisected specimen with the forceps and carefully dissect out one of the biramous walking legs, being sure to include the gills as well as the walking structures. Make a careful, detailed drawing of the complete leg and name the parts as in Fig. 6.4. Keep the specimen at your workstation until you have completed Question 4.

QUESTION 3

Examine a specimen of *Limulus* (horseshoe crab), which is a representative chelicerate. First, place the specimen with the dorsal side up and note its division into carapace-covered prosoma, fused opisthosoma, and telson. Then, turn it over so that the mouth and uniramous appendages are visible. With reference to Fig. 6.2, list the appendages, give the number of pairs of each type, and give at least one apparent function:

	Appendage	*Number of pairs*	*Location (prosoma or opisthosoma)*	*Apparent function*
Anterior				
Posterior				

QUESTION 4

Compare the number of appendages on the cephalothorax of the crustacean with the number on the prosoma of the horseshoe crab. Which has the greater degree of differentiation among these appendages?

____

____

In what ways are the pedipalp and walking legs of the horseshoe crab designed to assist in feeding?

____

____

QUESTION 5

An actual specimen or replica of a eurypterid is provided for study. Compare this specimen with the *Limulus* studied for the previous two questions, and fill in the table that follows. Eurypterid specimens are commonly incomplete or have critical information not visible, which may result in the necessity to write NA (for not available) in some places in the following table.

	Limulus	*Eurypterid specimen*
1 Shape of anterior outline of prosoma	______	______
2 Ratio prosoma : opisthosoma lengths (not including telson)	______	______
3 Number of segments in opisthosoma	______	______
4 Opisthosoma fused, or segmented	______	______
5 Telson present, or absent	______	______
6 Number of pairs of prosomal appendages	______	______
7 Ratio width:length (not including telson)	______	______

QUESTION 6

Your instructor will provide ostracodes from two or more orders. Each should be examined at ×20–×40 magnification through a stereomicroscope. For each specimen or suite of specimens in a single species, list the visible characteristics that are consistent with the characteristics of the order to which it belongs.

Genus	*Order*	*Characteristics of order shown by specimen*
______	______	______
______	______	______
______	______	______
______	______	______
______	______	______

QUESTION 7

The columnar section on the following page is taken from Wilson (1975) and represents two series of deposits from the Devonian Duperow Formation of the Williston Basin in North Dakota. Each series of deposits represents a single cyclic progression of environmental conditions. By comparison with the distribution of organisms as shown in Fig. 6.8, draw horizontal lines to the left of the columnar section where organisms in the deposits represent a change from one environment to another, and label the environments represented. Does the cycle represent progressive shoaling, or progressive deepening as one moves upward across younger deposits?

QUESTION 8

Examine the barnacle-encrusted specimen provided. Are all individual barnacle skeletons bilaterally symmetrical?

______

If not, what is the cause of the asymmetry?

______

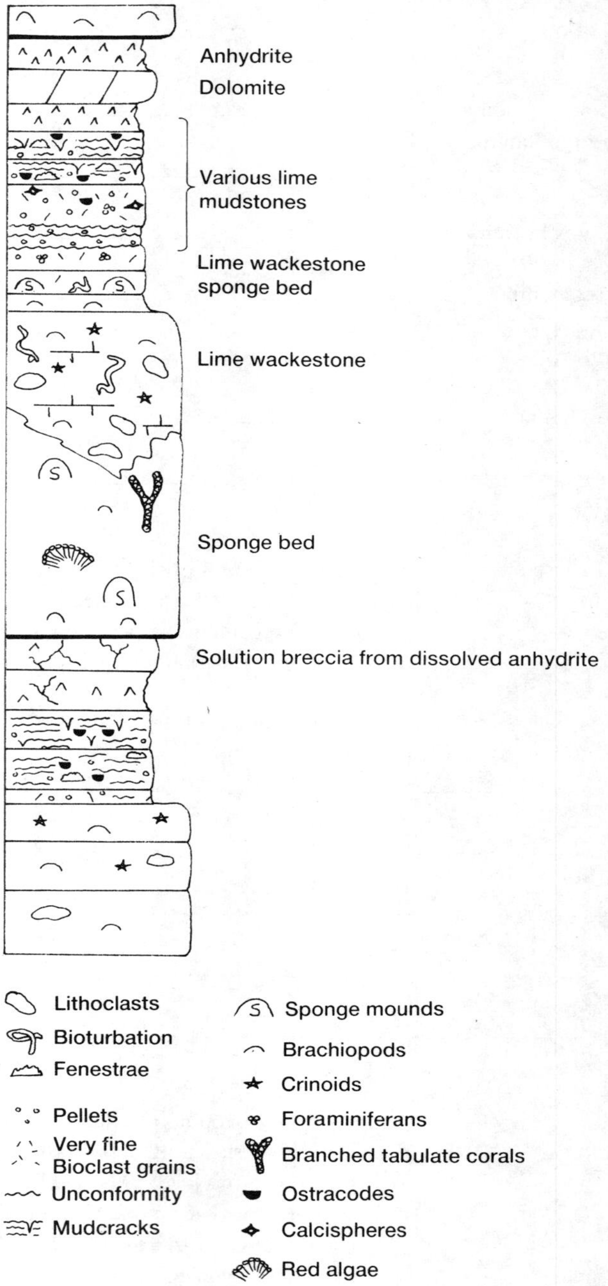

Is there a perceptible pattern to the distribution of the barnacles on the substratum? (e.g. apparently "random," in rows, obviously clumped, uniformly spaced)

Distributional patterns vary depending on microenvironment, density of recruitment, and tendency of individuals to settle nearby or spaced away from individuals already present. What seems to have been the determining factor(s) that caused the pattern in the specimen that you are examining?

EXERCISE 7

Arthropoda II: Trilobita

7.1 INTRODUCTION

Trilobites are the most common megascopic arthropod fossils in Paleozoic marine rocks. (Ostracodes are actually more widespread and abundant, but they usually are not conspicuous and are overlooked because of their size.) The organization of the exoskeleton into three longitudinal lobes gives the group its name. The three lobes are an **axial lobe** centered over the digestive tract and on each side a single **pleural lobe** that covers the appendages.

Trilobites are the most abundant of all easily visible fossils in Cambrian rocks, and they are diverse and abundant in Ordovician rocks as well. However, they began a decline by the Silurian Period and by Late Carboniferous time were represented by very few genera. They apparently went extinct near the end of the Permian Period, although more than one marine biologist has hoped to be the one to find living trilobites in some deep-ocean refuge as vigorous exploration of the deep-sea fauna has been carried out in the 20th century.

Unexcelled for Cambrian biostratigraphy, trilobites are also quite useful for correlation of Ordovician through Devonian rocks, but they are not much used for most of the Carboniferous and Permian rocks because of limited diversity and conservative morphology. Many pre-Carboniferous trilobites had highly derived morphologies, some of which have been quite instructive in studies of functional morphology and paleoenvironmental interpretation.

The purposes of this exercise are:

1 to generate familiarity with trilobite morphology;
2 to examine representatives of major groups of trilobites;
3 to infer general stratigraphic position based on morphology seen in a trilobite assemblage; and
4 to demonstrate a powerful biostratigraphic correlation procedure, using a trilobite-rich fauna in the example.

7.2 MORPHOLOGY

Trilobites are complex animals, and an extensive set of morphological terms has been applied to them. The most widely used terms for trilobites as seen in dorsal view are illustrated in Fig. 7.1. All illustrations in Fig. 7.1 clearly show the axial and two pleural lobes, which extend along the entire length of the body. In addition, all show the clearly differentiated fused **cephalon**, the articulated segments of the flexible **thorax**, and the **pygidium**, which consists of a single segment in *Olenellus* but is extended by fusion of several segments derived from the posterior of the thorax in all others illustrated.

Within the cephalon, the axial lobe comprises the **glabella**, which is primitively parallel-sided with clear remnant indications of segmentation. More highly derived trilobites may have the traces of segmentation in the glabella reduced substantially, and the glabella may either expand or taper toward the anterior edge.

There is a symmetrical line of weakness, the **suture**, present in the cephalon of trilobites, apparently designed to provide a line along which the exoskeleton could split during molting. The suture may be absent, as in Fig. 7.1(a), marginal, as in Fig. 7.1(d), or it may extend across the dorsal surface of the trilobite (Figs 7.1b and c). If the rear end of the suture crosses the margin on the anterior side of the corner **(genal angle)** where the lateral and posterior edges of the cephalon meet, the suture is called **proparian** (Fig. 7.1b). If the rear end of the suture crosses the posterior margin of the cephalon, it is termed **opisthoparian** (Fig. 7.1c). **Gonatoparian** sutures (not shown) cross the margin of the cephalon at the genal angle.

The portion of the cephalon between the suture and the margin is the **free cheek**, also called the **librigena**. The part of the pleural lobe between the suture and the glabella is the **fixed cheek**, also called the **fixigena**. The glabella and the fixigenae together comprise the **cranidium**. The compound eyes may be situated either on the librigenae or on the fixigenae.

Various furrows occur on the cephalon, the most consistent and prominent being the furrow that bounds the glabella, and also the transverse furrow present in many that divides the posterior **occipital ring** from the remainder of the glabella. In addition, various border furrows may occur.

The genal angle is the pronounced bend at each of the two postero-lateral corners of the cephalon. It may be rounded or sharp or may be extended posteriorly as a spine (Fig. 7.1a). In some trilobites, the genal spine extends well beyond the posterior end of the pygidium.

The undersurface of the cephalic area is rimmed anteriorly and laterally by a **doublure**, the turned-under margin of the cephalon. A variably shaped **rostral plate** may be present, centered against the doublure or anterior edge of the cephalic margin, and there is usually a hinged **hypostome** as well (Fig. 7.2). The hypostome abutts against the posterior edge of the rostral plate (if both are present) and usually is hinged near the forward edge at a pair of anterior wings, or the rostral plate and hypostome are fused together into a single structure. Apparently the hypostome covered the mouth and served some function in feeding.

The thorax consists of two to many articulating segments. In the trilobites that enroll, each segment is shaped so that when the trilobite is enrolled the lateral **pleurae** can overlap, especially along their outer ends. However, when the trilobite is stretched out in a flat plane, the segments slide apart so they are adjacent but not as overlapped as when the animal is enrolled. Ends of the pleurae were extended primitively as posteriorly directed spines, which were of varying length in some (Fig. 7.1a). Eventually pleural ends became rounded, although secondary spinosity developed in a few.

In the apparently most primitive trilobites, the pygidium consisted of only the posterior terminal segment or telson. However, there was a tendency for the posterior thoracic segments to fuse into the pygidium. Size of the pygidium is given in reference to size of the cephalon. A **micropygous** trilobite has a pygidium that is tiny, a **heteropygous** form has a pygidium that is smaller than the cephalon but not tiny, an

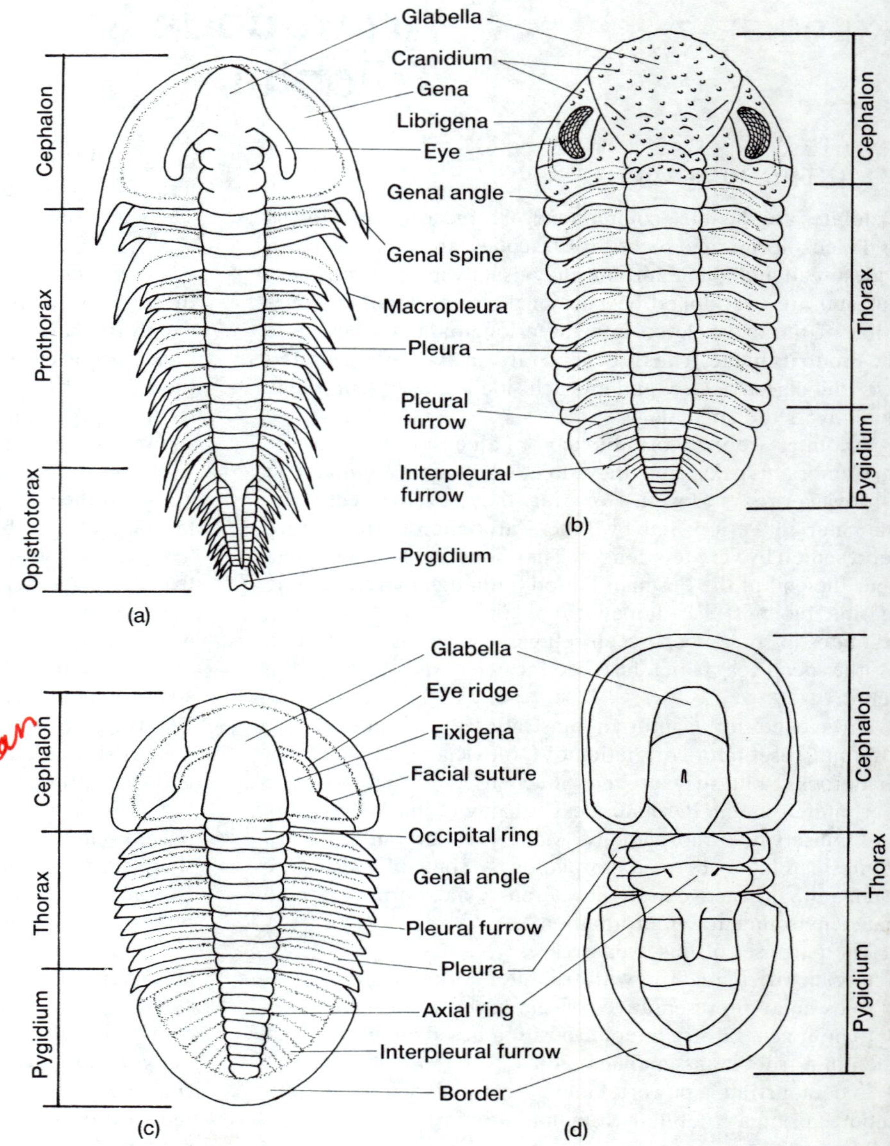

Opistoparian

Fig. 7.1 Some descriptive terminology applied to trilobites. Adapted from Harrington in Harrington *et al.* (1959) with permission from the Geological Survey of America.

isopygous form has a pygidium that is approximately the same size as the cephalon, and a **macropygous** trilobite has a pygidium larger than the cephalon. In most trilobites the fused segments can be seen clearly as axial rings and interpleural furrows. Some large post-Cambrian clades of trilobites, however, lost the furrows that bounded structures both in the cephalon and in the pygidium so that the dorsal exterior surface is smooth, with minimal evidence of original segmentation or features of any kind.

The surface texture of trilobite exoskeletons may be microscopically pitted, finely ridged, deeply pitted in certain regions, tuberculate, or spinose. Because of high organic content of the exoskeleton, it is common for trilobite fossils to be dark relative to the surrounding rock. Most are fragments of moulted skeletons, with a smaller percentage derived from dead animals.

7.3 CLASSIFICATION

As for the entire arthropod group, there is no universal agreement among specialists on trilobite classification and relationships. Early classifications emphasized some particular feature, such as position of the primary cephalic suture. Current attempts at classification utilize as many meaningful characteristics as possible. These include such criteria as cephalic sutures, structure of the eyes, shape and extent of the glabella, distinctive surface textures, number of thoracic segments, marginal structure of the cephalon, placement of the eyes, presence and shape of distinctive mouthparts, distinctiveness of furrows reflecting cephalic and pygidial segments, etc.

Two classification schemes are presented in Tables 7.1 and 7.2. The first represents an attempt to recognize monophyletic

Table 7.1 Orders and characteristics of orders of the Trilobita. Adapted from Clarkson (1986).

Order Agnostida	Small, with subequal cephalon and pygidium Thoracic segments number only two or three Hypostome small, with ribbon-like wings Eyes and sutures absent in most Lower Cambrian–Upper Cambrian
Order Redlichiida	Cephalon large, semicircular, with strong genal spines Thoracic segments numerous, usually extended as posteriorly oriented spines at lateral margins Pygidium tiny Lower Cambrian–Middle Cambrian
Order Corynexochida	Glabella usually parallel-sided or expanded anteriorly Cephalic sutures opisthoparian with subparallel anterior branches Rostral plate fused with hypostome Thorax segments number seven or eight Some macropygous Lower Cambrian–Upper Cambrian
Order Ptychopariida	Cephalon moderate-sized to large Glabella commonly simple, forwardly tapered but swollen in some, not reaching front of cephalon, furrows straight and simple or variously developed and curved Fringe may occur around anterior and lateral margins of cephalon, and genal angles may be pronounced Eyes small and simple to huge and multifaceted Sutures opisthoparian or marginal Thorax with six to many segments Pygidium small in primitive and some highly derived forms, others isopygous Some isopygous forms with smooth pygidia Subdivided into five rather diverse suborders Lower Cambrian–Upper Devonian
Order Proetida	Glabella large, vaulted, well defined Cephalon opisthoparian, usually with genal spines Eyes holochroal, usually large Rostral plate narrow and backwardly tapered, hypostome long Thoracic segments numbering eight to 10 Isopygous, pygidium furrowed but not spiny Ordovician–Permian
Order Phacopida	Glabella tapered or expanded forwardly, commonly with four to five glabellar furrows Normally proparian or gonatoparian Eyes holochroal or schizochroal Rostral plate present or absent Thorax with eight to 19 segments Pygidium smaller than cephalon in most, rounded, subtriangular, or lobed and spiny Subdivided into three important suborders Lower Ordovician–Upper Devonian
Order Lichida	Glabella very broad, commonly fused with lateral lobes Opisthoparian Isopygous or macropygous, with three pairs of blade-like pleurae Highly tuberculate Usually rare, but quite distinctive Lower Ordovician–Upper Devonian
Order Odontopleurida	Glabella with three pairs of lateral lobes, with separated ocular ridge on fixigenae Opisthoparian Rostral plate short and wide, hypostome small Thorax with eight to 10 segments, spinose Pygidium short, spiny Highly spinose overall ?Upper Cambrian, Lower Ordovician–Upper Devonian

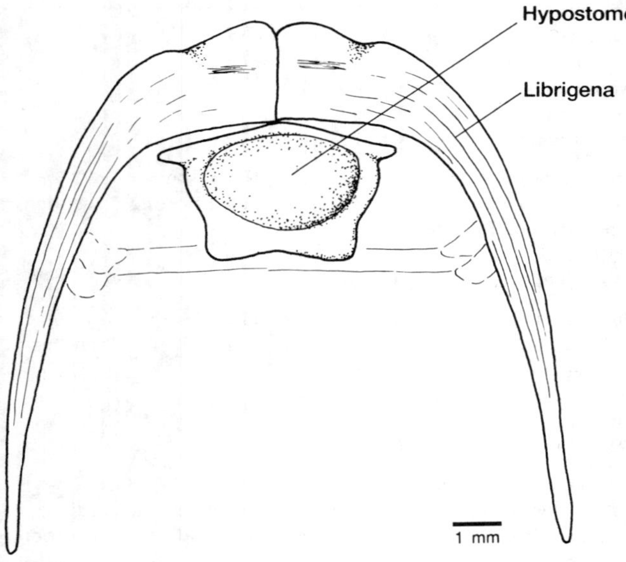

Fig. 7.2 Ventral side view of hypostome and doublure surface of librigenae of the Ordovician trilobite *Remopleurides caphyroides*. Drawn after Whittington (1959).

orders based on current knowledge; the second groups the trilobites into a single large order, excepting one highly distinctive group that constitutes a second order. The relationship between the two classifications and the ranges of the groups are shown in Fig. 7.3.
(Question 1)

7.4 MORPHOLOGIC TRENDS

Trilobites as a whole show remarkable stability in body plan throughout their stratigraphic range. However, when representative trilobites are arrayed stratigraphically as in Fig. 7.4, some general temporal patterns or evolutionary trends become apparent. This temporal pattern carries across even the highly specialized morphologies shown by many trilobites that invaded particular habitats such as reefs and muddy bottoms or niches such as epipelagic browsers, or scavengers on concentrations of floating graptolites.

Lower Cambrian trilobites commonly have no facial sutures, may exhibit clear pseudosegmentation in the glabella, may have long genal spines, typically have a large number of segments in the thorax, have thoracic segments terminating laterally in sharp angles or spines, and generally are micropygous with segments poorly fused in the pygidium. Others are agnostids, which are small, blind, lack facial sutures, have only two or three segments in the thorax, and are isopygous.

Agnostids are also common in Middle Cambrian trilobite faunas. Other Middle Cambrian trilobites tend to retain pseudosegmentation in the glabella, pronounced genal spines, and a large number of terminally angled thoracic segments. Otherwise they commonly differ from Lower Cambrian trilobites in that they have opisthoparian facial sutures, are heteropygous, and pygidial segments are more-or-less lightly furrowed in the pleural lobes but clearly scored in the axial lobes.

Upper Cambrian faunas still include some agnostids. Other Upper Cambrian trilobites, however, resemble Middle Cambrian taxa in most respects but may differ in having relatively shorter glabellas so that there is a wide preglabellar border, and genal angles and terminations of thoracic segments are rounded rather than pronouncedly angular.

Ordovician trilobites that occur in pure limestones commonly are large, relatively flattened, isopygous, and almost smooth with little or no external indication of segmentation in cephalons and pygidia. Others occurring in limestones to terrigenous rocks encompass an exhuberantly diverse set of morphologies: the glabella may be strongly inflated or greatly tapered anteriorly, glabellar segmentation is absent or marked only by lateral furrows, eyes may be huge or very small, facial sutures may be opisthoparian or proparian, thoracic segments are less numerous than in most Cambrian forms, terminations of thoracic segments are rounded or ornately spinose, they may be heteropygous or isopygous, and the surface texture may be smooth, pitted, tuberculate, or spinose.

Silurian through Permian trilobites in general resemble those of Ordovician faunas, but with progressively less diverse morphology. By the end of Early Carboniferous time, trilobites basically were of one conservative type, existing as only a few genera.
(Question 2)

7.5 FINE-SCALE STRATIGRAPHIC CORRELATION

The preceding section addressed approximation of geologic age based on how well the general aspect of a fauna matches

Table 7.2 Orders and characteristics of orders of the Trilobita. Adapted from Robison in Boardman *et al.* (1987).

Order Agnostida	Small Eyeless, lacking facial sutures, glabella well separated from anterior border furrow Hypostome saddle-shaped with large inwardly directed processes Two thoracic segments, the more anterior lacking articulating half ring Isopygous, pleural regions of pygidium unsegmented Lower Cambrian–Upper Ordovician
Order Polymerida	Small to large Cephalon with eyes, facial sutures Thorax with five or more segments, anterior segment always with articulating half ring Pygidium variable in size, with pleural regions commonly segmented Lower Cambrian–Upper Permian

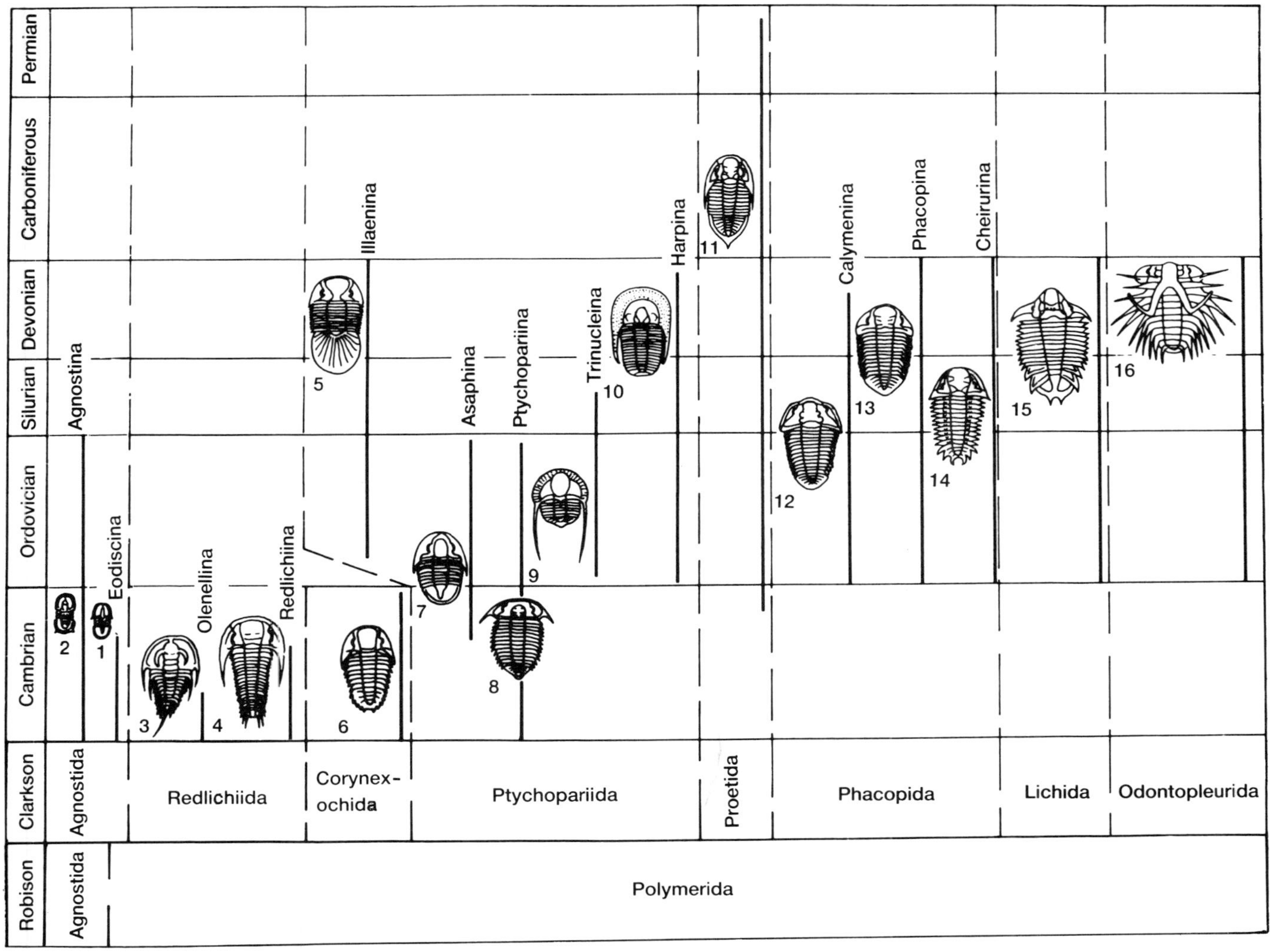

Fig. 7.3 Stratigraphic distribution of the major groups of the Trilobita. Modified from Clarkson (1986).

well-established evolutionary trends. More commonly, geologists establish the age of rocks by a more detailed analysis of the known ranges of genera and species present. In this procedure, the age of the rock is determined as within the time during which the various fossils present are known to have co-existed, that is, the age is determined by **concurrent range zones**. Use of concurrent range zones is included in Exercises 15 and 16.

A more precise method of biostratigraphic correlation has been devised in order to establish finer-scaled correlations for detailed analysis of geologic structures, paleoenvironmental and paleogeographic reconstructions for small increments of time, and studies in evolution and migration of faunal elements. This method requires carefully detailed and extensive collection and analysis of fossils so is rather time-consuming; however, the information produced can be enormous, critical for many investigations, and unavailable from any other physical or paleontological source. This method of analysis was devised by Alan B. Shaw for his work in the petroleum industry and will be referred to as **Shaw's method**.

From measured **stratigraphic sections** (local exposures through an interval of sedimentary rock that are examined bed-by-bed with at least the rock type and thickness of each successive bed recorded), large collections of fossils should be obtained at closely spaced intervals. Of course, for such detailed biostratigraphic work, each of the measured stratigraphic sections should encompass rocks known or suspected to represent the same approximate span of time. The stratigraphic distance (measured perpendicular to bedding of the rock) used for the spacing between successive collections is determined by the scale needed to achieve the purposes of the study and by availability of fossils, and each collection should include at least hundreds of specimens where possible. (There are statistical *rarefaction* methods that help with the decision of when to stop collecting because of decreasing rate of encounter of additional species.)

In the laboratory, specimens are identified to species, and their stratigraphic ranges, from first or stratigraphically lowest to last or stratigraphically highest occurrences, are plotted beside a **columnar section**, which represents the measured stratigraphic section from which they were collected (Fig. 7.5). This diagrammatic representation of rock types

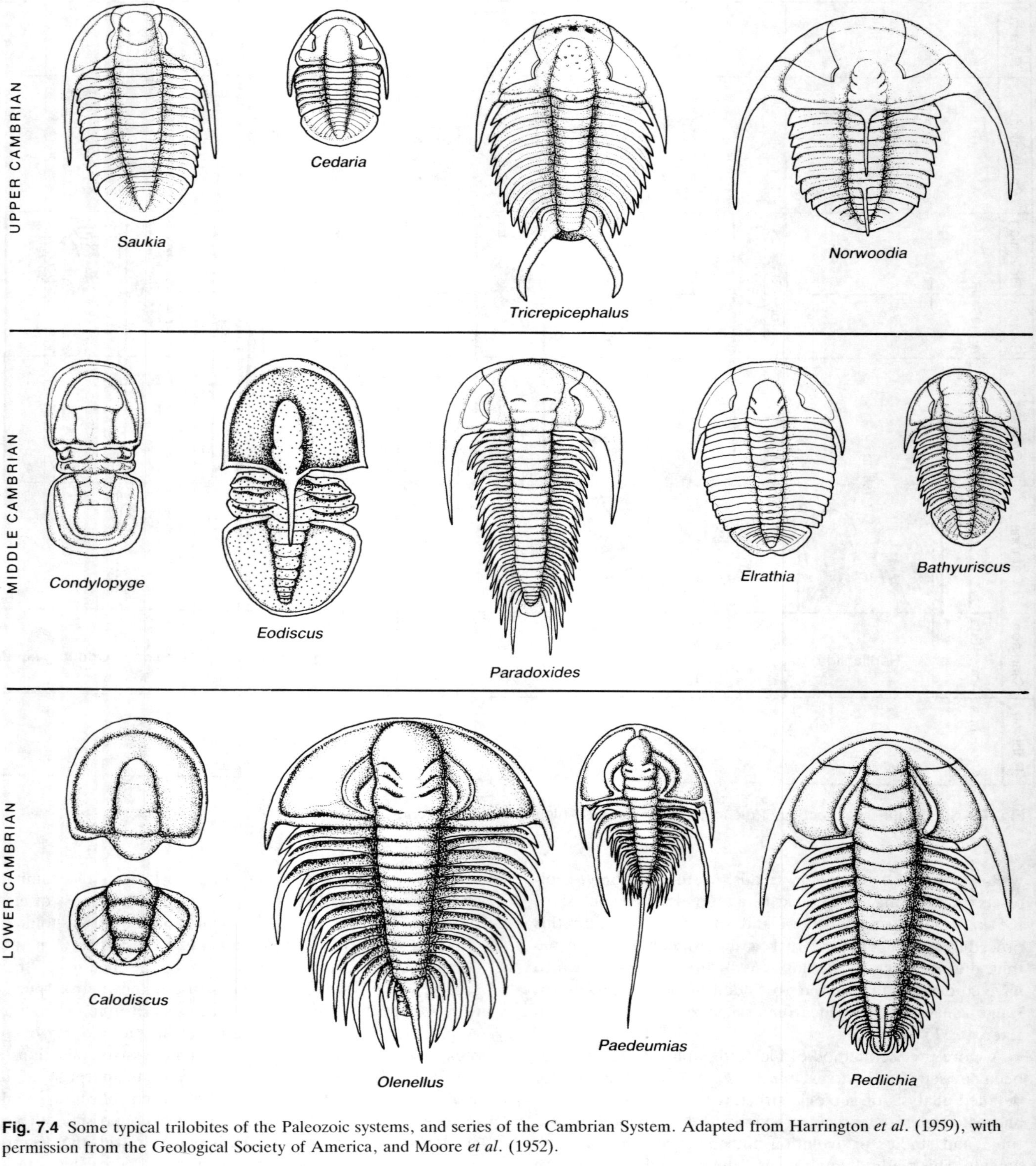

Fig. 7.4 Some typical trilobites of the Paleozoic systems, and series of the Cambrian System. Adapted from Harrington *et al.* (1959), with permission from the Geological Society of America, and Moore *et al.* (1952).

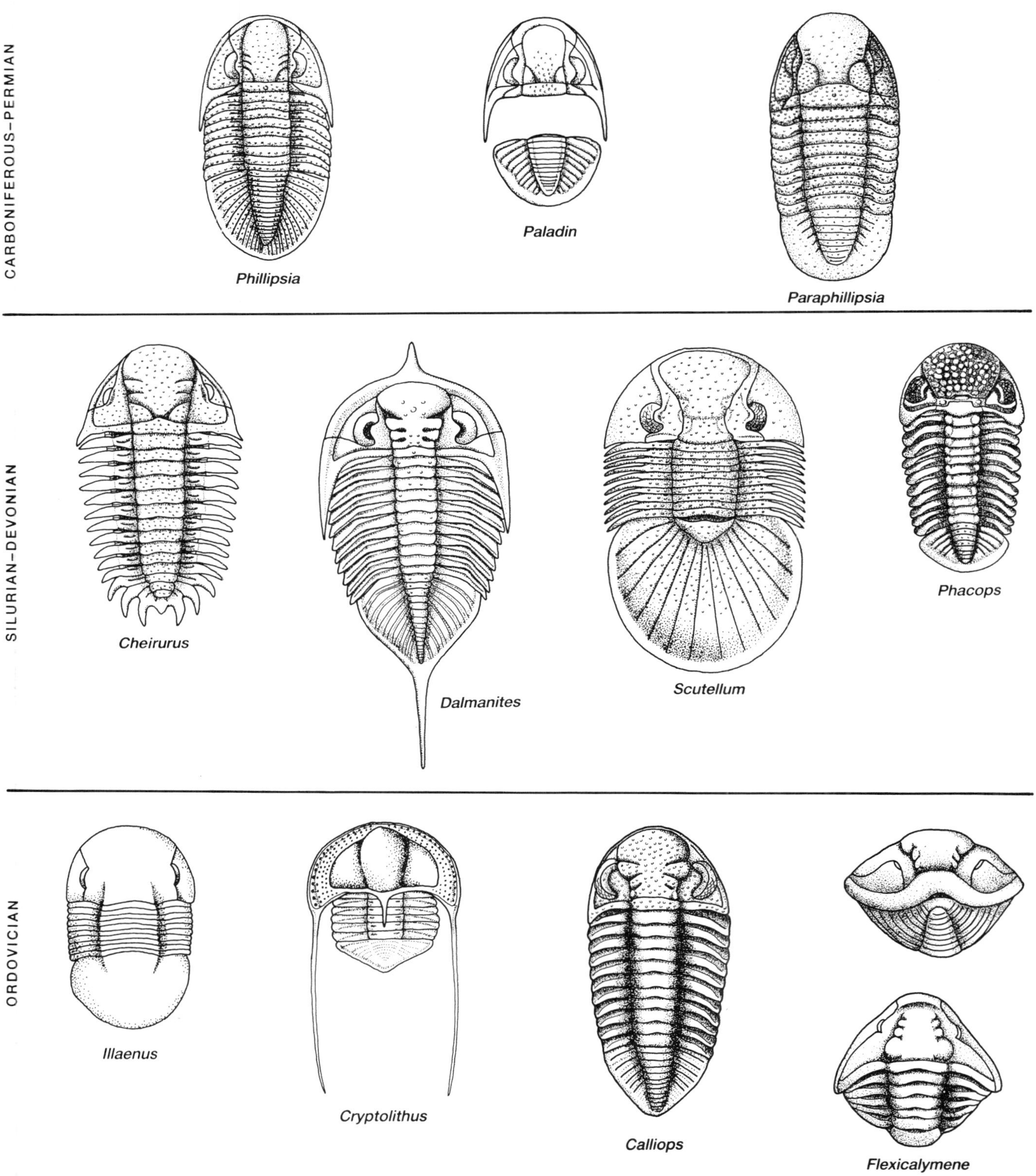

Fig. 7.4 *(Continued)*

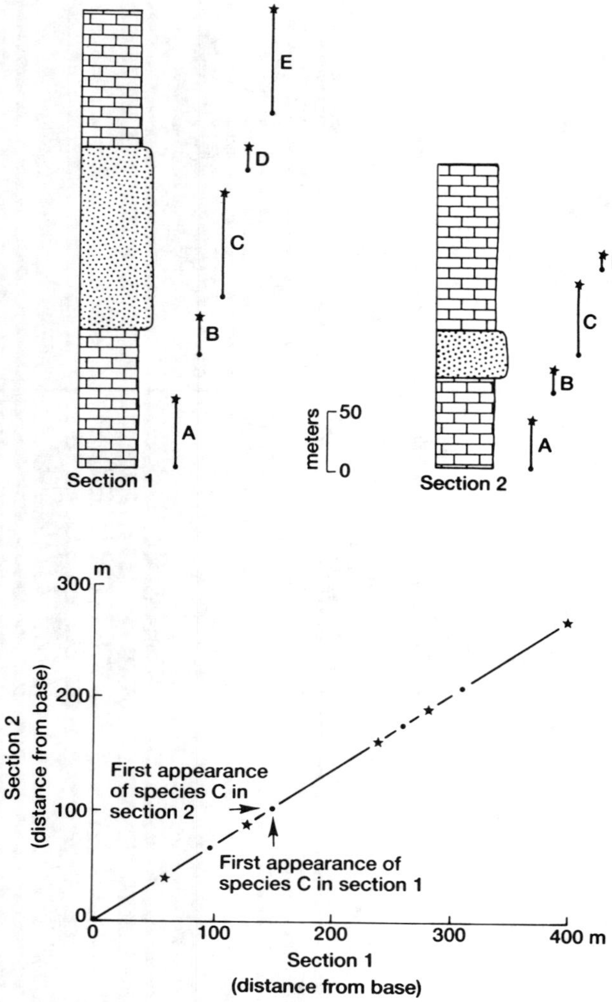

Fig. 7.5 Two hypothetical columnar sections and species distributions within them (*top*), and correlation between the two sections using Shaw's method (*bottom*).

and fossil ranges is largely to assist in conceiving patterns and relationships and in communicating with others rather than being essential for actually establishing correlations. A columnar section and plot of associated ranges of fossils is drawn up for each stratigraphic section to be used in the study. Only the moderately common species are plotted, because subsequent comparisons of first and last appearances of a species are much less precise for rare than for common species. Inclusion of rare species would therefore introduce unnecessary and confusing "noise," obscuring the statistical patterns that Shaw's method was devised to clarify.

In order to correlate between two stratigraphic sections such as those depicted at the top of Fig. 7.5, an $X-Y$ graph is drawn, with one section represented on the X axis and the other on the Y axis. The position of each end of a species' range in the two stratigraphic sections is used as co-ordinates to plot a point on the graph, as shown in the lower portion of Fig. 7.5.

Critical assumptions are made about each species used in correlation by Shaw's method: the first appearance of the species within a stratigraphic section closely approximates the time of evolution or of migration into the entire geographic area encompassed by the study, and the last appearance closely approximates the time of extinction or of extermination within the area encompassed. (First or last appearances at the very bottom or very top of a section probably do not represent an evolutionary event but may be highly useful in interpretation of depositional history of the rock.) If there is a reasonable suspicion that first or last appearance within a section does not correspond with evolution of the species, the point should be purged from the set of data.

The most common problem is appearance and disappearance of species (**facies fossils**) that lived only in particular limited environments, so that they appear within a local stratigraphic section only when the environment to which they were restricted migrated over the spot and resulted in accumulation of rock. If the environment was rapidly migrating so that it reappeared rapidly and frequently throughout the region being studied, then the problem of miscorrelation is minimal and can be ignored. This usually is not the case. One indication that a species is a facies fossil is its appearance within the stratigraphic section where a particular rock type begins and its disappearance where the rock type ends. If the same pattern holds for the species in several sections, its interpretation as a facies fossil is virtually assured.

First and last occurrences of the species in Fig. 7.5 are in the same sequence in both stratigraphic sections, and none correspond exactly with bottom and top of a particular rock type. Therefore, none appear to be strongly controlled by facies. The ranges are tentatively accepted as reflecting evolutionary or migratory origins and terminations of species and are therefore retained in the graph at the bottom of Fig. 7.5.

A regression line can be drawn through the points in the graph, using the formula given in Exercise 2 for calculation of lines of regression. Calculation is not necessary in this example, however, because all points fall on a straight line. (There is a *statistical correlation* of 1.0). The regression line plots a highly accurate line for **stratigraphic correlation** between the two sections. This is the primary purpose of the exercise. For example, once the regression line through the points is plotted, if one wishes to know which level in section 2 correlates with a particular level in section 1, one locates the appropriate point for section 1 on the X axis, projects straight up to intersect the regression line, and from the point of intersection then projects horizontally to intersect the Y axis. The value on the Y axis is the level within measured section 2 that correlates with the starting point within section 1.

Probably there are no actual cases where comparison of ranges of species between two separate stratigraphic sections would result in a precisely straight alignment as in Fig. 7.5. Instead, each species is subject in some degree to environmental limitations that cause some disparity in the times that it was present at two different localities. In addition fossils may be destroyed or inaccessible locally, or they may be available but simply overlooked and not collected. All these vagaries introduce scatter into a plot of first and last occurrences compared between two stratigraphic sections.

The Upper Cambrian Riley Formation in the vicinity of the Llano Uplift, central Texas, provides a good example. The ranges of species in the trilobite-dominated fauna are known in great detail. On Figs 7.6 and 7.7 are plotted ranges of the

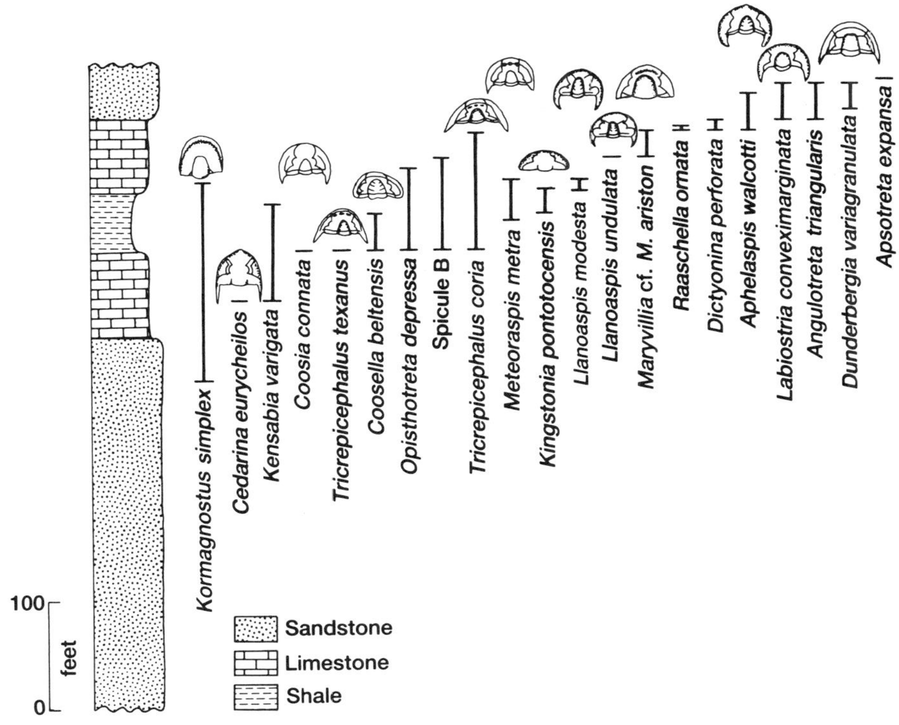

Fig. 7.6 Columnar section through Upper Cambrian Riley Formation at Morgan Creek, with distribution of fossils co-occurring in the White Creek section. After data from Palmer (1954) as given in Shaw (1964).

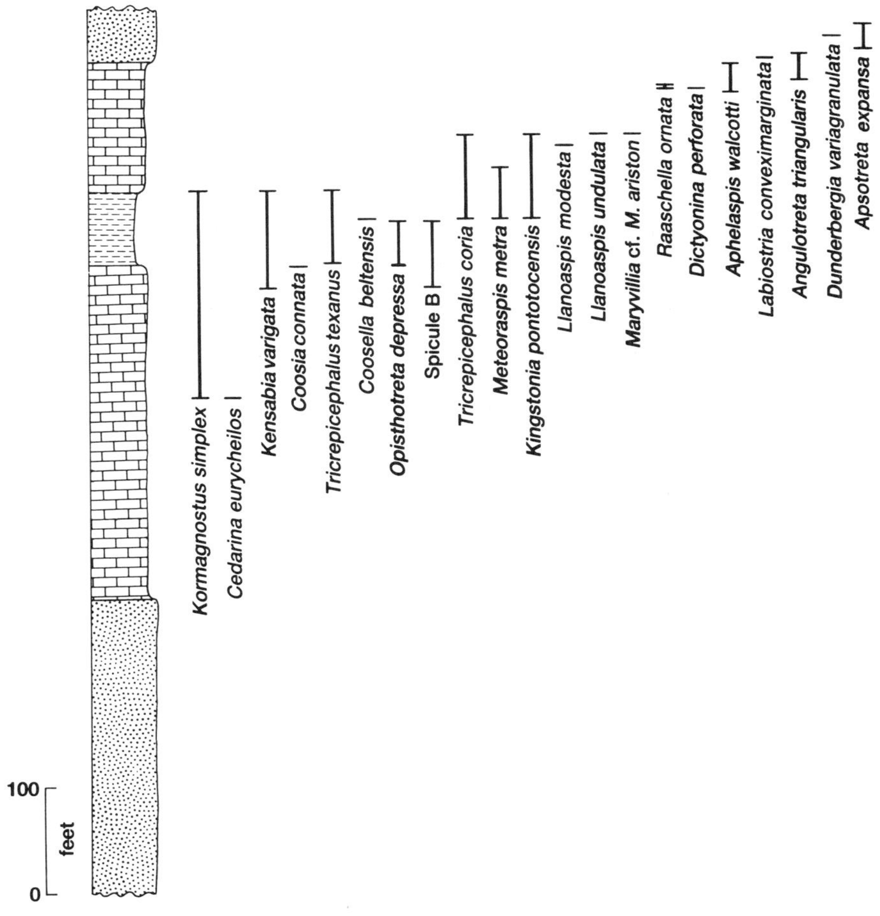

Fig. 7.7 Columnar section through Upper Cambrian Riley Formation at White Creek, with distribution of fossils co-occurring in the Morgan Creek section. After data from Palmer (1954) as given in Shaw (1964).

Table 7.3 Ranges of species that occur in both the Morgan Creek and White Creek sections of the Riley Formation in the Llano Uplift, Texas. Numbers given are distances in feet above the unconformable base of the formation. Data from A. R. Palmer as tabulated in Shaw (1964).

Species	Morgan Creek		White Creek	
	First	Last	First	Last
Kormagnostus simplex	299	485	460	655
Cedarina eurycheilos	(373)	(373)	(460)	(460)
Kinsabia varigata	373	464	561	653
Coosia connata	419	419	582	582
Tricrepicephalus texanus	419	(419)	586	(655)
Coosella beltensis	419	453	628	628
Opisthotreta depressa	419	494	582	725
Spicule B	419	504	561	725
Tricrepicephalus coria	419	(529)	628	(706)
Meteoraspis metra	446	485	628	677
Kingstonia pontotocensis	453	475	628	706
Llanoaspis modesta	472	483	695	695
Llanoaspis undulata	504	504	706	706
Maryvillia cf. *M. ariston*	504	(529)	706	(706)
Raaschella ornata	529	532	750	751
Dictyonina perforata	529	538	750	750
Aphelaspis walcotti	530	561	744	771
Labiostria conveximarginata	538	570	775	775
Angulotreta triangularis	538	570	756	779
Dunderbergia variagranula	547	570	795	795
Apsotreta expansa	574	574	784	808

species that co-occur in the stratigraphic sections measured at Morgan Creek and White Creek. Species that occur in only one of the sections have been omitted from the list. A preliminary scan of Figs 7.6 and 7.7, in which species ranges are presented in the same sequence away from the columnar sections, indicates that the species appear and disappear progressively in the same approximate order. However, it is also evident that the ranges do not match exactly, as seen, for example, in the second species occurring within the range of the first species at Morgan Creek but at the base of the range of the first species at White Creek. This variation by itself would destroy exact alignment of a graphic comparison of first and last appearances between the two sections.
(Questions 3–7)

The slope of the regression line is a measure of the relative rates at which the rock in the two stratigraphic sections being compared actually accumulated. (This is not the same thing as sediment accumulation rate, because some sediments such as shale compact appreciably during lithification while others such as quartz sand hardly compact at all.) If the coefficient of X is 1.0, the rock accumulation rate in the two sections was equal. Any other value means that the accumulation rates were unequal.

A change in rock accumulation rate in either section should show up as a reorientation in the trend of points that plot first and last appearances in the comparison of the sections. Separate regression lines for each of the regions of different trend should be calculated and plotted instead of forcing a single straight line from end-to-end of the complete set of points. This results in a "dog-leg" line that graphically represents the change in accumulation rates that the different coefficients of X specify.

A plateau in the trend of points may indicate one of three things:

1 a previously unsuspected change in environment represented in one of the sections by the abrupt appearance or disappearance of several species;
2 a fault that omits part of a section or duplicates a segment of the other section; or
3 a local unconformity where no rock accumulated in one section while deposition continued at the other.

Where a plateau occurs, close inspection of the field data for that part of both stratigraphic sections is needed.

By successively incorporating data from all the pertinent stratigraphic sections followed by readjustments of species ranges within the area, and reiterating the process a few times, a composite standard section for a region can be

produced. This allows a better approximation of the full range of each species and commonly results in a greater number of species being included than can be used in a single comparison of any two stratigraphic sections. The composite standard section can then be used within the region or for correlation with sections in distant regions.
(Question 8)

FURTHER READING

Edwards, L. E., 1989. Supplemented graphic correlation: a powerful tool for paleontologists and non-paleontologists. *Palaios* **4**:127–143.

Harrington, H. J. *et al.*, 1959. Arthropoda 1 (Arthropoda — general features; Protarthropoda; Euarthropoda — general features; Trilobitomorpha), in R. C. Moore (ed.), *Treatise on invertebrate paleontology*, Part O. Geological Society of America and University of Kansas Press: Lawrence, K. S.

Levi-Setti, R., 1975. *Trilobites*. University of Chicago Press: Chicago, Il.

Martinsson, A. (ed.), 1975. Evolution and morphology of the Trilobita, Trilobitoidea and Merostomata. *Fossils and Strata*, No. 4. Universitetsforlaget: Oslo, Norway.

Shaw, A. B., 1964. *Time in stratigraphy*. McGraw-Hill: New York.

EXERCISE 7 | **Worksheet** Name

QUESTION 1

Specimens of several genera of trilobites are provided for study. The order and geologic range of each may be given; if not, look up the order and range in *Treatise on invertebrate paleontology*, Part O. Record genus, geologic range of genus, order, and states of the morphological features that are listed in the table on p. 108. It is unusual to have only complete trilobites available for laboratory study, so for partial specimens, write NA in the spaces where information is not available from the specimens provided.

QUESTION 2

A set of trilobites collected from a single formation or a single stratigraphic series is provided. Characterize the stratigraphically important morphology of each species present, and determine the probable stratigraphic interval from which the assemblage was collected.

Species	*Characteristics*		
	Cephalon	*Thorax*	*Pygidium*

Probable stratigraphic interval: ____________________

QUESTION 3

Values for first and last appearances of the co-occurring species in the Morgan Creek and White Creek stratigraphic sections through the Riley Formation are given in Table 7.3. Plot the points for each mutual first and last appearance onto the graph that is provided on p. 109. Note that the Morgan Creek section is represented by the X axis and the White Creek section is represented by the Y axis. Each time one of the numbers circled in Table 7.3 is used to plot a point, draw a circle around the point.

QUESTION 4

Note that each of the circled points is not well aligned with the general trend of the other points. There is some reason for each why it is "suspect" and should not be used in detailed correlation between the two sections. Examine Figs 7.6 and 7.7 as well as the graph that you plotted and give the reason why each should probably be omitted from calculation of a regression line through the data.

Top and bottom of range of *Cedarina eurycheilos*:

____

Genus	Range	Order (Table 7.1)	Cephalon				Thorax		Pygidium	
			Shape of glabella	Glabella to anterior border	Eyes present or absent	Surface pattern	No. of segments	Shape of ends of pleurae	Size relative to cephalon	Distinctiveness of segmentation

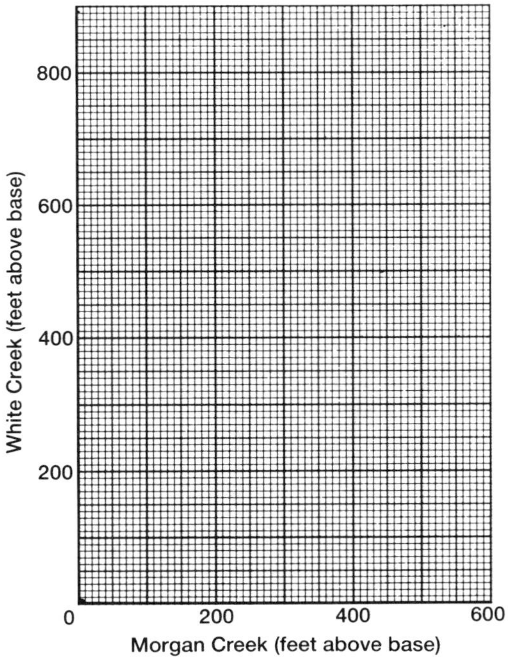

Top of range of *Tricrepicephalus texanus*:

____

Top of range of *Tricrepicephalus coria*:

____

QUESTION 5

Omitting the three points discussed in Question 4, the regression of Y on X for the data is calculated to be:

$Y = 70 + 1.27X$.

(If the laboratory is equipped with computers and appropriate software, the formula can be checked quickly for correctness based on the data in Table 7.3, but be sure to omit the circled points.)
Plot the regression line on the graph.

QUESTION 6

What stratigraphic level and rock type in the White Creek section correlates with the base of the lower limestone in the Morgan Creek section?

_______ feet; _______________ (rock type)

QUESTION 7

Projection of a regression line beyond the points on which it is based is always subject to errors because of the possibility of changed conditions about which we have no information. However, projection of the line is tempting, and in such enterprises as estimating size of future populations is the major purpose of determining the line in the first place. Assume that projection of the regression line is valid in the comparison of the Riley Formation in the Morgan Creek and White Creek sections. At what level does the line intersect the *Y* axis?

________ feet

What does this mean?

____

____

____

QUESTION 8

Interpret the graph below, explaining the meaning or possible meaning(s) of each segment or change in the line plotted.

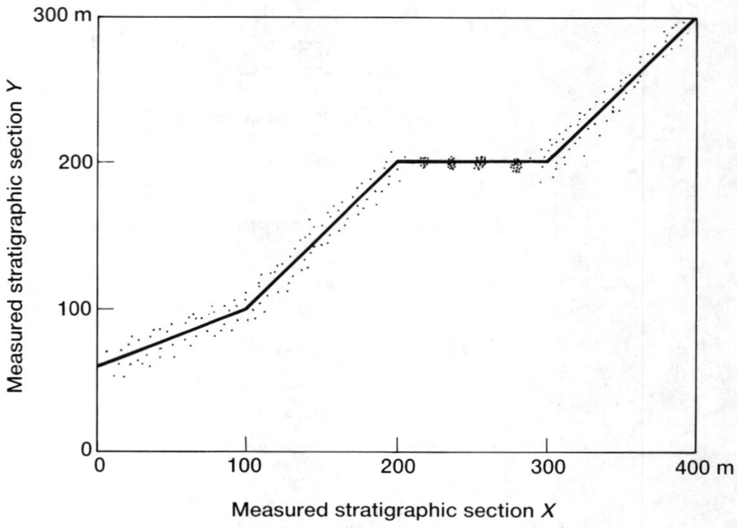

____

____

____

____

____

____

EXERCISE 8

Mollusca I: Gastropoda and Small Classes

8.1 INTRODUCTION

The phylum Mollusca is organized into subphyla and classes that contain radically different-appearing animals. The features that unite the mollusks are found entirely in the soft tissues and organs: a muscular **foot**; a **visceral mass**, which is the region that includes the digestive tract and associated organs; and a cavity that contains ambient fluid (water or air) and that is surrounded by a sheet of tissue termed the **mantle**. Mollusks lack segmentation and typically carry a calcareous exoskeleton on their dorsal side. Modifications of the foot and mantle, and concomitant changes in skeletal geometry, have been important major themes in molluskan evolution. Skeletons of mollusks such as clams, snails, ammonites, and nautiloids have accumulated abundantly in the fossil record.

The purposes of this exercise are:

1 to introduce the basic soft structures of mollusks as seen in a dissected marine gastropod (different expressions of many of the same structures will be seen in a dissected clam in the following exercise;

2 to examine representatives of some of the less diverse classes;

3 to examine various representative gastropods; and

4 to determine the basic shapes of gastropod shells and fit them within a framework of a theoretical range of morphologies of coiled shells.

8.2 DIVERSE MOLLUSKS

Each class of mollusks developed evolutionary innovations, which often enabled wholly new modes of life for members of the class. The various mollusks range from swimming carnivores (cephalopods), to crawling benthic grazers and carnivores (gastropods), to infaunal deposit and suspension feeders (pelecypods), among other modes of life. Many of the innovations leading to new modes of life involved the foot and mantle; mollusks are commonly cited as a superb example of adaptive radiation because of the immense diversity that resulted. The mollusk classes are characterized in Table 8.1, and their typical morphologies are illustrated in Fig. 8.1.
(Question 1)

8.3 GASTROPODA

Basic features of the soft parts of gastropods are shown in Fig. 8.2. Note that the specimen extracted from the shell exhibits an inner spiralled, tapered portion that contains the stomach and gonads and occupied the entire shell cavity. The *Buccinum* illustrated in Fig. 8.2 is a type of gastropod that contains gills in the mantle cavity; it also belongs to a subgroup that has evolved a **siphon**, which is an extended, enrolled part of the mantle used to draw in relatively clean water from above the substratum. Not all gastropods have gills and siphons.
(Question 2)

The Gastropoda (snails and slugs) constitute the largest of the molluskan classes in number of species, abundance, and diversity of morphologies. Variations in basic shell morphology and surface sculpture are enormous within the class (Figs 8.3 and 8.4). Some of the common shell forms and attributes include:

Biconical — both spire and base shaped as moderately steep cones (see *Conus*).

Conoidal — sides of spire curve in toward apex so that the shape is gently curved in profile but retains a pointed apex (see *Aplexa*).

Discoidal — both spire and base low, flat, or concave so that shell has shape of disk (see *Lecanospira*).

Obconical — similar to biconical except that cone of spire is nearly flat (see *Maclurites*).

Ovoid — both apex and base of shell are rounded so that profile has approximate shape of an oval (see Fig. 8.3 and *Lunatia*).

Patelliform — low cap-shaped form with broadly flared aperture that is oriented against substratum in life (see *Crucibulum* and *Fissuridea*).

Pupaeform — high ovoid form rounded at apex and base, having general resemblance to the pupal instar of an insect (see *Gastrocopta*).

Siphonostomatous — margin of aperture interrupted anteriorly by an indentation or extended canal-shaped area where mantle siphon is extended (see Fig. 8.3 and *Busycon* and *Ecphora*).

Trochiform — apex is evenly tapered and base is flat, so that profile is triangular (see Fig. 8.3 and *Calliostoma*).

Turbinate — conical apex of evenly stepped whorls and base rounded (see *Lophospira*).

Turreted — very high conical apex with flat base (see *Turritella*).

Morphology of soft parts divides the class into three distinct subgroups; the subclass Prosobranchia is the one most commonly encountered as fossils. Classification and characterization of gastropods to order level is given in Table 8.2, and representatives of the major subdivisions of the class are illustrated in Fig. 8.4.
(Question 3)

8.4 THEORETICAL MORPHOLOGY OF COILED SHELLS

As an alternative method to the terms that describe shell form as given in Section 8.3, the basic morphology of gastropods and other coiled shells may be described by specifying four parameters: the cross-sectional shape of the chamber (S, also known as the shape of the generating curve), the relative distance of the generating curve from the axis of rotation (D), the rate of increase in the width of the generating curve per 360° rotation (W), and the curve's rate of translation along the axis (T). (This method of analysis was worked out largely by David Raup in papers published in 1962, 1965, and 1966).

Table 8.1 Characteristics, mode of life, and geological range of major mollusk taxa. After Pojeta in Boardman *et al.* (1987).

Subphylum and class	Characteristics (distinguishing features in italics)	Structural innovations	Inferred primitive mode of life	Geologic range
S. Amphineura	Radulate; mouth and anus at opposite ends of body; prominent primitive spicular skeleton			Late Cambrian–Recent
C. Aplacophora	Wormlike body with no head or shell; cuticle over body contains aragonitic *spicules*; *no shell*	Radula	Uncertain	Recent
C. Polyplacophora	Elongate body with head; *eight articulated valves make dorsal shell*, surrounded by muscular *mantle girdle*; *spicules limited to mantle girdle*'s cuticle	Flexible multivalved shell, large foot, and mantle girdle allowing tight clinging to substratum	Benthic algal grazers in high-energy, shallow marine environment	Late Cambrian–Recent
S. Cyrtosoma	Generally with conical univalved shell, commonly spiralled; single shell opening typically small; gut typically bent into U-shape so mouth and anus close; radulate			Early Cambrian–Recent
C. Monoplacophora	Some organs in series; some with gut bent into U-shape, but anus not over head (i.e. *untorted*); bilateral symmetry is typical; *univalved shell with single aperture* and apex pointing anteriorly over head; *no operculum*	Univalve shell that can be pulled down to cover entire animal; large foot	Benthic algal grazers in shallow marine environment of varying energy.	Early Cambrian–Recent
C. Gastropoda	Most with body in asymmetric, *helically coiled univalve shell* with apex pointing away from head; distinct head can be moved independently; foot solelike, creeping; visceral mass *torted*, i.e. rotated ~ 180° so that anus is above head; *single aperture* in shell closed by *operculum*	Helical, torted shell with columellar muscle for retraction of head and foot into shell and operculum for closing aperture	Benthic algal grazers in shallow marine environment of varying energy	Early Cambrian–Recent
C. Cephalopoda	Bilaterally symmetrical; *external shell when present commonly straight or coiled*, rarely helical; *septa and siphuncle* in shell; head well developed, with brain and eyes; foot modified into circle or tentacles around mouth; locomotion by mantle serving as hydraulic chamber, emptied through funnel in foot; visceral mass *not torted*; body dorsoventrally elongated with inverted U-shaped gut placing anus near mouth; *single aperture in shell*	Siphuncle that controls gas and fluid levels in chambers to adjust buoyancy	Swimming and/or crawling carnivores or carrion feeders	Late Cambrian–Recent
S. Diasoma	Univalved, pseudobivalved, or bivalved shells that may gape at anterior and posterior ends; gut straight, mouth and anus widely separated; head poorly developed or absent; radula present or absent; body laterally compressed primitively			Early Cambrian–Recent
C. Rostroconchia	*Pseudobivalved* shell; perhaps radulate; anterior plate (*pegma*) primitively connects the valves and was site of muscle attachment; ventral *gape*, greatest at posterior and anterior ends	Pseudobivalved, laterally compressed gaping shell and probably anterior structures for ciliary deposit feeding	Infaunal ciliary deposit feeders	Early Cambrian–Recent
C. Scaphopoda	*Univalved shell* is tapering *tube closed dorsally and ventrally but open at anterior and posterior ends*; radulate; anterior feeding structures (captacula) present; *not torted*	Tubular shell open at both ends	Infaunal ciliary deposit feeders including shelly protists as food	Middle Ordovician–Recent
C. Pelecypoda	*Bivalved*; dorsal shell margins typically with interlocking projections and indentations (teeth and sockets); dorsal *ligament* opens shell, closure is by *adductor muscles*; head and radula absent	Bivalved shell with complex musculature, ligament, and large foot allowing rapid burrowing	Infaunal ciliary deposit and suspension feeders	Early Cambrian–Recent

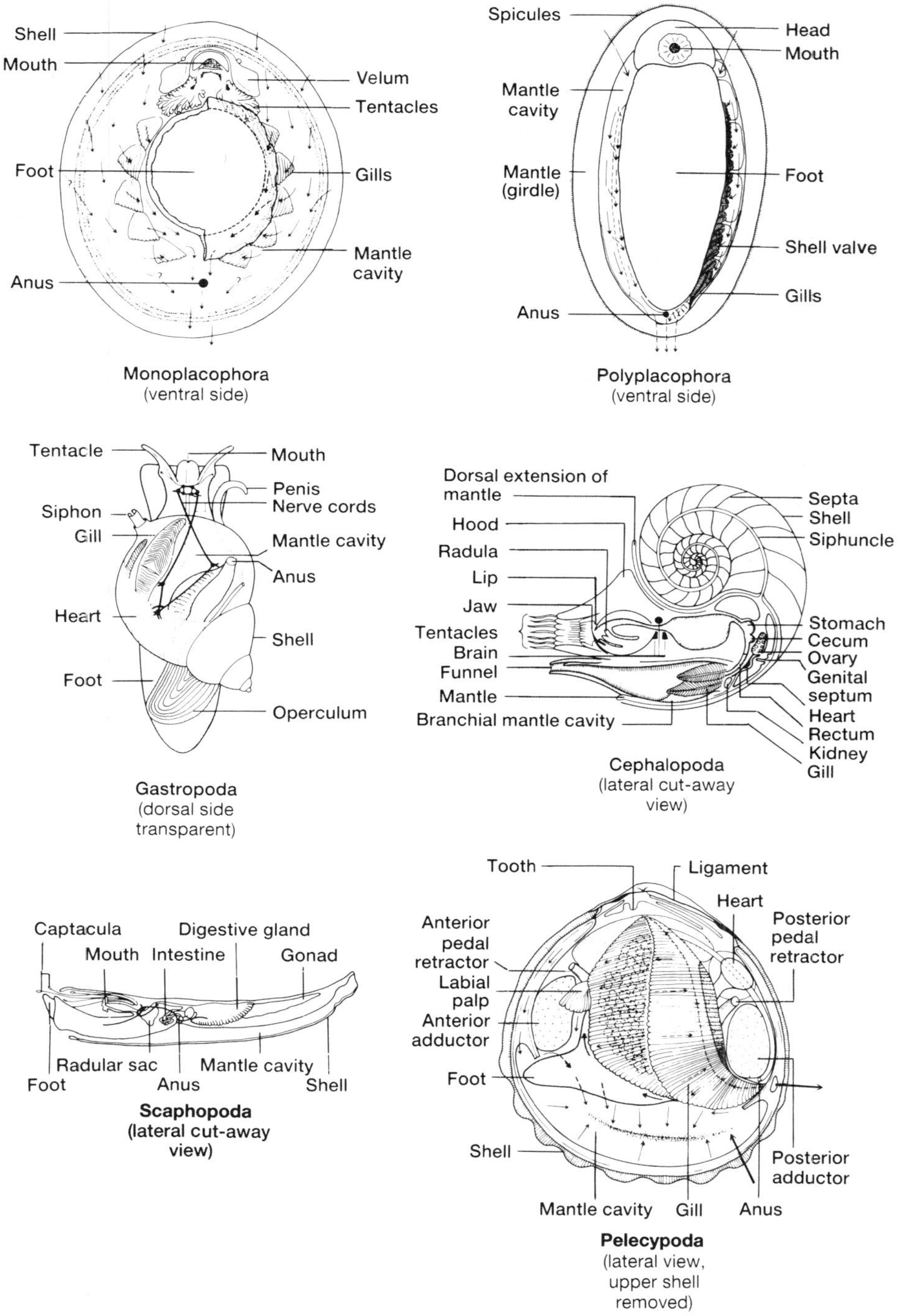

Fig. 8.1 Representatives of the classes of mollusks (excepting Aplacophora and Rostroconchia), showing organs and relationship of the soft body to the shell: Monoplacophora, a cap-shaped univalve; Polyplacophora, cap-shaped with eight plates in the shell; Gastropoda, a coiled and spiralled univalve; Cephalopoda, a coiled univalve with skeletal chambers; Scaphopoda, an elongate conical shell open at both ends and Pelecypoda, a bivalve. After various sources, from Pojeta in Boardman *et al.* (1987).

Figure 8.5 shows methods by which each of these may be determined for any coiled shell.

In Fig. 8.5, *S* is reduced to its simplest possible expression and omits indication of marginal indentations, crenulations, notches, etc. *S* does not equal shape of the aperture except for those gastropods in which the plane of the aperture may be projected through the axis of coiling.

Each of the four parameters defined above may vary due to ontogeny within an individual or may correspond with environmental changes. In some, such as the sedentary gastropod *Vermicularia* (Fig. 8.6), within a short interval of ontogeny, *D* and *T* may cease being predictable altogether, with *W* being reduced simultaneously to 1.0.
(Questions 4 and 5)

An understanding of the subtleties of gastropod shell form has barely begun developing. There have been some important papers published on the topic, some of which are listed below. However, there is much interesting detailed work and synthesis of patterns yet to be done in this area.

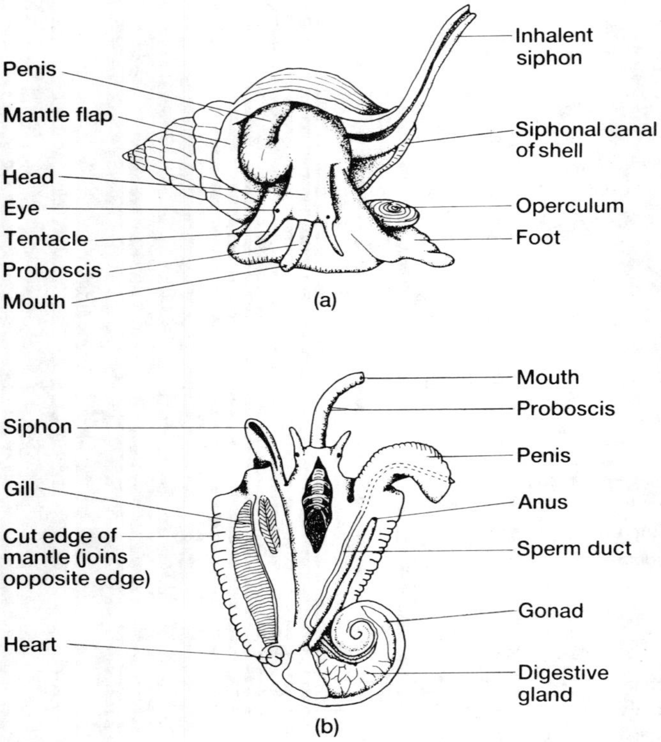

Fig. 8.2 *(Left)* Diagrammatic drawings of the gastropod *Buccinum*. (a) Terminology of protrusible soft parts and their relationship to the shell, and (b) various organs named; a semitransparent diagram with shell removed so that coiled visceral mass is visible and with the flesh pulled away from incisions along the dorsal side. After Cox *et al.* (1969) with permission from the Geological Society of America.

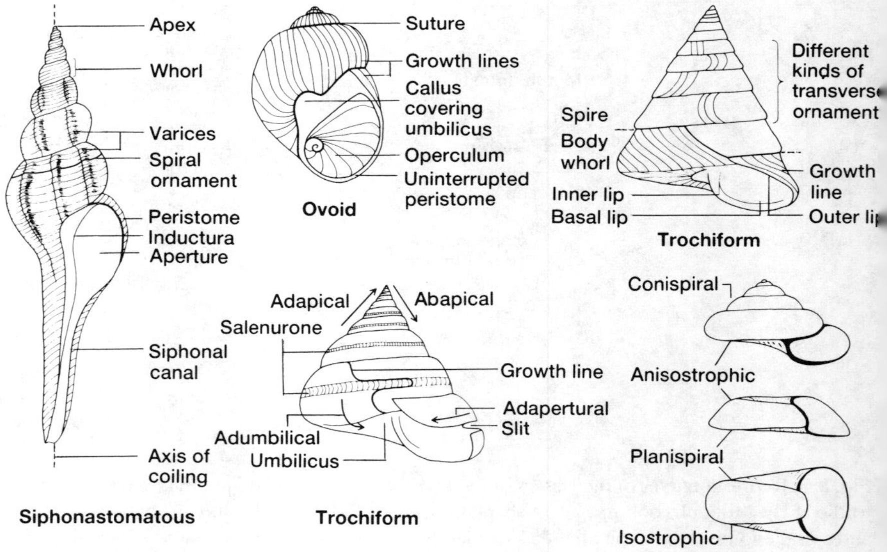

Fig. 8.3 Some basic terminology of gastropod shells. From Pojeta in Boardman *et al.* (1987). Other common shell forms are shown in Fig. 8.4.

Table 8.2 Characteristics of subclasses and orders of the Gastropoda.

Subclass Prosobranchia	Shells are cap-shaped or helical, with mantle cavity and gills at the anterior end Major nerve loop, which is located on the dorsal side, is **torted** (twisted into a figure eight) Early Cambrian–Recent
Order Archaeogastropoda	Primitively, two subequal gills are each provided with a double row of leaflets, but most forms with only one such gill that may have only a single row of leaflets Many exhibit a **selenizone**, which is a trace of the path of an indentation in the margin of the aperture where the anus trailed out (an improvement over having it drop feces directly on top of the head) An **operculum**, which is an organic or mineralized structure on the foot that is used to seal the aperture when the animal is withdrawn into the shell, is commonly present Exclusively marine Early Cambrian–Recent.
Order Mesogastropoda	Heart with single auricle, only one gill present, with a single row of leaflets No selenizone Some with **siphonal notch** which is an indentation in the anterior margin of the aperture where the siphon extends out of the mantle cavity Marine, freshwater, and terrestrial representatives Late Carboniferous–Recent
Order Neogastropoda	Heart and gill structure like that of mesogastropods Aperture with anterior siphonal notch or canal Radula and mouth at end of retractile **proboscis** Almost exclusively marine Cretaceous–Recent
Subclass Opisthobranchia	Reduced shell commonly internal or absent Mantle cavity and gill, where present, are posterior as result of detorting from prosobranch condition About 12 orders in most classifications, half of which lack shell Mostly marine Late Carboniferous–Recent
Order Pteropoda	Slender conical shell present or absent, distinct head lacking, foot modified as paired winglike fins for swimming Cretaceous–Recent
Subclass Pulmonata	Presumably detorted Conispiral shell present in most but may be reduced and concealed in mantle Gills absent, but mantle cavity altered into lung Operculum generally lacking Freshwater and terrestrial, rarely marine Late Carboniferous–Recent
Order Stylommatophora	Eyes at tips of the posterior of two pairs of tentacles Dominantly terrestrial Late Carboniferous–Recent
Order Basommatophora	Eyes at base of the single pair of tentacles Dominantly freshwater Late Jurassic–Recent

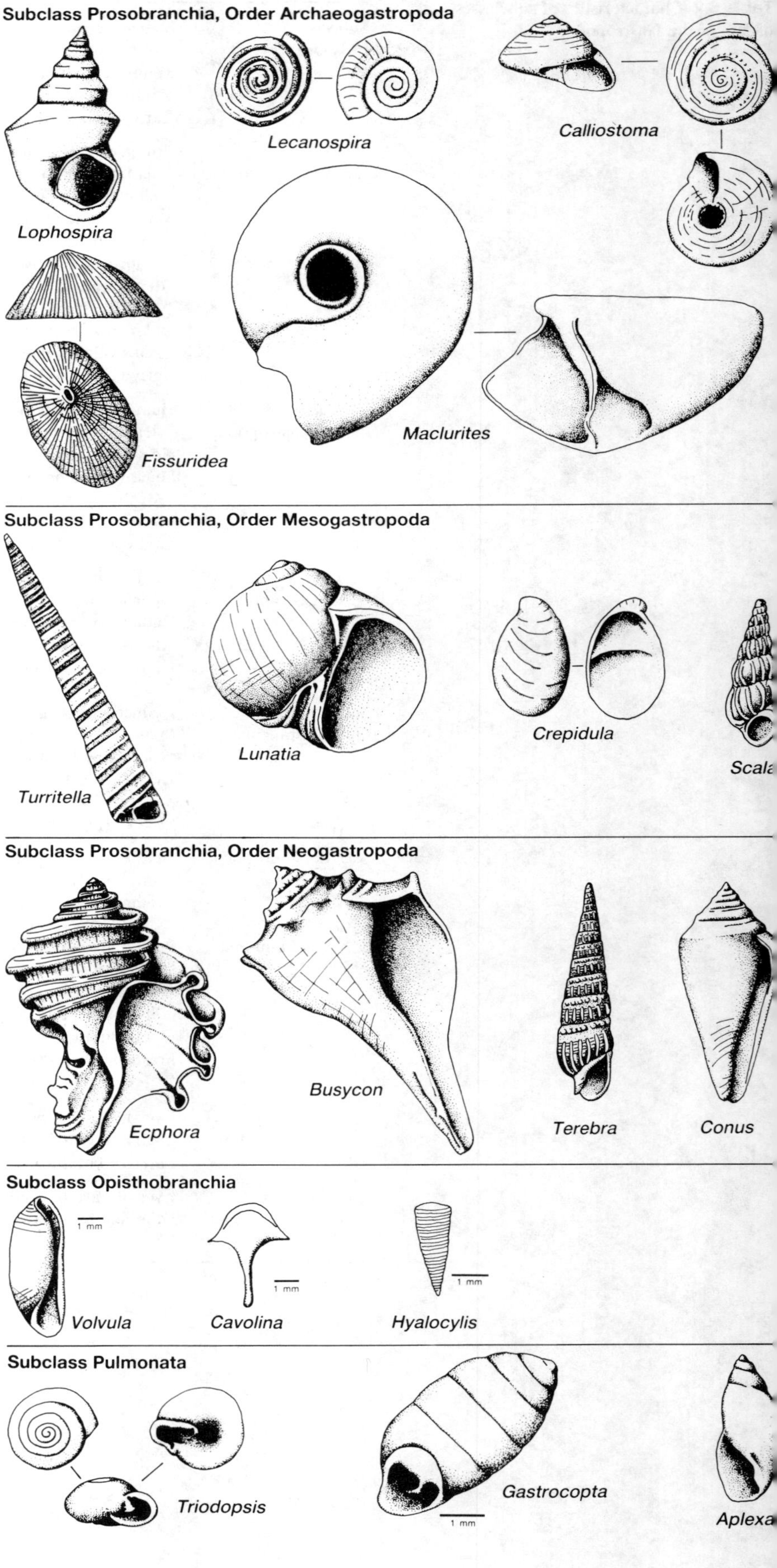

Fig. 8.4 Representative shells of the major subdivisions of the Gastropoda. Except where scale is given, shells are approximately natural size. From Clark (1904), Shattuck (1906), Butts (1926, 1941, with permission from the Virginia Division of Mineral Resources), and Leonard & Frye (1960, with permission from Illinois State Geological Survey).

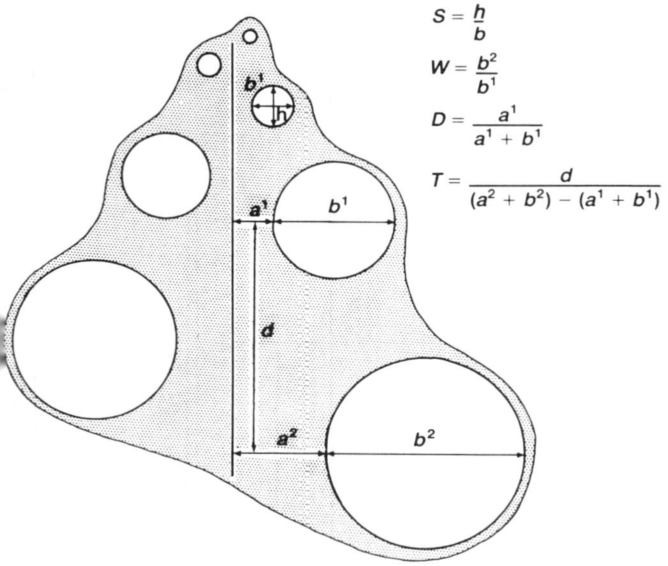

Fig. 8.5 Coiling parameters of shells used in this exercise.

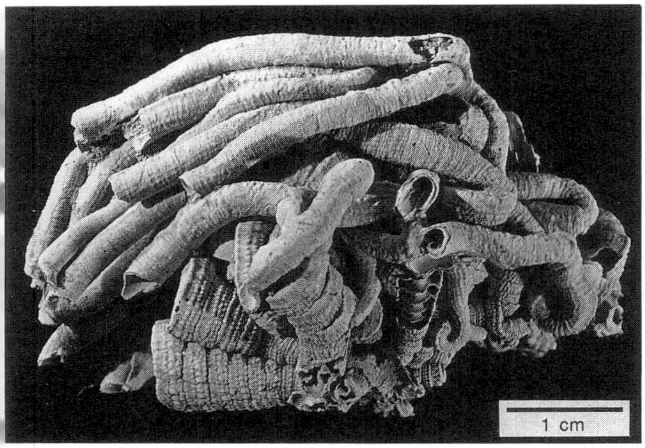

Fig. 8.6 The sedentary mesogastropod *Vermicularia* (Pliocene, North Carolina), in which coiling parameters are variable during ontogeny.

FURTHER READING

Graus, R. R., 1974. Latitudinal trends in the shell characteristics of marine gastropods. *Lethaia* **7:**303–314.

Linsley, R. M., 1977. Some "laws" of gastropod shell form. *Paleobiology* **3:**196–206.

Linsley, R. M., 1978. Shell form and the evolution of gastropods. *American Scientist* **66:**432–441.

Knight, J. B. *et al.*, 1960. Mollusca 1, in R. C. Moore (ed.), *Treatise on invertebrate paleontology*, Part I. Geological Society of America and University of Kansas Press: Lawrence, KS.

McNair, C. G., Kier, W. M., LaCroix, P. D. & Linsley, R. M. 1981. The functional significance of aperture form in gastropods. *Lethaia* **14:**63–70.

Morton, J. E., 1967. *Molluscs*. Hutchinson University Library: London.

Raup, D. M., 1966. Geometric analysis of shell coiling: general problems. *Journal of Paleontology* **40:**1178–1190.

Vermeij, G. J., 1971. Gastropod evolution and morphological diversity in relation to shell geometry. *Journal of Zoology, London* **163:**15–23.

Vermeij, G. J., 1978. *Biogeography and adaptation: patterns of marine life*. Harvard University Press: Cambridge, MA.

EXERCISE 8

Worksheet

Name

QUESTION 1

What features are exhibited by the polyplacophoran, monoplacophoran, scaphopod, and rostroconch specimens that fit their respective class description in Table 8.1? For each, what characteristics of the class may not be seen in the specimen? List several characteristics of the specimen that are not diagnostic for its class but that may be useful in assignment to lower-level taxa within the class.

Polyplacophora. Genus:________________________

Features indicative of class:

Class characteristics not seen:

Other characteristics visible:

Monoplacophora. Genus:________________________

Features indicative of class:

Class characteristics not seen:

Other characteristics visible:

Rostroconchia. Genus:________________________

Features indicative of class:

Class characteristics not seen:

Other characteristics visible:

Scaphopoda. Genus:____________________

Features indicative of class:

Class characteristics not seen:

Other characteristics visible:

QUESTION 2

Draw the partially dissected gastropod that has been extracted from its shell. Label the foot, head, eyes, mantle, gill, and digestive gland.

QUESTION 3

Examine the gastropods provided and enter the information required for each on the table on pp. 122–123.

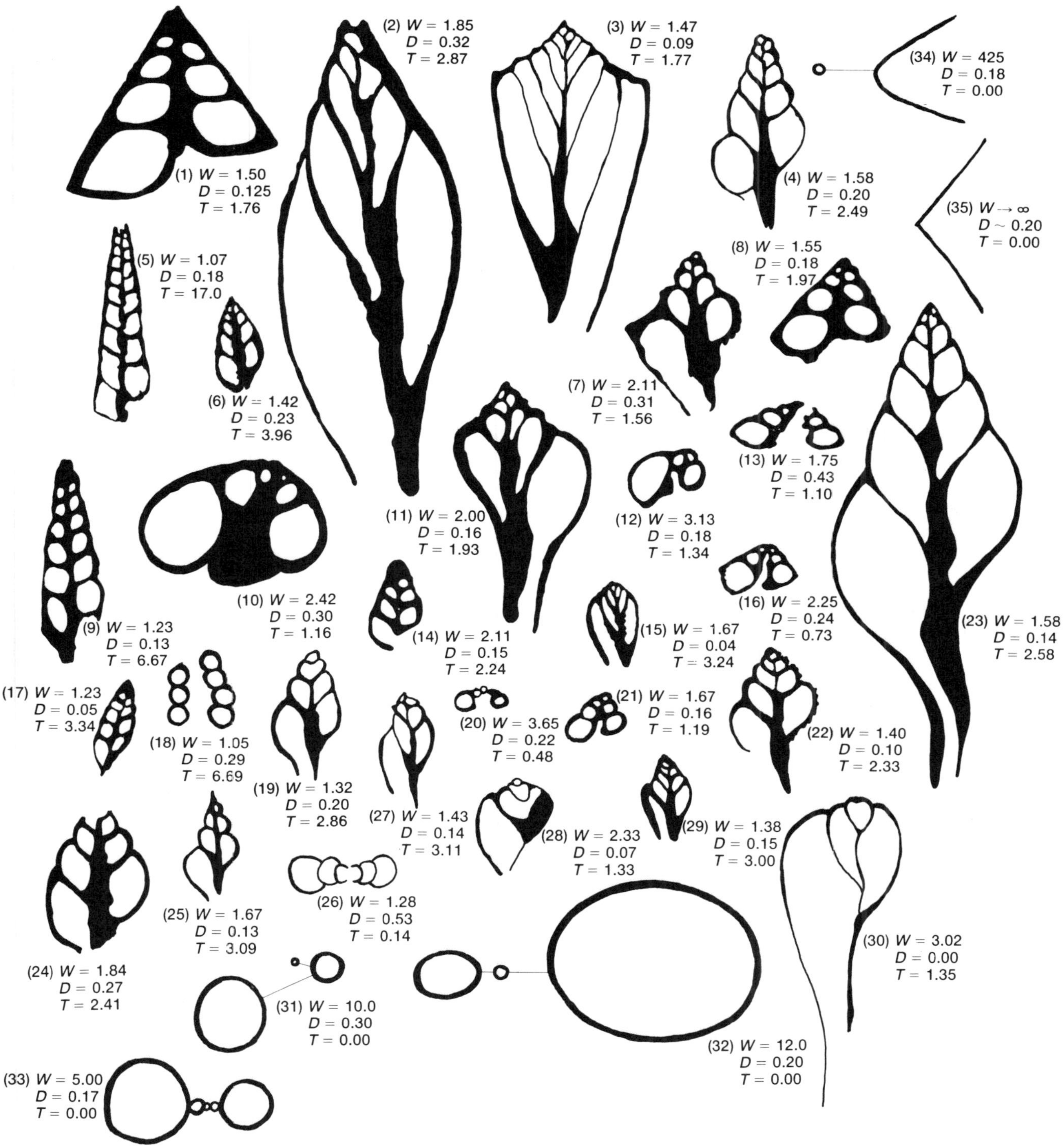

Fig. 8.7 Axial sections through gastropods (1–30) and monoplacophorans (31–35) for determination of *D*, *W*, and *T* as defined in Fig. 8.5. Sections 1–30 are stamped from inked surfaces of actual sections; 26–30 are constructed from imaginary sections through photographs of specimens.

Taxonomic assignment	*Range of genus*	*Shape* (Figs 8.3 and 8.4)	*Surface sculpture* (Fig. 8.3)	*Selenizone* (*present/absent*)	*Siphonal notch* (*present/absent*)	*Siphonal canal* (*present/absent*)	*Shape of aperture* (*approximately equidimensional, wedge-shaped, slit-like*)
Subclass: ______							
Order: ______							
Genus: ______	______	______	______	______	______	______	______
Subclass: ______							
Order: ______							
Genus: ______	______	______	______	______	______	______	______
Subclass: ______							
Order: ______							
Genus: ______	______	______	______	______	______	______	______
Subclass: ______							
Order: ______							
Genus: ______	______	______	______	______	______	______	______
Subclass: ______							
Order: ______							
Genus: ______	______	______	______	______	______	______	______
Subclass: ______							
Order: ______							

Subclass: ____________

Order: ____________

Genus: ____________ ____________ ____________ ____________ ____________ ____________ ____________ ____________

Subclass: ____________

Order: ____________

Genus: ____________ ____________ ____________ ____________ ____________ ____________ ____________ ____________

Subclass: ____________

Order: ____________

Genus: ____________ ____________ ____________ ____________ ____________ ____________ ____________ ____________

Subclass: ____________

Order: ____________

Genus: ____________ ____________ ____________ ____________ ____________ ____________ ____________ ____________

Subclass: ____________

Order: ____________

Genus: ____________ ____________ ____________ ____________ ____________ ____________ ____________ ____________

Subclass: ____________

Order: ____________

Genus: ____________ ____________ ____________ ____________ ____________ ____________ ____________ ____________

QUESTION 4

Fig. 8.7 shows axial sections through 30 gastropods and five monoplacophorans. *D*, *W*, and *T* are given for each. Plot scatter diagrams showing relationships between *D* and *T* and between *W* and *T* for each species, entering blue points for gastropods and red points for monoplacophorans on both scatter diagrams. After the points are completely plotted, tightly encircle each cloud with a line of the same color.

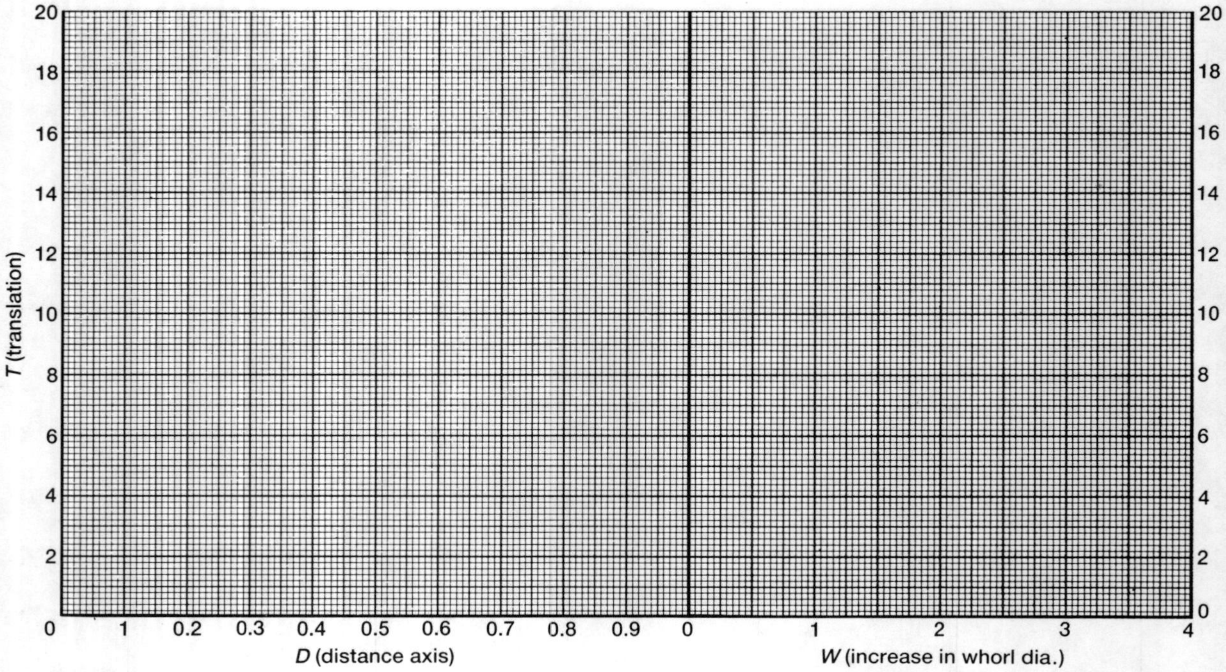

QUESTION 5

Characterize the differences in the cloud of points that represents the gastropods and the cloud of points that represents the monoplacophorans. What might be a cause of these differences in basic shell forms of gastropods and monoplacophorans?

EXERCISE 9 Mollusca II: Pelecypoda or Bivalvia

9.1 INTRODUCTION

This exercise concerns the class of mollusks that has a shell consisting of two dorsally hinged valves that enclose the visceral mass. These animals are characterized by a relatively straight digestive tract, a single pair of elongate gills, and, in most, a muscular foot opposite the hinge. The two valves are mirror images of one another except for those in which the original shell symmetry has been lost, generally through evolution of sedentary or swimming life habits, although symmetry has also been lost in some deposit feeders.

This is one of the groups of organisms that unfortunately have received several names that gained wide currency. At present three (of the approximately 15 that have been proposed) are employed: Lamellibranchiata is the least widely used of the three, and the current focus of dissention is whether to call them Pelecypoda or Bivalvia. The taxonomic name that was first given has priority and usually must be used in the lower-level categories (species through family). But priority does not apply to higher-level categories (order through phylum). Having no particular axe to grind on this matter, I lean slightly toward calling the class Pelecypoda since there are numerous other invertebrates that are also bivalved.

Pelecypods are superb organisms for the study of functional morphology. Many of their adaptations to various life habits (such as burrowing to various depths, free-lying or attached in various ways to different substrates, and swimming) are directly expressed in the overall shape of the shell and also are indirectly expressed in details of shell morphology that reflect adaptive modification of certain organs. The strong relationship between function and morphology in pelecypods is illustrated through frequent convergence in shell form among different lineages that evolved similar life habits.

The purposes of this exercise are:

1 to introduce the basic morphology of, and some of the morphological variation within, Pelecypoda;
2 to understand the extent to which soft-part morphology is reflected in shell morphology;
3 to study common or typical representatives of each of the pelecypod subclasses; and
4 to determine life habits of a few contrasting shell forms.

9.2 SOFT ORGANS AND THEIR RELATION TO SHELL MORPHOLOGY

The basic features of a burrowing pelecypod may be seen in the edible clam *Mercenaria* (Fig. 9.1) and its relatives. The shell valves are connected dorsally, where a tough **ligament** connects them near a **hinge line** that usually has one or more **teeth** and corresponding **sockets** on the opposite valve. Within the shell cavity, two large, approximately equal-sized **adductor muscles** attach to and close the valves when they contract; the anterior adductor muscle is near the mouth, and the posterior adductor muscle is near the anus. (Presence of such roughly equal-sized adductor muscles is termed the *isomyarian* condition.) Contraction of these two muscles pulls the valves closed and puts the ligament, which in this case lies predominantly outside the hinge axis of the valves, under tension. Relaxation of the adductors allows the ligament to contract, thereby absorbing the mechanical energy stored in it when it is stretched, and causing the shell to open because the hinge operates as a fulcrum.

The space between the valves is filled dorsally by the **visceral mass**, and ventrally by a seawater-filled **mantle cavity**. Some of the organs, such as gills and the foot, extend from the visceral mass into the mantle cavity. The **mantle**, which is the thin sheet of tissue that:

1 secretes the shell (both the outer, thin, organic periostracum, and the thicker organic ligament and calcified shell);
2 surrounds the mantle cavity;
3 contains various sensory structures; and
4 may house parts of the reproductive organs or provide supplementary respiration.

The mantle in some pelecypods, such as the giant reef-dwelling *Tridacna*, also houses symbiotic algae. The two lobes of the mantle (one lining each valve) extend the shell by secretion along their outer edge. The mantle margins must be retracted from this vulnerable position when the valves are closed tight, so they contain radial **pallial muscles** around their periphery by which the edge of the mantle is retracted into the shell. The pallial muscles commonly attach along a line (the **pallial line**) roughly parallel with but interior to the margin. Posteriorly the line of attachment of pallial muscles may be indented into a **pallial sinus** where the mantle margins have fused and hypertrophied into elongate siphons that may be extended well beyond the shell margin when the pelecypod is feeding, but may also be partially or entirely retracted into the shell at other times. The greater the mass of siphon that is retracted into the shell, the deeper is the pallial sinus.

Leaf-like gills that consist of numerous, interconnected filaments extend into the mantle cavity from either side of the visceral mass. The foot is medially placed on the ventral side of the visceral mass and is well developed in burrowing forms. In others it may be atrophied, rudimentary, entirely absent, or modified in part for secretion of tough **byssal threads** that serve for attachment or for anchoring to coarse sediment or to hard substrata.

Shell structures that reflect size and location of organs include muscle scars, usually a pallial line, and, in some, a pallial sinus. In addition, some deep-burrowing forms have a heroic-sized foot extending out the anterior end of the shell and similarly outsized siphons extending from the posterior of the shell: these very large features cannot be fully retracted. Closure of the valves does not sever the portion that remains outside. Instead, the valves do not fully touch during closure where tissues remain outside, leaving a **gape**. Smaller gapes exist where byssal threads extend out for anchorage.
(Question 1)

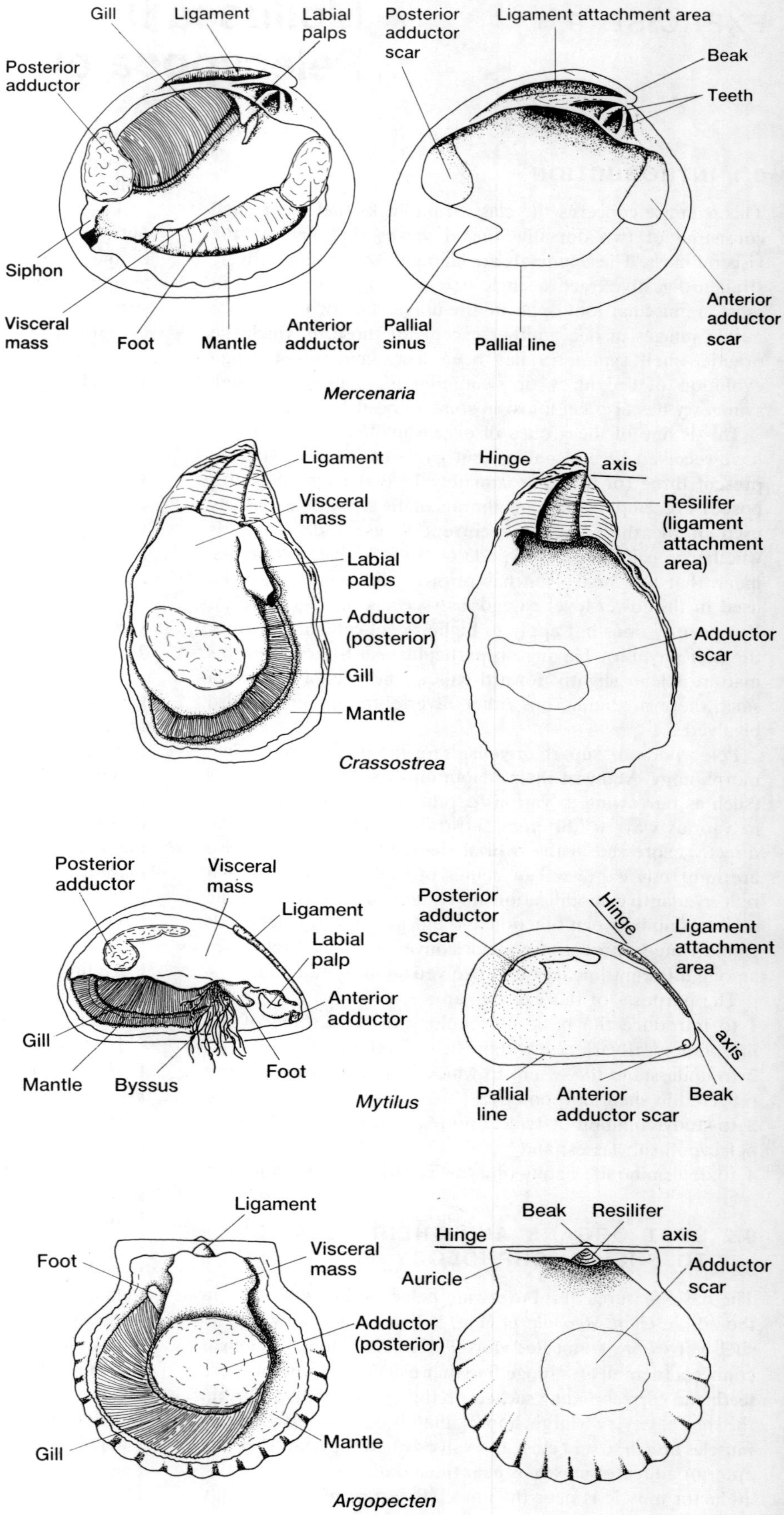

Fig. 9.1 Soft organs of various edible pelecypods and features of skeletal interiors that reflect soft parts.

9.3 VARIATIONS IN SOFT ORGANS AND SHELL FORM

Different adaptations in the gross anatomy of pelecypods are seen in *Mytilus*, *Argopecten*, and *Crassostrea* (Fig. 9.1), all of which along with their close relatives are also commonly eaten shellfish. *Mytilus* attaches ventrally to objects by byssal threads associated with a rudimentary foot, it tapers toward a pointed beak at the anterior end, its adductors are of conspicuously unequal size (therefore, **anisomyarian**), and its shell is very tough but flexible because it incorporates a high proportion of organic material. *Argopecten* swims by rapidly flapping its valves by contraction of its single, large adductor muscle. (Presence of only one adductor is termed **monomyarian**; the large muscle of pectenids is the seafood marketed as scallops.) The ligament is predominantly interior to the hinge and stores energy when compressed by the valves closing. When the adductor muscle relaxes, the ligament quickly springs out to the neutral condition, thereby pushing the valves apart. *Crassostrea* is cemented to a substratum by its larger, more convex valve, with the thinner, more planar valve away from the substratum. It exhibits no bilateral symmetry, but its shape is determined in part by the shape of the space into which it can grow. It, too, is monomyarian. *(Question 2)*

9.4 SUBCLASSES OF PELECYPODS

Many classification schemes have been proposed for pelecypods, each emphasizing differences in state of one particular character, or based on different combinations of characters. A recently proposed but unorthodox classification is given here (Table 9.1). Selected representatives of each subclass, with some morphology labeled, are illustrated in Fig. 9.2. A commonly used alternative subdivision into subclasses may be found in the *Treatise on invertebrate paleontology*, Part N or the paleontology text by Clarkson (1986).
(Question 3)

9.5 FUNCTIONAL MORPHOLOGY

Different pelecypods have evolved life habits that have allowed some members of the group to adapt to virtually all the marine and freshwater benthic habitats. Adaptive modifications of form that enable a pelecypod to live in a particular habitat are often conspicuous and commonly specific to one life habit. The outline that follows summarizes common morphological adaptations for each of the major life habits of pelecypods. The information is summarized in large part from Stanley (1970), supplemented by information from Pojeta in Boardman *et al.* (1987).

1 Byssal attachment.

(a) Commissure perpendicular to substratum: (i) equivalved; (ii) elongation of body and shell (length : height $\geqslant$ 1.35); (iii) reduction of the anterior end so that the shell beaks are terminal or are nearly terminal (although not true for many *Arca*); (iv) decrease in size of the anterior adductor muscle resulting in anisomyarian or even monomyarian condition; (v) presence of a byssal gape in the ventral commissure or a byssal sinus near the transition from the dorsal to the anterior margin of the shell; (vi) generally, the infaunal and semi-infaunal representatives have more similar-sized adductors, are less tapered anteriorly, and have maximum shell width medial or slightly dorsal; epifaunal representatives are more strongly anisomyarian, more strongly tapered anteriorly, and have maximum shell width near the ventral margin.

(b) Commissure non-perpendicular (horizontal or oblique to substratum); (i) inequivalved: (ii) commonly have a byssal sinus near the transition from the dorsal to the anterior margin of the shell, or a byssal notch or perforation in the lower valve; (iii) associated with byssal sinus, the hinge is elongated anteriorly into a pronounced auricle; (iv) posterior elongation of hinge in one group (the pteriids), possibly to separate inhalent and exhalent currents; (v) anisomyarian or monomyarian.

2 Attachment by cementation.

(a) Inequivalved.

(b) High external spinosity in tropical cemented species.

(c) Attached valved may be thick.

(d) Generally preserved *in situ*, attached to original substratum.

(e) Commonly monomyarian.

3 Reclining (unattached but resting on or partially buried in soft substrata).

(a) Almost always inequivalved, with lower valve convex and upper valve usually flat.

(b) Lower valve commonly very thick.

(c) Commonly monomyarian.

4 Intermittent swimming.

(a) Commissure non-vertical while swimming: (i) inequivalved; (ii) shells relatively thin, commonly radially ribbed or plicated near shore but smoother-shelled below normal wave base; (iii) anterior and posterior auricles, if present, about equal size (Fig. 9.3); (iv) umbonal angle large (Fig. 9.3); (v) permanent gapes adjacent to auricles in most (equivalent gapes absent or very narrow in bysally attached relatives); (vi) monomyarian.

(b) Commissure vertical while swimming: (i) equivalved; (ii) thin-shelled; (iii) anterior and posterior auricles, if present, about equal size; (iv) may occur rarely in thin-shelled, tubular burrowers; (v) isomyarian to monomyarian.

5 Burrowing in soft substrata.

(a) General features; (i) equivalved; (ii) most are isomyarian with distinct pallial line; (iii) those with circular shell outlines tend to rock back and forth through a large angle of rotation, penetrating substratum vertically downward. Moderately elongate species typically also have large angle of rotation and move forward somewhat during penetration of substratum. Highly elongate forms rock little if at all and penetrate substratum essentially parallel with long axis of shell (Fig. 9.4); (iv) some have distinct scars for pedal retractor muscles that are smaller than and near to the adductor muscles.

(b) Depth of burial: (i) highly sculptured shells generally result in shallow burrowing; (ii) most burrowing species with thick shells are shallow burrowers; (iii) marginal denticulation (along the commissure) is virtually restricted to shallow, near-spherical burrowers; (iv) deep burrowers are overwhelmingly thin-shelled with little surface texture; (v) depth of pallial sinus may indicate approximate depth of burial in a general way; longer siphons of deeper burrowers

Table 9.1 Characteristics of pelecypod subclasses. Largely after Pojeta in Boardman *et al.* (1987).

	Subclass				
Character	Palaeotaxodonta	Isofilibranchia	Heteroconchia	Pteriomorpha	Anomalodesmata
Overall shell form	Equivalved, beak displaced away from midline of valve and oriented toward either end; shell elongated posteriorly or anteriorly	Equivalved, beak strongly displaced from midline toward anterior end; highly elongate posteriorly	Variable, typically equivalved, beaks pointed anteriorly but not at end of shell	Variable, most with auricles, lobes, or wings, and anterior end reduced; many inequivalved; beaks may be terminal anteriorly	Short to elongate, most subequivalved, some equivalved or inequivalved; overlap of dorsal margin common; beaks may penetrate one another
Composition and microstructure of shell	Aragonite, often with inner nacreous layer	Aragonite or aragonite and calcite, inner layer commonly nacreous	Usually aragonite, ultrastructure variable	Aragonite, calcite, or both; ultrastructure variable	Aragonite, commonly nacreous inner and prismatic outer layers
Gapes	Lacking in most	Generally lacking except narrow byssal slit in some	Lacking in most	Byssal gape in many	Deep burrowers with pedal and siphonal gapes; byssal gape; or lacking
Hinge dentition	Most with many similar teeth (*taxodont*), some with short bladelike teeth radiating from beak and long teeth parallel with hinge axis (*actinodont*), a few lacking hinge teeth (*edentulous*)	Generally edentulous but may have one or more small teeth below beak or small posterior taxodont teeth	Edentulous space separates groups of a few large teeth (*heterodont*) in most, some actinodont, few taxodont	Variable; taxodont, actinodont, heterodont, or edentulous	Edentulous or with one large tooth; lateral teeth seldom present
Ligament	External, mainly posterior to beak, and attached in distinct ligamental grooves or internal, centered on beak and attached to spoon-shaped structure (*resilifer*)	External, centered on beak	Mainly external, centered on beak	Ligament external or internal, posterior to or centered on beak; resilifer present for some internal ligaments	Semicircular, centered on beak, primitively external, secondarily semi-internal, rarely internal; absent in a few

Pallial line and siphons	Lacking sinus in most	Sinus lacking, siphons absent	Sinus present or lacking, many with mantle fusion and siphons	Sinus lacking, siphons absent	Sinus lacking to enormous; siphons common, often with ventrally used manfle lobes
Foot and byssus	Foot well-developed; adults lacking byssus	Foot rudimentary, byssus generally well-developed	Foot well developed in most, reduced in a few, byssus present in some	Foot generally rudimentary or absent; byssus well developed to absent	Some with adult byssus; foot may be hypertrophied
Gills	Broad, leaf-like gill filaments either unconnected or attached by cilia tufts	Narrow filaments joined by cilia tufts (*filibranch*)	Filibranch in some; most with narrow gill filaments interconnected with themselves and neighboring filaments by connective tissue (*eulamellibranch*)	Filibranch in most, eulamellibranch in some	Eulamellibranch, absent in some
Muscles	Most isomyarian, some with posterior adductor reduced or lacking	Most anisomyarian, with anterior adductor reduced (absent in a few)	Usually isomyarian, some anisomyarian; some with internal shells lack adductor muscles	Commonly anisomyarian, some isomyarian or monomyarian	Most isomyarian, some anisomyarian
Life habits	Marine, infaunal, most living are detritus feeders	Mostly marine or brackish, mainly epifaunal and byssally attached, some semi-infaunal, some nestle, some boring	Marine to freshwater, most burrowing and suspension feeding; all life modes but swimming	Most marine, very few freshwater, attached epi- or semi-infaunal; some with byssus or cemented; all suspension feeders	Marine, most burrowing suspension feeders; some nestling, semi-infaunal, reclining, cemented, commensal
Geologic range	Early Cambrian–Recent	Early Cambrian–Recent	Early Early Ordovician–Recent	Late Early Ordovician–Recent	Late Middle Ordovician–Recent

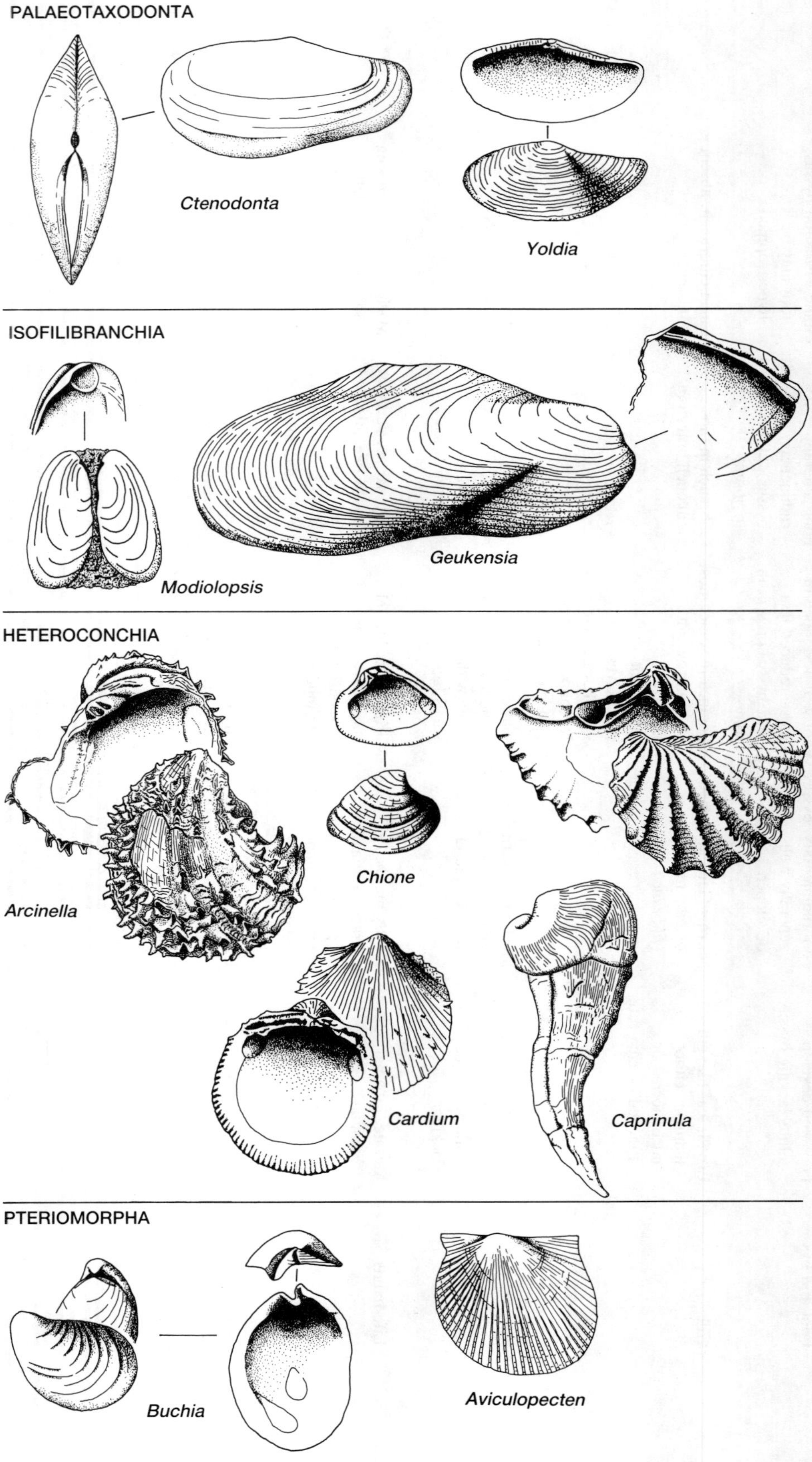

Fig. 9.2 Representative genera of the pelecypod subclasses. Approximately natural size or slightly reduced. Most after Zittel (1900), Clark (1904), Shattuck (1906), and Wade (1926).

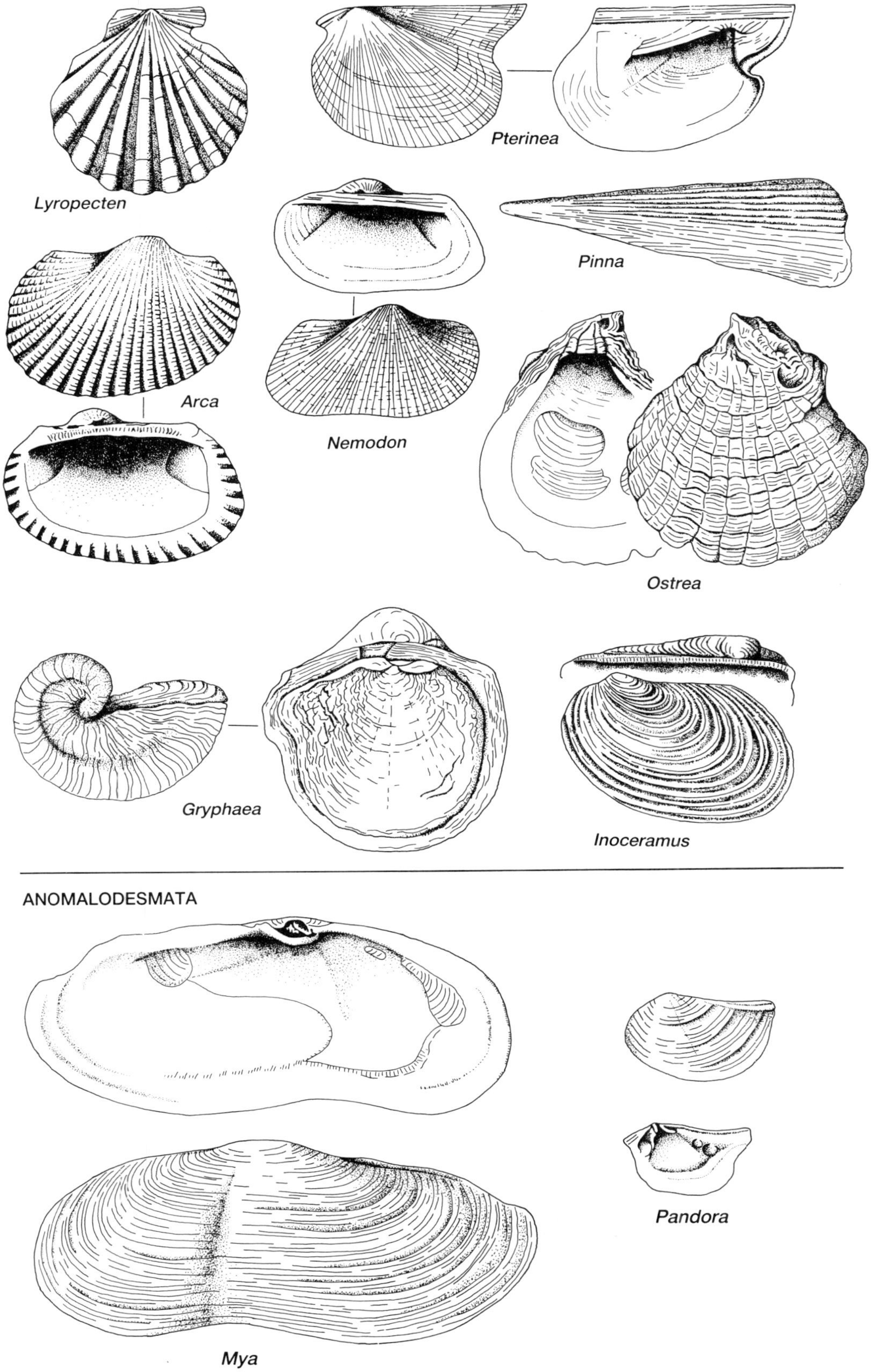

Fig. 9.2 *(Continued)*

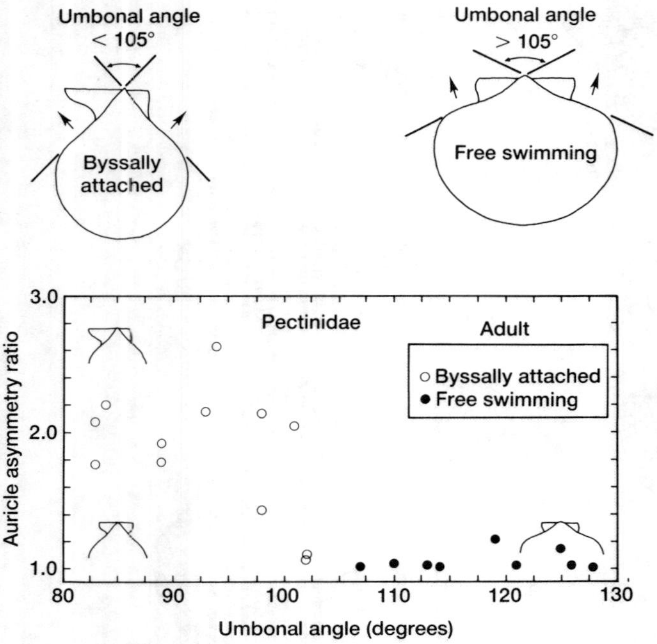

Fig. 9.3 Contrasting morphology of swimming and byssally attached species in the Family Pectinidae of the Subclass Pteriomorpha. Swimming pectinids have broader umbonal angles and typically more symmetrical hinge auricles than do byssally attached species. After Stanley (1970).

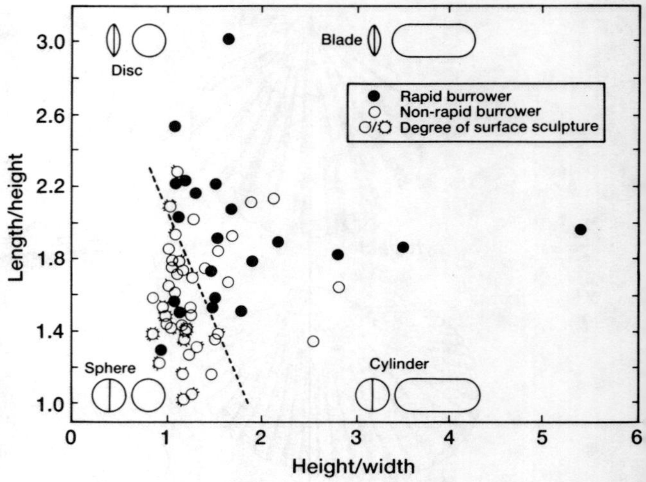

Fig. 9.5 Relationship between overall shell shape and burrowing rate in pelecypods. Both axes of the graph define shape by representing ratios of dimensions, and burrowing rate for various species is indicated by white versus black points plotted on the graph. To the left of the dashed line on the graph, most species are relatively slow burrowers and to the right, most species are relatively rapid burrowers. From Stanley (1970).

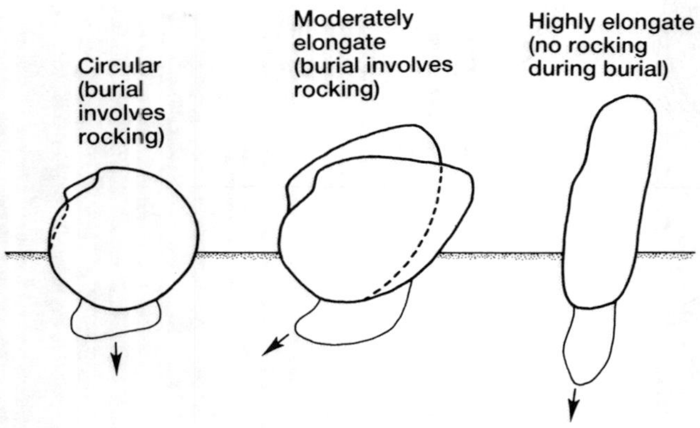

Fig. 9.4 Orientation, behavior, and angle of penetration of burrowing pelecypods that have different degrees of shell elongation. After Stanley (1970).

commonly require a deeper pallial sinus when they are withdrawn into the shell, but some deep burrowers with slender siphons do not require a deep pallial sinus (in any case, presence of a pallial sinus indicates a siphonate species); (vi) some deep burrowers have large pedal (anterior or antero-ventral) and/or siphonal (posterior) gapes.

(c) Rapidity of burrowing: (i) small animals often can burrow more rapidly than can large animals; (ii) species with relatively spherical shells most commonly are slow burrowers; those that are discoidal, bladed, or cylindrical most commonly are rapid burrowers (Fig. 9.5); (iii) highly textured shell surfaces generally result in slow burrowing, except that divaricate surface texture apparently aids burrowing; (iv) a tapered anterior edge and shift of maximum shell width towards the dorsal side are associated with rapid burrowing; (v) rapid burrowing is associated primarily with occupation of shifting substrata (where frequent reburial is required) or deposit-feeding (where food is encountered only by the animal's movement).

(d) Substratum texture and mobility: (i) rapid burrowers in coarse, strongly shifting substrata commonly have relatively thick, smooth shells — those in unstable but less vigorously shifting substrate are thinner-shelled; (ii) burrowers into soft, fine-grained substrata are typically small, and many are thin-shelled; (iii) stable, firm sedimentary substrata are inhabited by pelecypods exhibiting diverse morphologies; a common form has thick valves, elongate posterior, rounded anterior, and moderate inflation (e.g. *Mercenaria*).

6 Boring into solid substrata.
(a) Equivalved.
(b) Cylindrical cross-section.
(c) Many with short, external spines in shell.
(d) Commonly thin-shelled.
(e) Anisomyarian or monomyarian.

7 Nestling in cavities in solid substrata.
(a) See byssal attachment (1, above).
(b) Shells commonly distorted to fit available shape in cavity.
(Question 4)

FURTHER READING

Carter, R. M., 1972. Adaptations of British Chalk Bivalvia. *Journal of Paleontology* **46:**325–340.

Cox, R. L. *et al.*, 1969. Mollusca 6. Bivalvia, Vols 1 and 2, in R. C. Moore (ed.), *Treatise on invertebrate paleontology*, Part N. Geological Society of America and University of Kansas: Lawrence, KS.

Morton, J. E., 1967. *Molluscs*. Hutchinson University Library: London.

Stanley, S. M., 1970. Relation of shell form to life habits of the Bivalvia (Mollusca). *Geological Society of America Memoir* **125**.

Stenzel, H. B., 1971. Mollusca 6. Bivalvia, Vol. 3 (Oysters), in R. C. Moore (ed.), *Treatise on invertebrate paleontology*, Part N. Geological Society of America and University of Kansas: Lawrence, KS.

Thomas, R. D. K., 1975. Functional morphology, ecology, and evolutionary conservation in the Glycymeridae (Bivalvia). *Paleontology* **18**:217–254.

Waller, T. R., 1969. The evolution of the *Argopecten gibbus* stock (Mollusca; Bivalvia), with emphasis on the Tertiary and Quaternary species of eastern North America. *Paleontological Society Memoir* **3**:1–125.

EXERCISE 9

Worksheet

Name

QUESTION 1

A live or preserved whole specimen of *Mercenaria* or one of its close relatives is to be closely studied, then sketched. Refer to Fig. 9.1 and carefully note the positions of the anterior and posterior adductor muscles. If a preserved specimen is being used, insert a scalpel or knife and carefully cut through the adductor muscles adjacent to one of the valves. If the specimen is alive, heat the middle of one valve until the animal is steamed up and opens slightly, then cut the adductors as instructed above. Slowly open the shell by pulling the ventral edges away from one another. Be sure that the visceral mass, muscles, and mantle remain with the valve to which the adductor muscles are still attached. This may require some additional trimming with the scalpel or knife. Note the small muscles that connect the visceral mass to the shell near the hinge. Next, completely cut away the valve that has been separated from the adductors by cutting through the ligament.

From the mass of soft tissues, carefully cut away and discard the mantle that had lined the detached valve. This will give a clear view of the visceral mass, foot, adductor muscles, siphons, gill, and labial palps on the side that is uppermost, and the mantle that lines the valve that remains. Study the specimen until you are certain that you can identify each of the features that were just listed, then gently push the mantle up toward the visceral mass to see where the pallial muscles anchor along the pallial line.

Make a carefully drawn sketch of the soft parts and the valve in which they lie, labeling every organ and skeletal feature visible, including anterior, posterior, dorsal, and ventral margins of the shell. Then, draw the interior of the opposite, dissociated valve, labeling the hinge and all skeletal features that are indications of the size and location of soft structures. Along with the name of each of these latter skeletal features that you label, write the name of the organ(s) about which it provides information.

QUESTION 2

A mussel (*Mytilus* or *Geukensia*), scallop (*Argopecten* or a related type), or oyster (*Crassostrea* or *Ostrea*) is provided for study and sketching. Follow the directions in Question 1 for preparation, study, and sketching, being particularly careful to note where the muscle(s) are located before attempting to open the shell.

QUESTION 3

For each of the laboratory specimens available that represents the various pelecypod subclasses, enter the specimen's characteristics in the table below. The purpose of the assignment in this question is to have you make a detailed set of observations on several pelecypod shells that are distributed among the various subclasses. (Your instructor may have a different preference for subclass organization than the one given in this manual.) If any of the features that you record are not typical for the subclass as far as can be determined from Table 9.1, circle the symbol for the feature after you have entered it into the table for this question. Use the following symbols to complete the table.

Equality of valves: E = equivalved; I = inequivalved.

Muscles: I = isomyarian; A = anisomyarian; M = monomyarian; D = dimyarian, equality of muscles unclear.

Gapes: A = anterior gape; V = ventral gape; P = posterior gape.

Hinge dentition: E = edentulous, hinge teeth absent; T = taxodont, many similar teeth along hinge; H = heterodont, few teeth under beak separated from one or two additional teeth along more posterior portion of hinge; A = actinodont, narrow, elongate teeth radiating from under beak; O = other.

Ligament: EP = external, centered posterior to beak; EC = external, centered on beak; IR = internal, attached to resilifer; IN = internal, no resilifer.

Pallial line and sinus: N = none visible; L = pallial line, no sinus; SS = pallial sinus present, depth equal or less than width; SD = pallial sinus present, depth greater than width;

Thickness of shell: TN = thickness less than 1/20 length; M = thickness 1/20 − 1/10 length; TK = thickness 1/10 − 1/5 length; VT = thickness greater than 1/5 length.

First two letters of subclass name	*Genus*	*Geologic range of genus*	*Equality of valves*	*Muscles*	*Gapes*	*Hinge dentition*	*Ligament*	*Pallial line/ sinus*	*Thickness of shell relative to length and width*

QUESTION 4

Three specimens are provided that represent three different life habits. For each specimen, give the information requested and determine the life habit by referring to the outline of characteristics given in the text of this exercise.

Specimen 1. Genus: ____________________

Formation and age: ____________________ Locality: ____________________

Size. Length: ________ mm height: ________ mm width: ________ mm

Cross-sectional shape. Height : width ratio: ____________________

Location of maximum width: ____________________

Equivalved/inequivalved: ____________________

Lateral view outline. Length : height ratio (elongation): ____________________

Shape of anterior: ____________________

Shape of posterior: ____________________

Hinge elongation/asymmetry: ____________________

Commissure gapes (give location and size of each if present):

____

Thickness of shell relative to length and height: ______________

Texture of exterior of shell: ______________

Interior shell structures. Muscle scars: ______________

Pallial line/sinus: ______________

Relative length of pallial sinus if present: ______________

Ventral commissure denticulation: ______________

Inferred life habit: ______________

Specimen 2. Genus: ______________

Formation and age: ______________ Locality: ______________

Size. Length: ________ mm height: ________ mm width: ________ mm

Cross-sectional shape. Height : width ratio: ______________

Location of maximum width: ______________

Equivalved/inequivalved: ______________

Lateral view outline. Length : height ratio (elongation): ______________

Shape of anterior: ______________

Shape of posterior: ______________

Hinge elongation/asymmetry: ______________

Commissure gapes (give location and size of each if present):

____

Thickness of shell relative to length and height: ______________

Texture of exterior of shell: ______________

Interior shell structures. Muscle scars: ______________

Pallial line/sinus: ______________

Relative length of pallial sinus if present: ______________

Ventral commissure denticulation: ____________________

Inferred life habit: ____________________

Specimen 3. Genus: ____________________

Formation and age: ____________________ Locality: ____________________

Size. Length: ________ mm height: ________ mm width: ________ mm

Cross-sectional shape. Height : width ratio: ____________________

Location of maximum width: ____________________

Equivalved/inequivalved: ____________________

Lateral view outline. Length : height ratio (elongation): ____________________

Shape of anterior: ____________________

Shape of posterior: ____________________

Hinge elongation/asymmetry: ____________________

Commissure gapes (give location and size of each if present):

____

Thickness of shell relative to length and height: ____________________

Texture of exterior of shell: ____________________

Interior shell structures. Muscle scars: ____________________

Pallial line/sinus: ____________________

Relative length of pallial sinus if present: ____________________

Ventral commissure denticulation: ____________________

Inferred life habit: ____________________

EXERCISE 10 Mollusca III: Cephalopoda

10.1 INTRODUCTION

Cephalopods are a class of mobile mollusks, most (or perhaps all) of which have been nektic, i.e. free-swimming. Shelled cephalopods were important members of the marine fauna from latest Cambrian into the Cenozoic, although they are greatly reduced in diversity in modern seas: only two living genera have an external shell. During their long history, the class as a whole, and particularly certain subclasses and orders, experienced near-extinctions and subsequent radiations. Therefore, despite convergent evolution and common **homeomorphy** (great similarity in form), successive cephalopod faunas tend to be distinctive. In addition, cephalopod fossils commonly occur in more than one facies, probably because they were free-swimming above various types of sea floor conditions, and the shells of many probably tended to float for some time after death. Therefore, cephalopods hold a prominent place in biostratigraphy.

The purposes of this exercise are:

1 to introduce the basic morphology of cephalopods;
2 to examine representative fossils of the subclasses of cephalopods that have shells;
3 to focus on the patterns of junction of septa and the outer shell in ammonoid cephalopods; and
4 to determine the stratigraphic level of certain ammonoids.

10.2 BODY ORGANIZATION AND PROPULSION OF CEPHALOPODS

Squids (Fig. 10.1) and cuttlefish are members of the most recently evolved subclass, the Coleoidea. Coleoids have internal shells that may be reduced to no more than a tough, cuticular plate extending along the dorsal side. (This plate is homologous with the outer, chambered shell of *Nautilus*.) The cephalic region bears two well-developed eyes, as well as several forwardly-extending, sucker-bearing tentacles. The digestive tract is recurved, so that the anus empties into the mantle cavity just below the head. The mantle cavity consists of a ventral chamber, in which a pair of gills is located, and a narrow extension that encircles the head. The margin of the mantle is produced as a tube, known as the **hyponome**, on the ventral side. A slit-like opening into the mantle cavity extends along the entire anterior margin of the cavity and is designed so that when the cavity is dilated, water flows in. When muscles quickly deflate the mantle cavity, the slit-like opening locks shut so that the water that is pushed out escapes thorough the hyponome, producing movement of the squid by jet propulsion. This resulting movement of the squid is in the direction opposite that in which the hyponome points. More controlled swimming is generated by undulation of lateral fins.
(Questions 1 and 2)

The externally shelled cephalopod genus *Nautilus* is represented by several species that are broadly distributed across the western tropical Pacific Ocean. In many respects, *Nautilus*' anatomy (Fig. 10.2) is much like that of *Loligo* and its relatives. Differences are that:
1 the entire body is housed in the distalmost chamber (the **body chamber**) of the coiled shell;
2 there are two pairs of gills; and
3 many more tentacles are present.
An additional difference is the presence of a thin tissue strand that extends from the aboral end of the body through a narrow tube (the **siphuncle**) that penetrates the liquid- and gas-filled chambers that constitute the **phragmocone**. Body chamber, phragmocone and earliest-formed portion (**protoconch**) constitute the entire shell. The four extinct subclasses of cephalopods and all fossil nautiloids were characterized by chambered shells, and *Nautilus* is taken as generally representative of the basic soft-part morphology of them all. There is no guarantee that *Nautilus* is not aberrant in one or more ways, but on the other hand there is nothing at all to suggest that it is not typical, except that it may have more tentacles than many fossil relatives and that it continues to exist while other lineages have become extinct.
(Question 3)

Squid and *Nautilus* swim with the dorsal surface uppermost, and occasionally preserved color patterns in both straight and

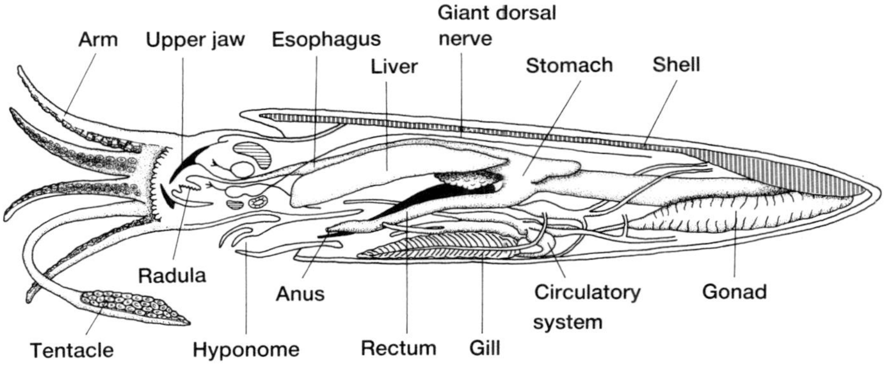

Fig. 10.1 Sagittal section (along the plane of symmetry) through the squid, *Loligo*. Modified from Barnes (1963) with permission from Saunders College Publishing, a division of Holt, Rinehart and Winston, Inc.

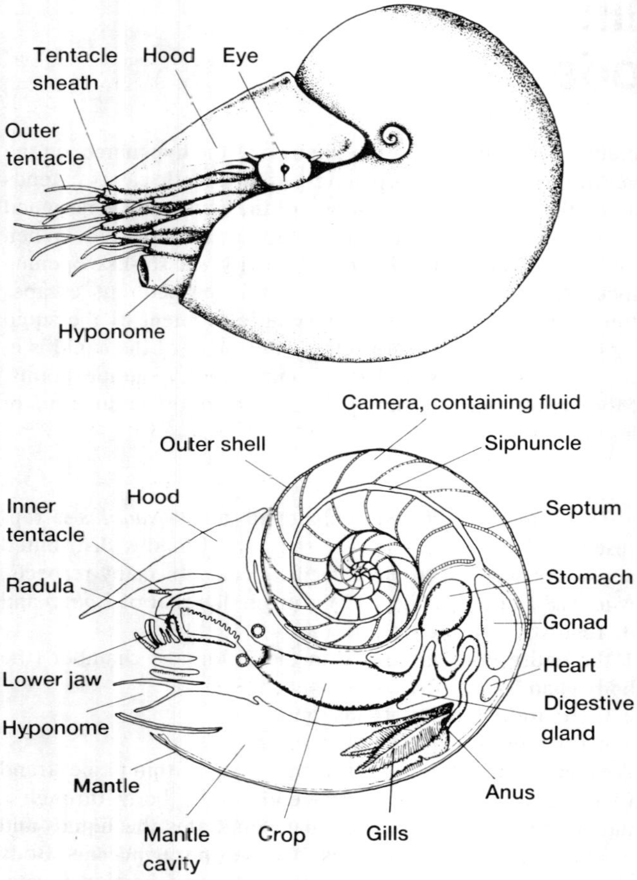

Fig. 10.2 Exterior and sagittal section of *Nautilus*. After Ward (1987).

coiled fossil cephalopods are consistent with the interpretation that this orientation has always been typical. The hydrodynamically streamlined shape of squids allows them to shoot rapidly in the reverse direction when water is ejected from hyponomes that point forward, and they are found at all levels through the water column, relatively independent of substratum. The coiled *Nautilus* is oriented in life position with the gas-filled chambers overhead and the hyponome on the bottom periphery of the shell. Consequently, when it swims the jet propulsion exerts a tangential force on the shell, causing the bottom of the shell to travel while the phragmocone tends to remain almost in place. The force is spent when the mantle cavity is depleted of water, and the undersurface of the shell swings back under the phragmocone, which has travelled only a short distance. This curious, inefficient seesawing motion during swimming may be one reason why *Nautilus* is a benthic carnivore. In fact, it is seldom found swimming at any great distance from the substrate (Saunders & Ward, 1987).
(Question 4)

10.3 CLASSIFICATION

Characteristics of the subclasses of cephalopods are summarized in Table 10.1, and representatives are illustrated in Fig. 10.3. Within most of the subclasses with external skeletons, a broad range of shell morphologies exists. Variations in overall shell form are shown in Fig. 10.4; in addition, some cephalopods had surface sculptures such as accentuated growth lines, annular pinching and swelling, longitudinal striations or ribbing, and various spines or nodes. Most, however, had relatively smooth shells with fine growth lines.

Sutures, which are the traces of the junctions between septa and outer shell, are of great importance in classification

Table 10.1 Characteristics of the subclasses of the class Cephalopoda.

Subclass Nautiloidea	Shells are small to large, primitively straight (**orthoconic**) or gently curved (**cyrtoconic**), but in derived forms they may be strongly coiled with the coils completely overlapping (**involute**) Deposits within the chambers (**cameral deposits**) are common in non-coiled forms, apparently functioning as ballast The siphuncle may be positioned medially or may be displaced toward or to the midventral margin Siphuncle may be slender to large and may have straight to slightly curved and swollen septal necks Deposits within the siphuncle may be present, but they are never endoconic or actinoceratid (see definitions below) Upper Cambrian–Recent
Subclass Endoceratoidea	Shells are medium to huge, usually straight and long (orthoconic), but some may be short (**breviconic**) to slightly cyrtoconic The siphuncle is medium to large and is generally located ventrally or subventrally Septal necks may be short or so long that they overlap Deposits within the siphuncle are closely packed cone-in-cone (**endoconic**) or are concentric longitudinal lamellae No cameral deposits are present Lower–Upper Ordovician

Table 10.1 *(Continued)*

Subclass Actinoceratoidea	Shells are medium to huge and are usually straight The siphuncle consists of septal necks that are recurved and slightly enrolled (**cyrtochoanitic**) and connecting rings that are inflated and typically are broader than they are long. If siphuncles are large, they typically contact the ventral wall, but smaller siphuncles are subventral to near-medial in position. Annular deposits are secreted in the siphuncle and in later stages may be actinoceratid, which means that they fill all but a central canal, radiating canals, and some space beside the connecting rings Cameral deposits may occur on either side of the septa or on the chamber side of the septal neck Middle–Upper Ordovician
Subclass Bactritoidea	Shell is straight to slightly curved, with a small, globular- to egg-shaped protoconch that is marked off from the succeeding shell by a slight to strong constriction, and a large body chamber. The septa are concave toward the distal end and are penetrated by a narrow, ventral siphuncle that has generally straight septal necks No siphuncular or cameral deposits The surface is typically smooth, with growth lines that have a ventral indentation (**hyponomic sinus**) and a dorsal flexure Sutures are simple, characterized by a ventral lobe and in some a dorsal saddle Bactritids apparently link the nautiloids with the ammonoids and with the coleoids Ordovician–Permian.
Subclass Ammonoidea	Most shells are planispiral and to some degree involute Sutures vary from simple gentle lobes and saddles to several superimposed orders of lobes and saddles that may be very tightly curved and bent (i.e. from goniatitic to ammonitic) Siphuncles are simple No siphuncular or cameral deposits Surface sculpture varies from smooth to complex, and some were subject to periodic constrictions in chamber width Lower Devonian–Upper Cretaceous
Subclass Coleoidea	Shells are internal or lacking The basic shell consists of the chambered portion (**phragmocone**) that may have a distal extension (the **pro-ostracum**) on the dorsal side The proximal end of the phragmocone may be extended by massive, concentric skeletal deposits that constitute the **rostrum** (alternatively termed the **guard**) The shell is reduced to varying degrees in all but the extinct order Belemnitida Living members have two gills, eight equal-sized, sucker-bearing tentacles and, in all but the octopus, two longer tentacles that are retractile Lower Carboniferous–Recent

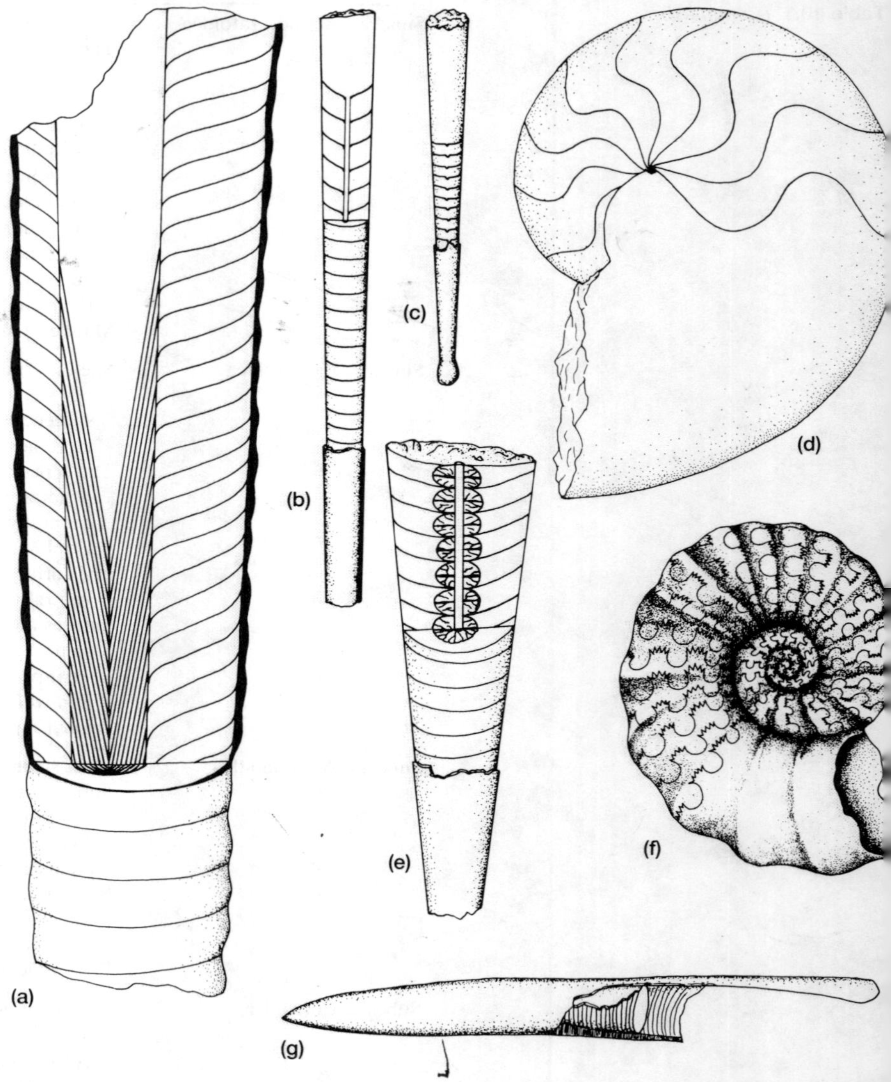

Fig. 10.3 Representatives of the subclasses of the Class Cephalopoda. (a) Endoceratoidea, (b) and (d) Nautiloidea, (c) Bactritoidea, (e) Actinoceratoidea, (f) Ammonoidea, (g) Coleoidea. Approximately natural size. Based in part on illustrations in Hall (1847), Nicholson (1872) and Flower (1962, 1968).

Table 10.2 Types of sutures present in cephalopods. After Pojeta & Gordon in Boardman *et al.* (1987).

Orthoceratitic	Lobes and saddles lacking or broadly undulating and gently rounded Late Cambrian–Recent
Agoniatitic	Characterized by narrow mid-ventral lobe, broad lateral lobe, and perhaps few additional undivided, simple lobes and saddles Early–Middle Devonian
Goniatitic	Several (typically eight) simple, undivided lobes and saddles Lobes rounded to pointed, and ventral lobe subdivided by median saddle Saddles usually rounded Upper Devonian–Upper Permian, uncommonly in Triassic and Cretaceous rocks
Ceratitic	Saddles undivided, but lobes subdivided by second-order crenulations Uncommonly Lower Carboniferous–Permian, common in Triassic, uncommonly Cretaceous
Ammonitic	Lobes and saddles subdivided by second- and perhaps third-order crenulations, producing highly complex sutures Permian–Cretaceous, most characteristically Jurassic and Cretaceous

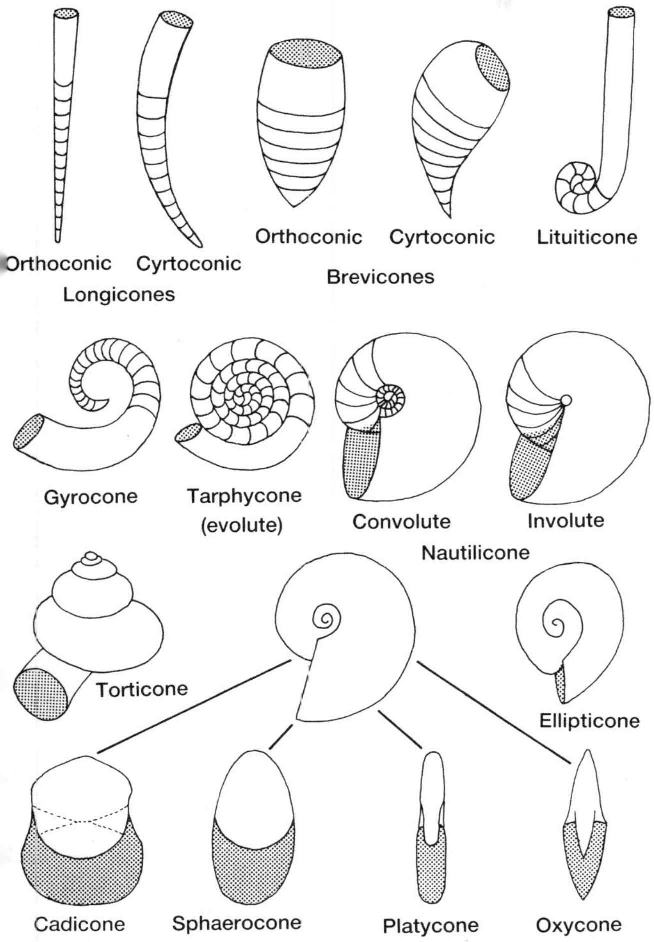

Fig. 10.4 Some common types of cephalopod shells. After Arkell *et al.* (1957) and Teichert *et al.* (1964) with permission from the Geological Society of America.

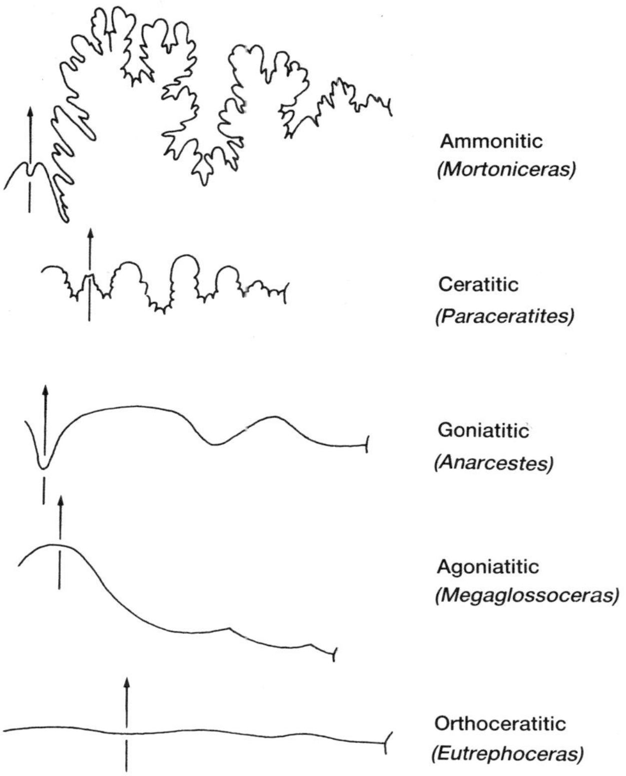

Fig. 10.5 Representatives of the five ammonitic suture types used in this manual.

of shelled cephalopods at all taxonomic levels down to species. However, sutures are more complex and are most important for classification at lower taxonomic levels (species to family) within the subclass Ammonoidea. While some cephalopods, generally some straight-shelled forms, have sutures that project onto a flat plane as a perfectly straight line, most septa are broadly to tightly flexed where they meet the outer shell, producing gently sinuous to finely crinkled projections onto a flat plane. Sutures are classified into categories defined by complexity. Five categories are used here; they are described in Table 10.2 and illustrated in Fig. 10.5.
(Question 5)

10.4 CEPHALOPOD BIOSTRATIGRAPHY

Ranges and diversities of the cephalopod subclasses are given in Fig. 10.6. Width of the spindles for any geological series or set of stages indicates number of genera of the subclass recorded for that interval, except that the spindle for ammonoids plots numbers of families. In addition to differing relative diversities of subclasses up through the geologic column, different types of sutures dominated at different levels within the column. Information on the history of suture types is summarized in Table 10.2.
(Question 6)

The ammonite suborder Goniatitina ranges from Middle Devonian through Upper Permian rocks. Shell form varies from thin discoidal to globular, and whorls range from partially to completely overlapping. Most have goniatitic sutures, some are ceratitic, and a very few have ammonitic sutures. The basic number of lobes is eight, although some have fewer and others have more. As is true for any of the ammonoid groups, goniatites are quite useful for correlation when they are found, and some are so short-ranging that their occurrence precisely pinpoints the stratigraphic level down to a single stage or less. Sutures characteristic of the fat and globular goniatite genera are illustrated in Fig. 10.7, and ranges of the genera are plotted in Fig. 10.8.
(Question 7)

FURTHER READING

Arkell, W. J. *et al.*, 1957. Mollusca 4. Cephalopoda Ammonoidea, in R. C. Moore (ed.), *Treatise on invertebrate paleontology*, Part L. Geological Society of America and University of Kansas Press: Lawrence, KS.

House, M. R. & Senior, J. R. (eds), 1981. *The Ammonoidea*. Academic Press: London.

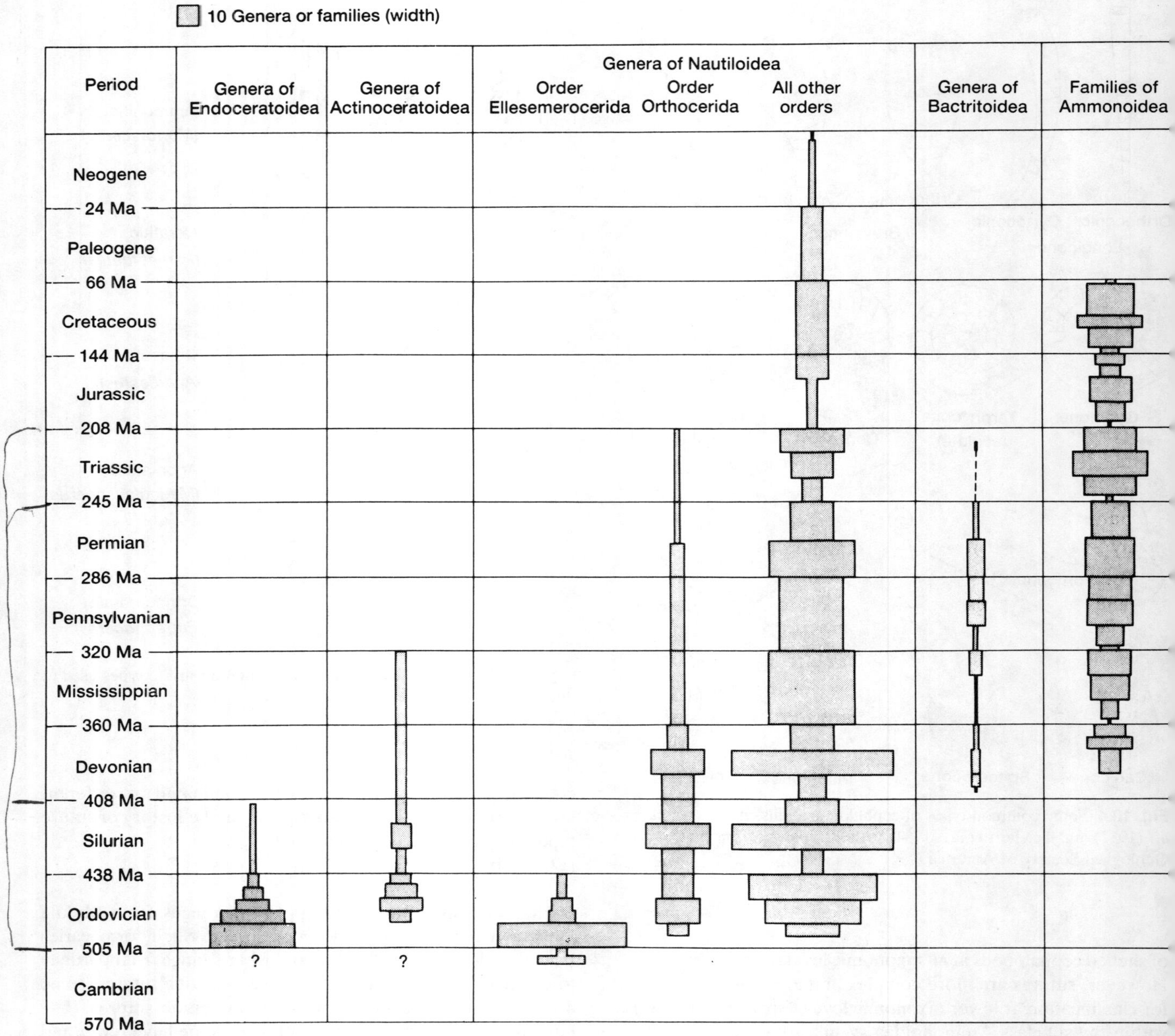

Fig. 10.6 Diversity of shelled cephalopods through time. After Pojeta & Gordon in Boardman *et al.* (1987).

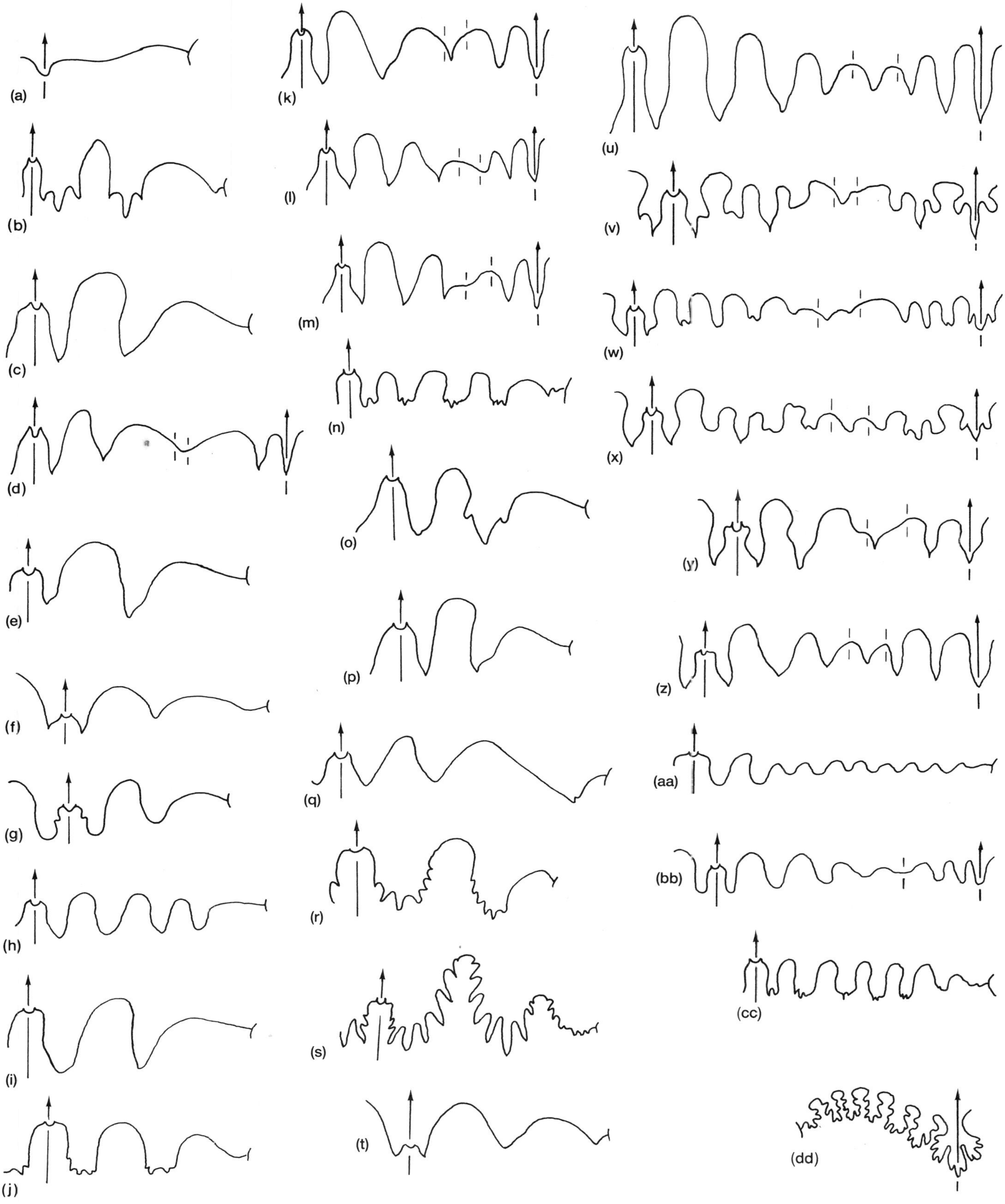

Fig. 10.7 Sutures found in fat to globular genera of the ammonite suborder Goniatitina. (a) *Ramondiceras*, (b) *Delepinoceras*, (c) *Eoasianites*, (d) *Goniatites*, (e) *Cravenoceras*, (f) *Lyrogoniatites*, (g) *Neodimorphoceras*, (h) *Agathiceras*, (i) *Bisatoceras* (smooth) or *Pennoceras* (transversely ribbed), (j) *Eothalassoceras*, (k) *Gastrioceras*, (l) *Paralegoceras*, (m) *Diaboloceras*, (n) *Peritrochia*, (o) *Neoschumardites*, (p) *Owenoceras*, (q) *Emilites*, (r) *Thalassoceras*, (s) *Epithalassoceras*, (t) *Neoglyphioceras*, (u) *Schistoceras*, (v) *Schumardites*, (w) *Vidrioceras*, (x) *Subschumardites*, (y) *Juresanites*, (z) *Melalegoceras*, (aa) *Adrianites*, (bb) *Texoceras*, (cc) *Stacheoceras* (may have 5–10 lobes), (dd) *Waagenoceras*.

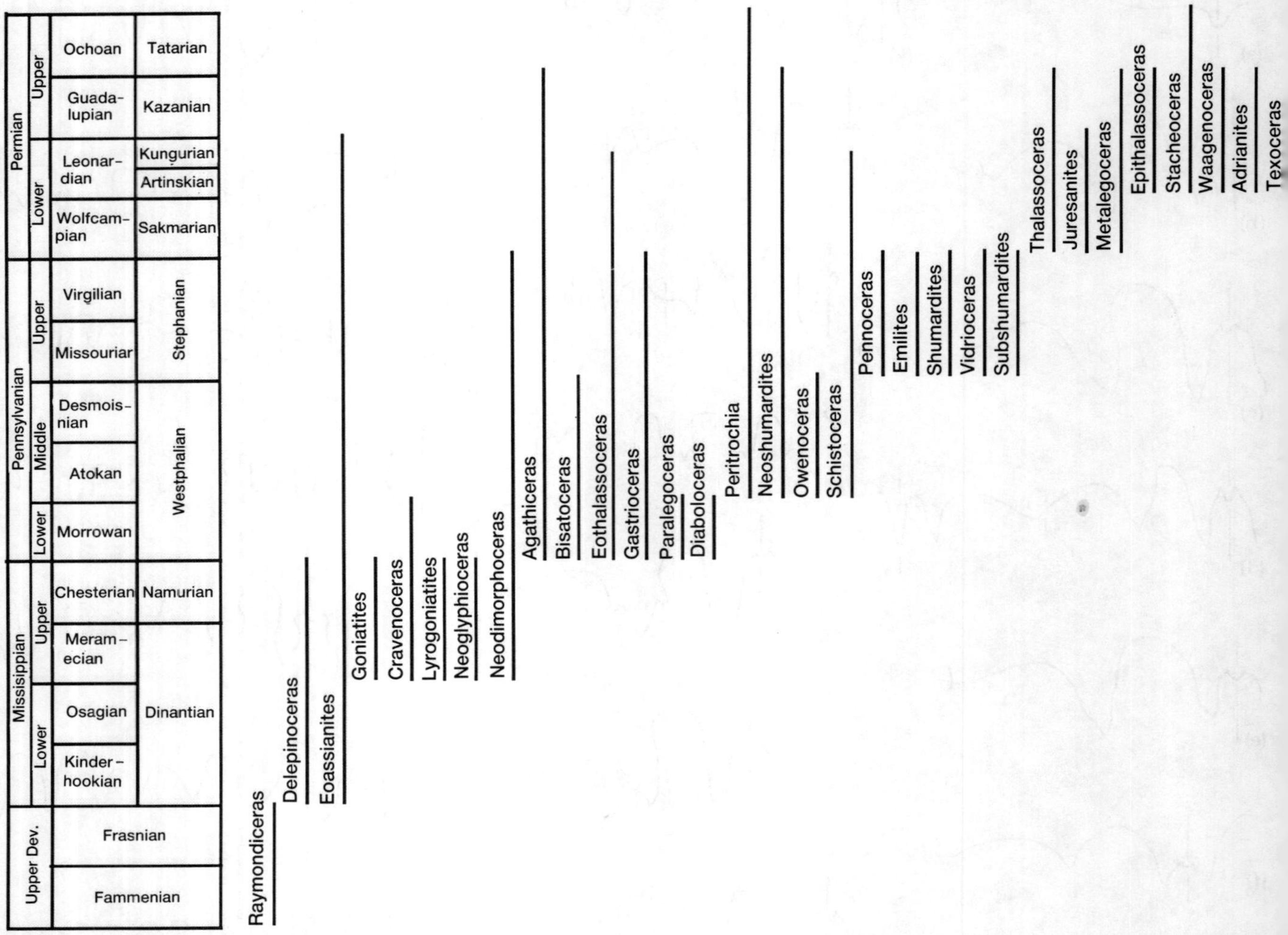

Fig. 10.8 Stratigraphic ranges of the fat and globular genera of the suborder Goniatitina. After information in Arkell *et al.* (1957).

Lehmann, U., 1981. *The ammonites: their life and their world*. Cambridge University Press: Cambridge.

Saunders, W. B. & Landman, N. H., 1987. *Nautilus – the biology and paleobiology of a living fossil*. Plenum Press: New York.

Teichert, C. *et al.*, 1964. Mollusca 3. Cephalopoda Nautiloidea, in R. C. Moore (ed.), *Treatise on invertebrate paleontology*, Part K. Geological Society of America and University of Kansas Press: Lawrence, KS.

Ward, P. D., 1987. *The natural history of Nautilus*. Allen & Unwin: Boston, MA.

EXERCISE 10

Worksheet

Name

QUESTION 1

Carefully dissect a preserved specimen of the squid *Loligo* or a related genus, observing each structure, particularly in the anterior end of the specimen, as you proceed. Make an incision along the plane of symmetry from the ventral side in shallow stages, scoring the specimen along its full length with each stroke. When the dorsal plate of cuticle is encountered, turn the specimen over without cutting through the cuticle. Make a new incision from the dorsal side down to the level where the cuticular plate is again encountered. When the incision is complete, pull the two halves apart so that the cuticular plate remains embedded in one of the halves. In the space below, draw the specimen as viewed toward the cut surface of this half of the dissected specimen. Label the hyponome, mantle cavity, gill, digestive tract, mouth, dorsal cuticular plate, and tentacles.

QUESTION 2

In the space below, make a drawing 3× natural size of the mantle cavity and adjacent tissues as seen in the cut surface, including the part that can be seen on the dorsal side of the head. Draw an arrow to the two tissues that act as a one-way valve into the mantle cavity and that separate from one another and allow water to flow into the cavity, but that make contact and prevent outflow around the mantle periphery when the mantle cavity is depressed.

QUESTION 3

Examine a shell of *Nautilus* that has been cut along the midplane. Within each chamber, the siphuncle consists of two parts: the **septal neck**, which is the skeletal sleeve around the proximally-directed, indented perforation in the **septum** (the plate that divides one chamber from the next), and the **connecting ring**, which is a calcified tube that connects from the septal neck back to the indentation centered on the septal neck of the next-proximal septum. What are the relative thicknesses of:

1 the outer shell: ______________________

2 the septum: ______________________

3 the septal neck: ______________________

4 the connecting ring (if present): ______________________

Determine the thickness of the outer shell at a position 2–2.5 whorls out from the protoconch, using callipers.

Thickness: ____________ mm.

Why is the outer shell thicker than each of the other shell structures?

____

Which part of the shell is thinnest? ______________________

Why? ______________________

QUESTION 4

Record the distribution of the relatively darker part(s) of the surface of the *Nautilus* shell versus the relatively lighter region(s). Taking the living orientation and location into account, what is the best explanation for such a color pattern?

Location of darker area(s): ______________________

Location of lighter area(s): ______________________

Probable reason(s): ______________________

____

QUESTION 5

For each of the study specimens provided, list in the following table the characteristics as displayed by the specimen itself. Ranges may be given on specimen labels or may be determined from the *Treatise on invertebrate paleontology*, Parts K and L. Use the following list as a guide for filling in the characteristics on the table:

Shell shape: see Fig. 10.4. Coleoids should be scored for the following two characteristics of the rostrum.

1 Cross-sectional shape: C = circular, G = indented by single groove, B = bilobed, O = other.

2 Longitudinal profile: F = flared distally, C = cylindrical except tapered proximal end, T = tapered distally from proximal bulge.

Suture type: see Fig. 10.5; O = orthoceratitic, Ag = agoniatitic, G = goniatitic, C = ceratitic, Am = ammonitic.

Siphuncle position: C = central, S = subcentral (PV = ventral, P = peripheral if on convex side of curved or coiled shell, PD = dorsal, if on concave side of curved or coiled shell).

Siphuncular deposits: P = present, A = absent.

Chamber deposits: P = present, A = absent.

Genus	*Subclass*	*Shell shape*	*Suture*	*Siphuncle position*	*Deposits*		
					Siphuncular	*Chamber*	*Rostral*

QUESTION 6

Examine the assemblage of cephalopods that were collected from a single locality, formation, or stratigraphic level. Determine the subclass to which each belongs and the type suture that it exhibits. Then, referring to Fig. 10.6 and Table 10.2, determine the most probable position in the geologic column from which the collection was made. Depending on the assemblage available, you may be able to suggest an interval of less than one system, or the assemblage may not be sufficient to specify an interval more closely than a span of 2–3 systems.

	Subclass	*Type of suture*
Specimen 1		
Specimen 2		
Specimen 3		
Specimen 4		
Specimen 5		
Specimen 6		

Probable position of the assemblage in the geologic column: ______________________

QUESTION 7

A sample of one or more specimens belonging to a single genus of one of the fat or globular Goniatitina is provided. Draw the suture pattern from the specimens in the space below. Then, using Fig. 10.7, identify the genus, and from Fig. 10.8 determine the geologic range.

Genus: ____

Geologic range: ____

EXERCISE 11 Brachiopoda

11.1 INTRODUCTION

Brachiopods are bivalved animals, which gives them a superficial resemblance to pelecypods although the bivalved state arose independently in the two groups. Brachiopods are among the most abundant fossils in Paleozoic rocks that accumulated in marine areas, and some are found also in apparent estuarine and other subsaline deposits. Although brachiopods are most abundant in shallow-water marine rocks, as a group they were adapted to a broad range of habitats. They inhabited substrata of virtually all textures, degrees of firmness, and compositions in shallow seas. This includes hardgrounds, sandy shores, and terrigenous and carbonate muds. In addition, they may be found *in situ* in basinal sediments where sufficient oxygen was present, and deep basinal anaerobic muds commonly contain diminutive, thin-shelled brachiopods that were epibionts on planktic organisms in the waters above. Brachiopods have not been as diverse and abundant since the end-Permian extinction; they still occupy habitats from the estuaries to the deep sea, but with local exceptions they generally are not abundant on shallow sea floors, where they formerly thrived worldwide.

Existence on substrates as different as hardgrounds, firm sands, and soft muds requires a large diversity of adaptations. Whether attached or free-lying, brachiopods exhibit many adaptations to the physical environment. Therefore, interpretations of their functional morphology, especially in combination with their salinity tolerances, make brachiopods useful in studies of ancient environments.

The purposes of this exercise are:

1 to examine representative specimens of the two major groups through careful drawing of soft tissues;
2 to study skeletal morphology in representative genera of various orders;
3 to determine the orders represented and the probable age of a brachiopod assemblage; and
4 to analyze adaptations to firmness of the substrate.

11.2 CHARACTERISTICS OF SOFT STRUCTURES

Two quite distinct groups of brachiopods came into existence during the Early Cambrian and continue to the present day. The two groups are the **Inarticulata** and the **Articulata**, and they exhibit several consistent differences in their shells and organs. The names refer to presence (Articulata) or absence (Inarticulata) of a hinge mechanism consisting of teeth and sockets. The hinge lies along the posterior of the shell, where a fleshy stalk emerges in most species, and the side opposite the hinge is anterior. Characteristics of the shells are summarized later in this exercise; the present section focuses on differences between organs of the two groups (Fig. 11.1).

The two valves of brachiopods are lined by sheets of tissue that constitute the two lobes of the **mantle**. Most organs occur in a visceral mass in the posterior portion of the space surrounded by the two valves. The anterior part of the space is termed the **mantle cavity**. Within the mantle cavity there is a **lophophore**, which consists of two symmetrical arms

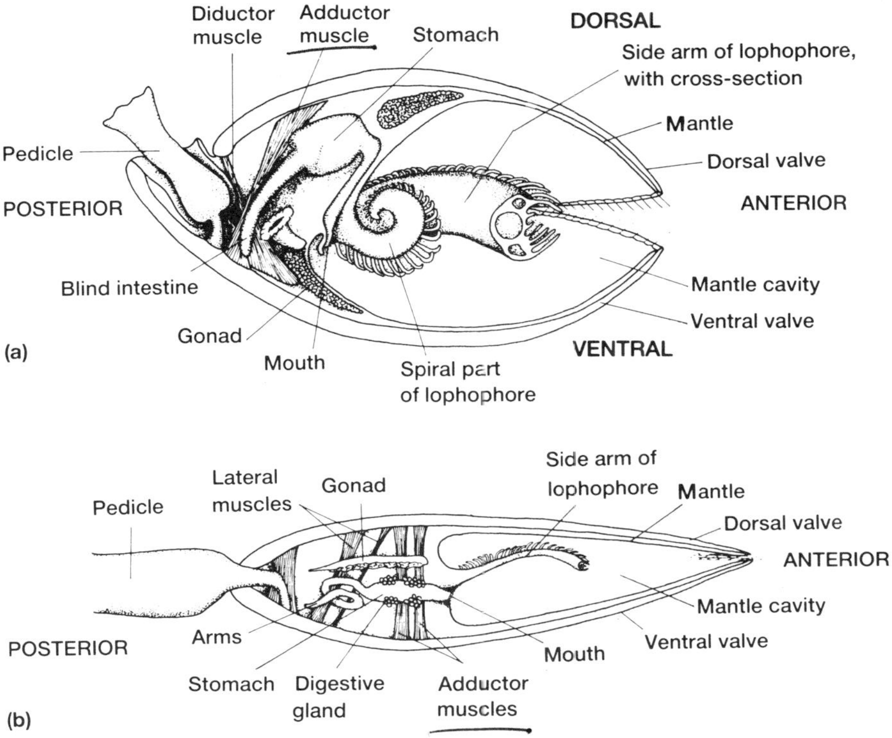

Fig. 11.1 Stylized saggital sections through (a) articulate and (b) inarticulate brachiopods. Part (a) after Williams *et al.* (1965) with permission from the Geological Society of America.

(**brachia**, which contribute to the name of the phylum) that bear many ciliated filaments for feeding. The brachia may be variously curved, looped, or spiralled. A ciliated groove extends along each and serves to transport food to the centrally placed, slit-like mouth. Skeletal supports for the lophophore, if present, extend from the **dorsal valve** (also called the **brachial valve**).

Most brachiopods have a fleshy stalk, the **pedicle**, that emerges from a central area between the valves along the hinge or from a hole present within one of the valves near the midpoint of the hinge. The pedicle is quite different in the two brachiopod groups. In the inarticulates, it is a hollow extension of the ventral body wall and contains an extension of the coelom. The pedicle of articulates, however, extends from the junction of the ventral and dorsal body walls and contains a mass of muscles and connective tissue that attach to the **ventral valve** (also called the **pedicle valve**), but it is not an extension of the coelom. The pedicle typically anchors the animal by cementation to solid substrata or by multiple strands that adhere to grains in sediment, although in some, such as the inarticulate genus *Lingula*, the pedicle adheres to sand grains by a mucus.

Other organs lie within the visceral mass or extend into the mantle lobes. The digestive tract has only one opening (the mouth) within the articulates so is said to terminate blindly because it lacks an anus. Both mouth and anus are present at opposite ends of the "complete" digestive tract of inarticulates.

Muscles that open and close the valves are quite simple and relatively small in articulates, presumably because the hinge articulations inhibit shearing motion of the valves as they open and close. Muscles of the articulates are concentrated near the hinge and extend almost directly across from one valve to the other (Fig. 11.2). Muscles of inarticulates, in contrast, are much more numerous, robust, and broadly distributed within the shells (Fig. 11.2). In addition, some of the muscles cross diagonally from one valve to the other. This musculature in inarticulates not only prevents or minimizes shearing motion that could be placed on the shell (because of the lack of hinge teeth) by water motion, predators, or the process of opening and closing the shell, but also is used in generating valve shear in lingulids in order to burrow.

Each mantle lobe consists of a connective tissue-filled envelope of epithelium around a ramifying coelomic space. The outer epithelium secretes the shell, and the epithelium that lines the cavity serves in gas exhange. **Mantle canals** are branched, tubular extensions of the coelomic cavity into the mantle lobes. In addition, gonads, the nervous system, and the circulatory system extend into the mantle lobes.
(Question 1)

11.3 SHELL CHARACTERISTICS

Symmetry distinguishes brachiopod shells from pelecypod shells. Within a brachiopod shell, the dorsal and ventral valves differ from one another. Therefore, the **commissural plane**, which passes between the valves, is not a plane of symmetry in brachiopods. Instead, all brachiopods have a plane of symmetry that is perpendicular to the valves, passing from the midpoint of the posterior hinge to the midpoint of the anterior edge. Each valve of a brachiopod shell is bilaterally symmetrical within itself.

In contrast, the plane of bilateral symmetry of pelecypod shells, if present, passes between the valves, along the commissural plane. Also, the hinge is dorsal in pelecypods, with anterior and posterior at opposite ends of the shell so that each valve develops an asymmetry: there is no fundamental plane of bilateral symmetry that divides a single pelecypod valve into two mirror-image halves.

Pelecypod shells are constructed only of a calcium carbonate (with a small amount of organic matter), but although most brachiopod shells are constituted of calcium carbonate some are made of calcium phosphate intermixed with much organic material (i.e. **chitinophosphatic**). Most inarticulate brachiopods have chitinophosphatic shells. In addition, pelecypod and brachiopod shells differ conspicuously in their microstructure and their structures associated with opening and closing.

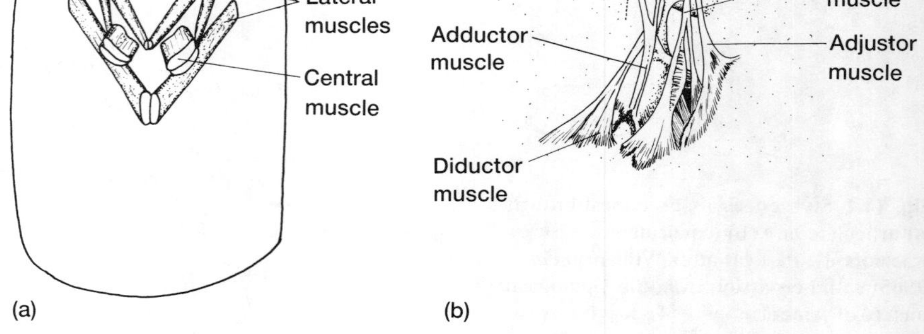

Fig. 11.2 Musculature of (a) the inarticulate brachiopod *Lingula* (an outline view of the ventral valve as seen from the anterior) and (b) the articulate brachiopod *Marandrevia* (an oblique close view of the complete musculature, which is concentrated in the middle of the hinge area). After Williams *et al.* (1965) with permission from the Geological Society of America, and Cooper (1975).

11.3.1 Shape

Both valves of brachiopods may be convex (**biconvex**), have a rounded outline, and be equal in degree of inflation, or their form may be distinctly different from this simple case (Fig. 11.3). At least one valve is always convex, but the other may be flat or even concave, resulting in **planoconvex** or **concavoconvex** shells if the dorsal valve is planar or concave. If the ventral valve is planar or concave, the shell is **convexo-plane** or **convexoconcave**, respectively.

The shell profile varies greatly among brachiopod taxa as viewed toward the broad surface of either valve, with the commissural plane perpendicular to the line of sight (Fig. 11.4). The variation is produced by differences in the rate (and direction) that the shell grows at different points around its periphery.

The hinge may be short or long. It if is short, the shell is typically circular, ovoid, rounded-polygonal or triangular in profile. Those with long hinges may be semi-circular, roughly parallel-sided with the greatest width at the hinge or anterior to the hinge, or slightly to highly **alate** in which the hinge is conspicuously the widest part of the shell.

If shell growth produces a space between the initial point of growth of the shell (the **beak**; centered on the larval shell, which commonly is near the posterior edge of the shell), that part of the shell is termed the **inter-area** (Fig. 11.5). Inter-areas may be present in both dorsal and ventral valves. Inter-areas may extend posteriorly from the beak, or they may be reflected under the beak so that the beak itself forms the most posterior portion of the valve, situated on a posterior projection termed the **umbo**.

11.3.2 Surface sculpture

The surfaces of brachiopods may be virtually smooth, with only concentric growth lines visible. Commonly, however, the surface is more conspicuously textured. Concentric texture may include frilled **lamellose** sculpture of highly pronounced growth lines, or **rugae**, which are concentric or even diagonal wrinkles.

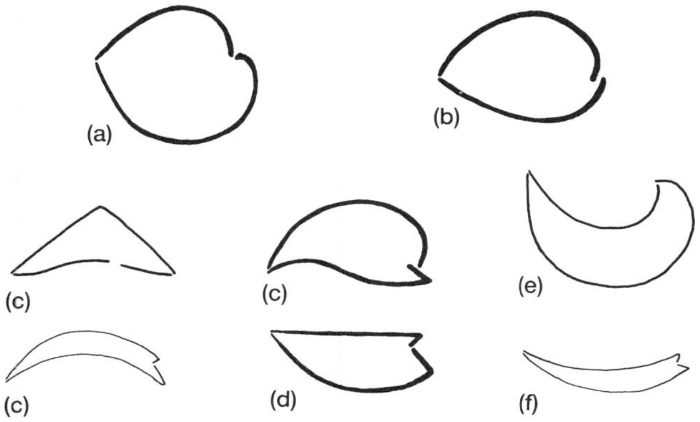

Fig. 11.3 Cross-sectional profiles of brachiopods along the plane of symmetry (beak to midpoint of commisure). The ventral valve is on the underside of each profile. (a) Equally biconvex; (b) unequally biconvex; (c) convexoconcave; (d) planoconvex; (e) deep concavoconvex; (f) shallow concavoconvex.

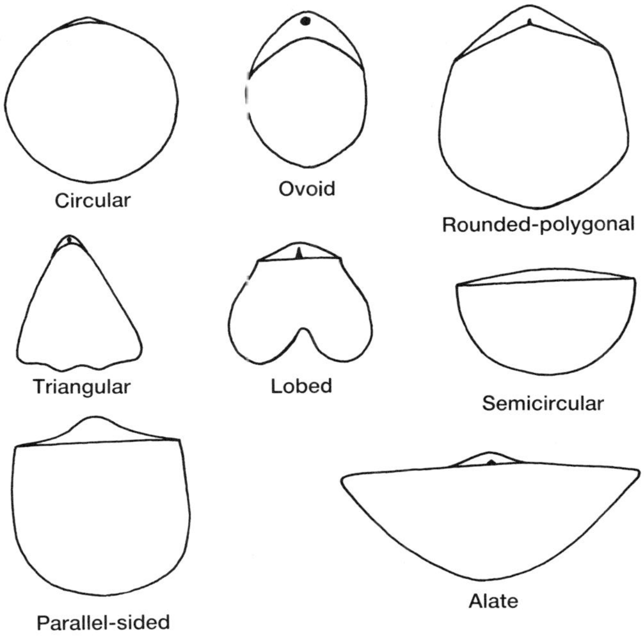

Fig. 11.4 Outlines of brachiopods as viewed from the dorsal or ventral sides.

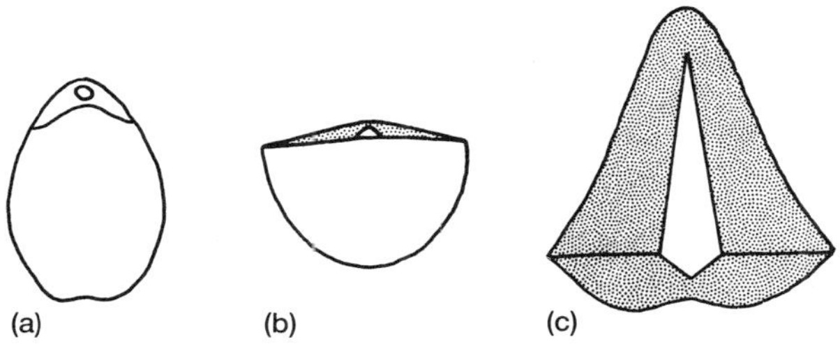

Fig. 11.5 Relative development of inter-areas between hinge and beak. (a) Inter-area minimal or lacking, (b) small inter-area on ventral valve but lacking or minimal on dorsal valve, (c) large inter-areas on both ventral and dorsal valves.

Radial surface features extend from the beak to the anterior and lateral shell edge. The most pronounced of these is seen in articulates as a gentle to deep median depression (a **sulcus**) in one valve and a corresponding elevated **fold** in the other valve. This structure is related to redirection of medially outflowing filtered water away from the lateral regions of inflowing food-bearing water. (Some inarticulates show the reverse: medial inflow and lateral outflow.) There may be multiple radiating **ribs** of various sizes that extend from the beak or that arise as the shell grows.

Both concentric and radial elements may be present. If the two have roughly equal amplitudes, intersection of them produces a **cancellate** surface.

11.3.3 Pedicle opening

A gap or opening through which the pedicle extends is typically present, centered along and located on or near the hinge. The part of the opening that is located in the ventral valve is termed the **delthyrium**, and the part that indents the dorsal valve is the **notothyrium**. If the space is not modified by plates that partially or entirely close it off, it produces a triangular (delta-shaped) notch in each valve. The space may be partially or entirely closed in either valve by a single plate that grows toward the hinge from the beak of the valve or by a pair of plates that close the space from the sides (which may leave a circular opening for the pedicle at the beak of the ventral valve). In some brachiopods the original pedicle opening is completely closed by plates, the pedicle has been lost, and the brachiopod lies free on the sea floor.

11.3.4. Interior structures

Four types of interior features may be found in brachiopods. These include:
1 hinge structures in the articulates;
2 muscle scars and platforms;
3 supports for the arms of the lophophore in articulates; and
4 radiating indentations formed especially by the mantle canals.

The first three types of interior features are summarized below. There is an immense and complex terminology for these features: only the barest introduction to necessary terms is given here.

A single pair of hinge **teeth** and corresponding **sockets** into which they fit are characteristic of the Articulata (Fig. 11.6). The teeth are situated immediately by the corners of the delthyrium or covering plates of the ventral valve, and they may be supported on small **dental plates**. The sockets are located immediately opposite in the dorsal valve. Elongate hinges may have crenulations distributed along them that may supplement the function but rarely replace the pair of true hinge teeth.

The hinge sockets of the brachial valve are part of a complex of structures termed the **cardinalia** (Fig. 11.6) that is associated with many functions. Muscles for opening and closing the valves attach to certain areas of the cardinalia, including various skeletal outgrowths called **cardinal processes**. There are corresponding muscle attachments and their resulting scars situated directly on the surface of the ventral valve.

A pair of narrow projections may extend anteriorly from the cardinalia (Fig. 11.6). Where these features of the dorsal valve appear to have been too short to reach the base of the lophophore arms in some early articulates, they are called **brachiophores**. If long enough that they are inferred to have provided support for the base of the lophophores, they are termed **crura**. More extensive development of lophophore supports (**brachidia**) formed simple or complex **loops** or were spiralled (**spiralia**).
(Question 2)

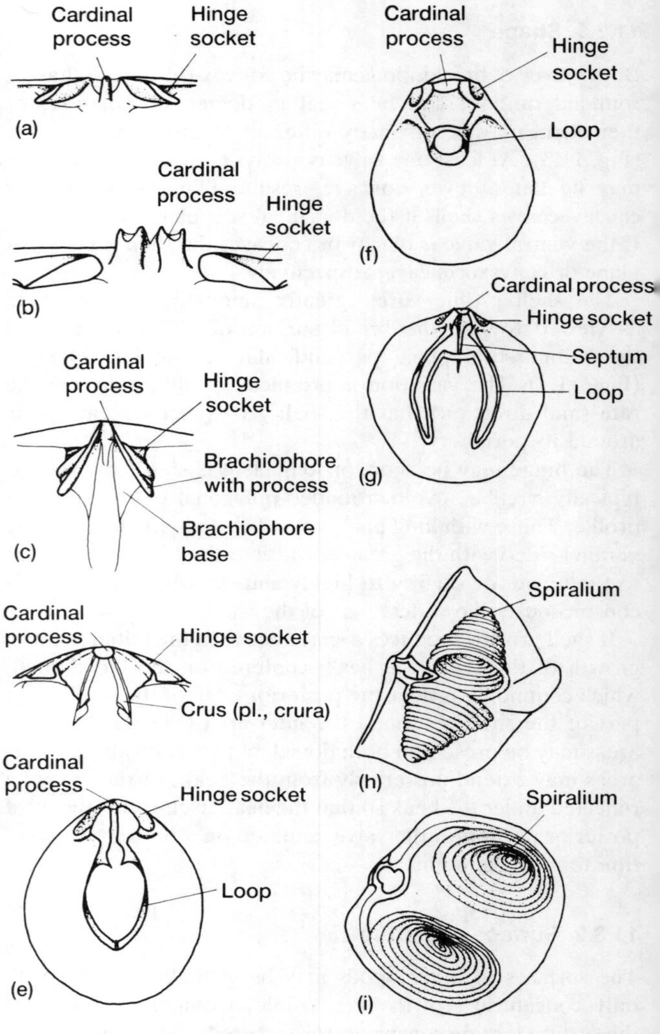

Fig. 11.6 Cardinalia (structures on inside of beak area of dorsal valve for articulation, muscle attachment, and lophophore support) and brachidia (extensive lophophore supports that extend away from cardinalia) in articulate brachiopods. (a) and (b) Lophophore supports not developed, (c)–(i) lophophore supports present: (c) brachiophore, (d) crura, (e)–(g) loop-shated brachidia, (h)–(i) spiralled brachidia (spiralia). After Williams *et al.* (1965) with permission from the Geological Society of America.

11.4 CLASSIFICATION

It has been argued by some that inarticulate and articulate brachiopods were independently evolved and constitute two separate phyla. This argument is based upon many of the contrasts between the two that were seen in Section 11.2 of this exercise, plus very different larvae and transitions from larval to adult states. However, they are treated here in the more standard way, as phylogenetically related subdivisions of a single phylum. Characteristics of the Articulata and Inarticulata and their constituent orders are summarized in Table 11.1. Representatives of the orders and some of the more important suborders are illustrated in (Fig. 11.7).
(Questions 3 & 4)

11.5 DISTRIBUTION AND RELATIVE ABUNDANCE THROUGH TIME

All but one of the orders of brachiopods were established by the end of the Ordovician, and the final order originated during the Devonian (Fig. 11.8). Knowing the relative abundance of the orders for each period, and when various orders became extinct, it is often possible to determine to which

Table 11.1 Characteristics of the classes and orders of the Brachiopoda. After Williams *et al*. (1965) and Rowell & Grant in Boardman *et al*. (1987).

Class Inarticulata	Most with shells of calcium phosphate intermixed with appreciable proportion of organic material, some with calcareous shells Hinge teeth and sockets absent Adductor and oblique muscles present, diductor muscles absent Digestive tract complete (anus present) Hollow pedicle is an extension of ventral body wall Ventral and dorsal body walls separated by a posterior body wall Coelomic space originates in embryos as a hollowing out of the mesoderm Development direct (without organ reorganization in late larval stages) No shelly lophophore supports Pedicle developed from ventral mantle Lower Cambrian–Recent
ORDER LINGULIDA	Most with calcium phosphate shells that are finely **punctate** (penetrated by tissue-filled perforations) or **impunctate** (lacking perforations) Shells gently biconvex with beak marginal in both valves Generally circular, triangular or elongate ovoid in outline Most bearing pedicle, which emerges posteriorly between valves Inter-area-appearing regions (**pseudointer-areas**) in some early forms Lower Cambrian–Recent
ORDER ACROTRETIDA	Shell phosphatic or punctate calcareous, generally circular or subcircular in outline Pedicle opening, if present, located in convex or subconical ventral valve Dorsal valve convex Calcareous forms typically attached by cementation of ventral valve Lower Cambrian–Recent
ORDER OBOLELLIDA	Small, calcitic, biconvex shells, subcircular in outline Beak marginal Pedicle emerging posteriorly between valves, through tube at apex of ventral valve, through resorbed tear-shaped opening anterior to beak, or absent Lower–Middle Cambrian
ORDER PATERINIDA	Small, biconvex, phosphatic shells, rounded or elliptical in outline, with straight posterior margins Pseudointer-areas divided by open delthyrium and notothyrium, commonly partially closed by outwardly convex triangular plates through which pedicle, if present, emerges Lower Cambrian–Middle Ordovician
Class Articulata	Shells calcitic, valves articulated along hinge axis, typically with a pair of ventral teeth and dorsal sockets Adductor and diductor muscles located near posterior midline of valve Digestive tract blind (anus absent) Pedicle solid or fibrous, locally separating ventral and dorsal body walls, which are fused and continuous along the posterior margin lateral to the pedicle Coelomic spaces originate in embryos as infolded pouches Mantle rudiment undergoes reversal in late larval stage Shelly lophophore supports present in many Pedicle develops directly from embryonic rudiment Lower Cambrian–Recent

Table 11.1 *(Continued)*

ORDER ORTHIDA	Shells punctate or impunctate, typically **strophic** (having an elongate hinge line), with hinge about equal to maximum width of shell Typically semicircular to truncated subcircular in outline Unequally biconvex with ventral valve deeper Usually with fine or medium ribs Inter-areas well developed in both valves Delthyrium and notothyrium open or less commonly closed by plates Cardinalia variable Blade-like brachiophores present in some Lower Cambrian–Upper Permian
ORDER STROPHOMENIDA	Characteristically **pseudopunctate** (with laminae of shell interrupted by solid granular rods), strophic, hinge forming or nearly equal to maximum width of shell Shells plano- to concavoconvex, less commonly biconvex or other Inter-areas lacking to prodigious Delthyrium and notothyrium usually closed Pedicle foramen rarely open in adult shells Cardinal process usually present as partition or pair of lobes, socket ridges commonly vestigial or absent Ventral muscle field commonly expanded well beyond umbonal cavity Ordovician–Triassic
Suborder Strophomenidina	Vast majority plano- to concavoconvex, semicircular or subquadrate in outline Ordovician–Triassic
Suborder Chonetidina	Resembling semicircular, concavoconvex strophomenidines, but with a few well-developed tubular spines along posterior margin of ventral inter-area Ordovician–Permian
Suborder Productidina	Inner fibrous and outer lamellar layers in shell, generally strongly concavoconvex shape that may be irregular or extended as conical ventral valve Spines elaborate, not confined to hinge Cardinal process lobate Pedicle absent, but ventral valve attached by ring structure in early stage Devonian–Permian
Suborder Oldhaminidina	Highly derived strophomenids with dorsal valve restricted to small triangular plate from which a large, generally lobate plate extends Mantle cavity not closed dorsally Upper Carboniferous–Permian
ORDER PENTAMERIDA	Generally biconvex, strophic or **astrophic** Ovoid, circular, triangular, trapezoidal, rhombic or pentameral in outline Delthyrium generally unmodified Brachial processes simple, commonly supported by plates at posterior end Dental lamellae fused in apical region of ventral valve, forming spoon-shaped **spondylium** Shell impunctate Middle Cambrian–Devonian
ORDER RHYNCHONELLIDA	Generally impunctate, biconvex, astrophic, triangular to circular in outline, with dorsal fold and ventral sulcus Delthyrium partially closed, but functional pedicle present Highly branched mantle canals Median septum, commonly supporting hinge plates in dorsal valve Dental plates generally present in ventral valve Lophophore spiral Middle Ordovician–Recent

Table 11.1 *(Continued)*

ORDER SPIRIFERIDA	Punctate or impunctate, biconvex, astrophic to highly strophic, circular to alate in outline Delthyrium open or closed Spiral brachidium, first whorl of arms connected by a bridge (**jugum**) Middle Ordovician–Jurassic
Suborder Atrypidina	Impunctate, astrophic Primary whorl of spiralia commonly parallel to plane of commissure Middle Ordovician–Upper Devonian
Suborder Spiriferidina	Impunctate or punctate, strophic Ventral inter-area well developed Jugum simple or absent Primary whorl of spiralia approximately parallel to plane of symmetry of shell Silurian–Lower Jurassic
Suborder Athyridina	Impunctate, astrophic, generally smooth shells intererrupted only by lamellae edges Inter-areas reduced Jugum commonly elaborate Primary whorl of well-developed spiralia parallel to plane of symmetry of shell Upper Ordovician–Jurassic
Suborder Retziidina	Generally punctate, astrophic, strongly ribbed, resembling rhynchonellides Inter-areas reduced Jugum simple Primary whorl of spiralia parallel to plane of symmetry of shell Upper Silurian–Triassic
ORDER TEREBRATULIDA	Punctate, biconvex, typically astrophic, most commonly circular to ovoid in outline Inter-areas confined to ventral valve Pedicle opening restricted by plates Brachidium loop-shaped Lower Devonian–Recent

system of rocks a brachiopod assemblage belongs even without identifying the assemblage below the level of order or suborder.

On average, Cambrian brachiopod faunas are dominated by inarticulates and orthids. Ordovician brachiopods are mostly orthids and gently concave strophomenids, with fairly common inarticulates. Orthids are greatly reduced in Siluran faunas, in which they are slightly less common than gently concave strophomenids and nonalate spiriferids. Spiriferids greatly dominate Devonian brachiopods, and they are progressively overtaken by strongly concave productid strophomenids through the Carboniferous and Permian. Strophomenids, orthids, and spiriferids died out at the beginning of or early during the Mesozoic, so that post-Triassic faunas are dominated by terebratulids and secondarily by rhynchonellids.
(Question 5)

11.6 ADAPTIVE MORPHOLOGY

Brachiopods may be cemented or attached by the pedicle to a hard substrate (including other brachiopods), rooted by the pedicle, or free-lying. The morphology of the shells and their position of burial often indicate which of these modes of life fossil brachiopods had. As a demonstration of adaptive morphology of brachiopods, this part of the exercise focuses on the contrasts in shell characteristics possessed by two brachiopods, one adapted to life on soft mud bottoms and the other ill adapted to soft mud but capable of living on silts and sands swept by moving water.

Benthic invertebrates exhibit a range of adaptations to soft substrates (Thayer, 1975). The various adaptations depend on decreasing the stress that the organism places on the substrate on which it lies. For any object denser than a surrounding fluid medium, the static stress (σ) that it places on the underlying bearing surface is

$$\sigma = (\rho - \rho_f)\left(\frac{S_2}{n\, S_1}\right) g,$$

where ρ is the density of the object, ρ_f is the density of the fluid (1.028 g/cm^3 for 35‰ seawater), S_2 = a shape factor for volume of the object, S_1 is a shape factor for maximum horizontal area of the object, n is the proportion of the horizontal area in contact with the bearing surface, and g is

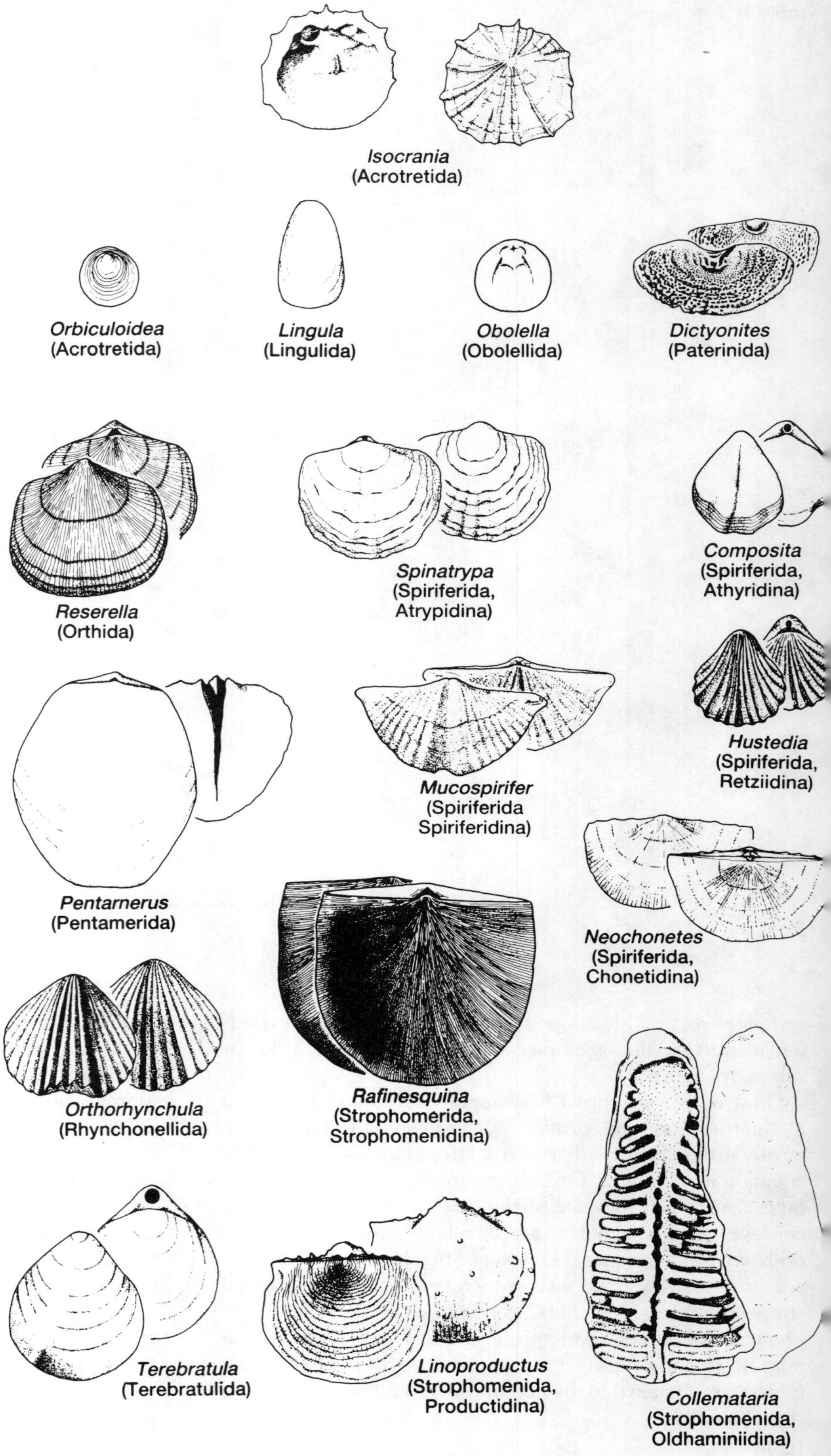

Fig. 11.7 Representative examples of the orders and selected suborders of the Brachiopoda. Class Inarticulata: *Lingula*; *Isocrania*, interior and exterior of ventral valve; *Orbiculoidea*, dorsal valve; *Obolella*, interior of ventral valve; *Dictyonites*, ventral and partial dorsal valves. Class Articulata: (f) *Reserella*, ventral and partial dorsal valves; *Spinatrypa*, dorsal and partial ventral valves; *Composita*, ventral and partial dorsal valves; *Mucrospirifer*, ventral and partial dorsal valves; *Hustedia*, ventral and partial dorsal valves; *Pentamerus*, dorsal valve and partial interior mold of ventral valve; *Rafinesquina*, dorsal and partial ventral valves; *Neochonetes*, dorsal and partial ventral valves; *Linoproductus*, dorsal and partial ventral valves; *Collemataria*, dorsal and partial ventral valves; *Orthorhynchula*, dorsal and partial ventral valves; *Terebratula*, ventral and partial dorsal valves. Approximately natural size or slightly enlarged.

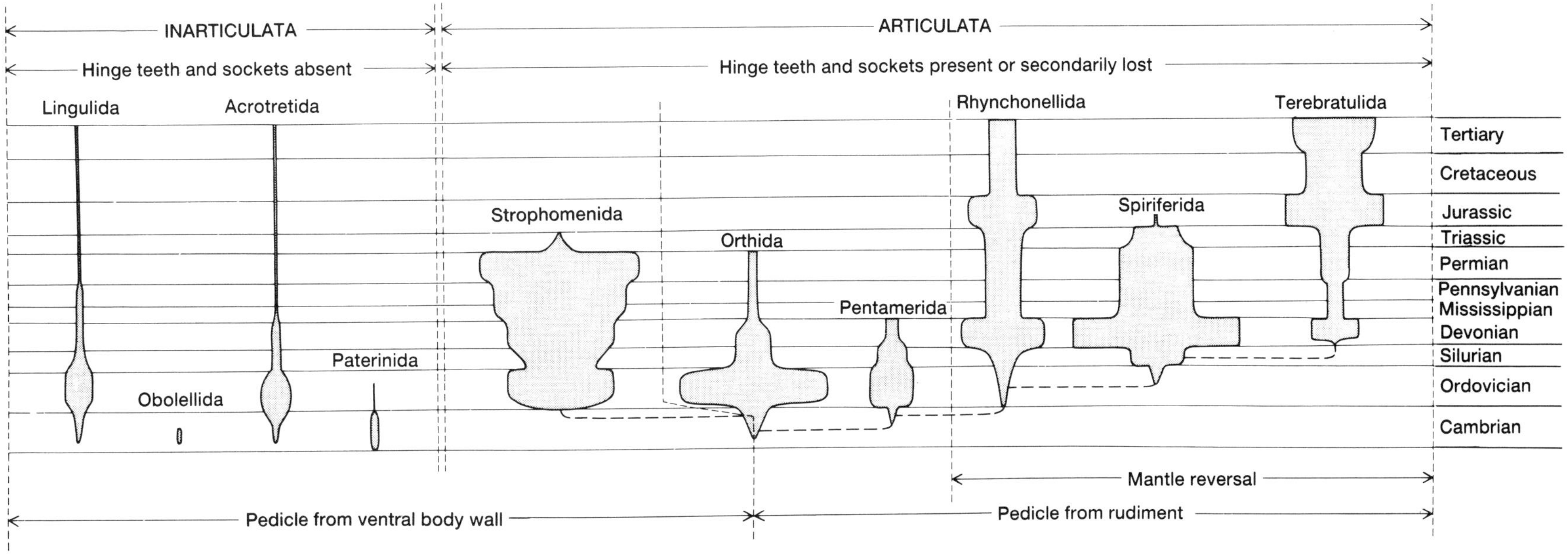

Fig. 11.8 Stratigraphic ranges of most orders of brachiopods. Width of spindles represents approximate relative diversity of genera within the order. From Rowell & Grant in Boardman *et al.* (1987).

the acceleration of gravity. The avenues open to organisms for reducing stress on the substrate include:

1 reduce ρ;

2 increase the bearing area relative to volume or mass, that is, increase nS_1 more rapidly than S_2. Thayer's term for this is the **snowshoe effect**;

3 remain small, that is, limit the increase in r;

4 increase overall ρ_f by sinking the lower parts of the organism into a more dense fluid while the upper portions remain in the less dense medium. Thayer's term for this is the **iceberg effect**; and

5 any combination of (1)–(5).

Where $\rho > \rho_f$, then $\sigma > 0$, and the object will sink into the fluid. But if the fluid increases in density downward, as it does from seawater into underlying sediments, then the difference between ρ and ρ_f diminishes. Where $\rho = \rho_f$, then $\sigma = 0$, and the object will float.

The transition in density from seawater ($\rho_f = 1.028$ g/cm^3) and well-packed sandy sediment is relatively abrupt, occurring at a clear boundary between the water and the grain-supported sediment (even though water fills the pore space between the grains). The transition from water to muds, however, is more gradual as the proportion of clay or mixed clay and silt gradually increases and the proportion of water gradually diminished downward (summarized in Thayer, 1975). Many silts and clays contain more than 75% water by weight and are permanently fluid, and clay-rich muds with 50–75% water become fluid when disturbed. Many of these muds have densities between 1.14–1.25 g/cm^3 in the top 2–3 cm, and density may be as low as 1.3 g/cm^3 even at 30 cm depth. *(Questions 6–8)*

FURTHER READING

Dutro, J. T. & Boardman, R. S. (eds), 1981. *Lophophorates – notes for a short course*. University of Tennessee Department of Geological Sciences Studies in Geology 5.

Rudwick, M. J. S., 1970. *Living and fossil brachiopods*. Hutchinson University Library: London.

Thayer, C. W. & Steele-Petrovic, H. M. 1975. Burrowing of the lingulid brachiopod *Glottidia pyramidata*: its ecologic and paleoecologic significance. *Lethaia* **8**:209–221.

Williams, A. *et al*. 1965. Brachiopoda, in R. C. Moore (ed.), *Treatise on invertebrate paleontology*, Part H. Geological Society of America and University of Kansas Press: Lawrence, KS.

EXERCISE 11

Worksheet

Name

QUESTION 1

Two specimens of Recent brachiopods with preserved soft tissues are provided. In each, the organs remain attached to one shell valve, and the other valve is free of tissues. Part of the tissues that were against the unattached valve have been removed so that more organs can be seen. In the spaces provided below, draw the outline of both valves of each specimen as viewed from the interior, with the apex toward the top of the page. Label the valves (dorsal, ventral). For each specimen, carefully draw the organs that are visible and label any skeletal indication of the organs that can be seen in the unattached valve. Look for the features that are listed below. Not all can be seen, but find and label all that are available in the study specimens.

1 Inarticulate specimen
- (a) Soft parts
  - (i) mantle
  - (ii) lophophore
  - (iii) filaments
  - (iv) mouth
  - (v) stomach or intestine
  - (vi) anus
  - (vii) posterior and anterior adductor muscles
  - (viii) oblique muscles
  - (ix) pedicle
- (b) Shell
  - (i) muscle scars
  - (ii) mantle canals
  - (iii) pedicle groove

2 Articulate specimen
- (a) Soft parts
  - (i) mantle
  - (ii) lophophore
  - (iii) filaments
  - (iv) mouth
  - (v) blind intestine
  - (vi) adductor muscles
  - (vii) pedicle
- (b) Shell
  - (i) muscle scars
  - (ii) cardinalia
  - (iii) delthyrium or notothyrium
  - (iv) mantle canals

Inarticulate brachiopod: ______________________________

Valve interior *Soft tissues*

Articulate brachiopod: ______________________________

Valve interior *Soft tissues*

QUESTION 2

Examine the interior of the dorsal valve of the articulate brachiopod provided. Using a binocular microscope, carefully draw the hinge area of the interior surface, including all the cardinalia, and label the parts that can be seen.

QUESTION 3

For each specimen designated by the instructor, list the following information in the table opposite. Give *only* the information that is available from examination of the skeleton, and write NA wherever information is not available. (For example, if both valves are present and the interior cannot be seen, no information on the cardinalia is available from your specimen.) Use terminology from the following list.

Outline: see Fig. 11.4.
Hinge: short, long, alate.
Inter-area: ventral valve — absent, present; dorsal valve — absent, present.
Pedicle opening: absent, or describe shape of delthyrium as triangular, slit-like, or circular.
Valve convexity: see Fig. 11.3.
Surface sculpture; smooth, growth lines, fold and sulcus, ribbed, reticulate, spinose (give their location).
Cardinalia: cardinal process, brachiophores, crura, loops, spiralia (give orientation of their axis, which will be parallel with or perpendicular to the commissural plane).

QUESTION 4

Briefly re-examine the specimens for question 3, keeping in mind which are articulates and which are inarticulates. Is there a greater proportion of relatively dark (dark red or black) specimens in one or the other?
If so, which?

____

If there is a difference in color, why might this be?

____

____

QUESTION 5

For the assemblage provided, assign each species to order (and suborder in rhynchonellides and spiriferides) and determine the probable system from which it was drawn.

Order (or suborder) of species 1: ____________________________

Order (or suborder) of species 2: ____________________________

Genus	*Order*	*Outline*	*Hinge*	*Inter-area*		*Pedicle opening*	*Valve convexity*	*Surface sculpture*	*Cardinalia*
				Ventral	*Dorsal*				

Order (or suborder) of species 3: ____________________

Order (or suborder) of species 4: ____________________

Order (or suborder) of species 5: ____________________

Order (or suborder) of species 6: ____________________

Order (or suborder) of species 7: ____________________

Order (or suborder) of species 8: ____________________

System from which assemblage was drawn: ____________________

QUESTION 6

The overall density of an invertebrate animal depends on the density of its soft tissues, which is near that of seawater, and density of its skeleton, generally some combination of aragonite, calcite or silica with organic material. As one step in contrasting the mode of life of the brachiopods *Rafinesquina* and *Orthorhynchula*, calculate the approximate ρ, S_1, and S_2 for each. For the purposes of calculation, geometry will be simplified, but the error introduced is small in comparison with the differences in values between the two brachiopods.

Effective density of soft tissues (ρ_{st}) = 1.03−1.03 g/cm^3 = 0 g/cm^3

Effective density of shell (ρ_s) = 2.6−1.03 g/cm^3 = 1.57 g/cm^3

Rafinesquina — modeled as a tabular hollow semicylinder
Measure from shell:

r = ______ cm

$\bar{h}$ = average of 10 h scattered across shell = 0.______ cm

t = thickness of shell = 0.______ cm
Calculations:

$S_1 = 1/2\ \pi\ r^2 =$

$S_2 = \bar{h}\ S_1 =$

$$\sigma = \left(\frac{2\ S_1 t \rho_s + (S_2 - 2S_1 t)\ \rho_{st}}{S_1}\right) g =$$

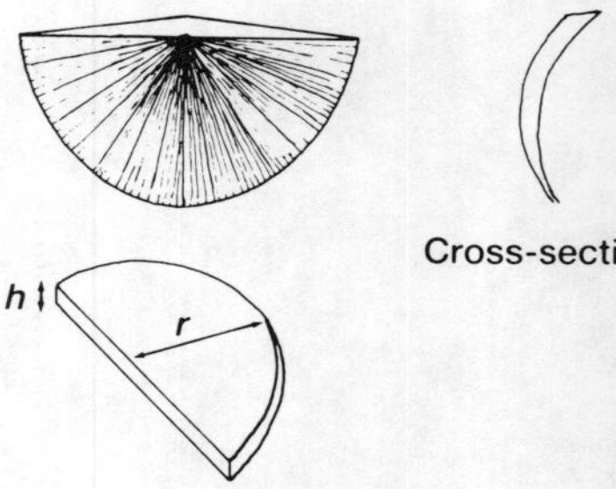

Cross-section
Model

Orthorhynchula — modeled as a hollow sphere with uniform wall thickness
Measure from shell:

r_2 = ______ cm

$\bar{t}$ = average shell thickness = 0.______ cm

$r_1 = r_2 - \bar{t}$ = ______ cm

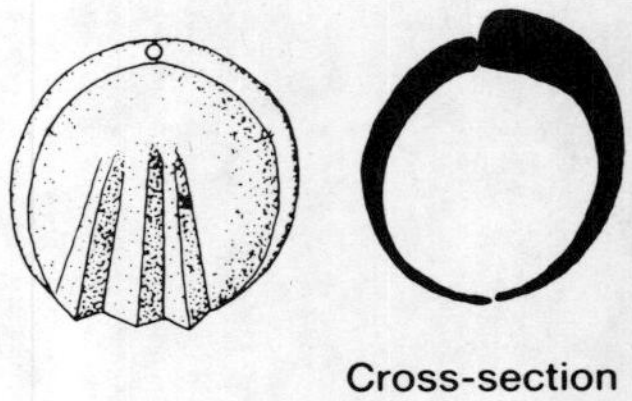
Cross-section

Calculations:

$S_1 = \pi r_2^2 =$

$S_2 = 4/3\ \pi r_2^3 =$

$$\sigma = \left(\frac{(S_2 - 4/3\ \pi r_1^3)\ \rho_s + (4/3\ \pi r_1^3)\ \rho_{st}}{S_1}\right) g =$$

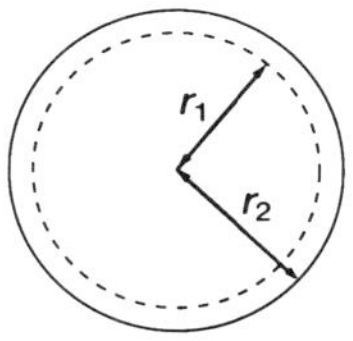

Model

QUESTION 7

Which of the two brachiopods is adapted to lying free on soft substrates?

What specific adaptation(s) to soft substrates does it exhibit?

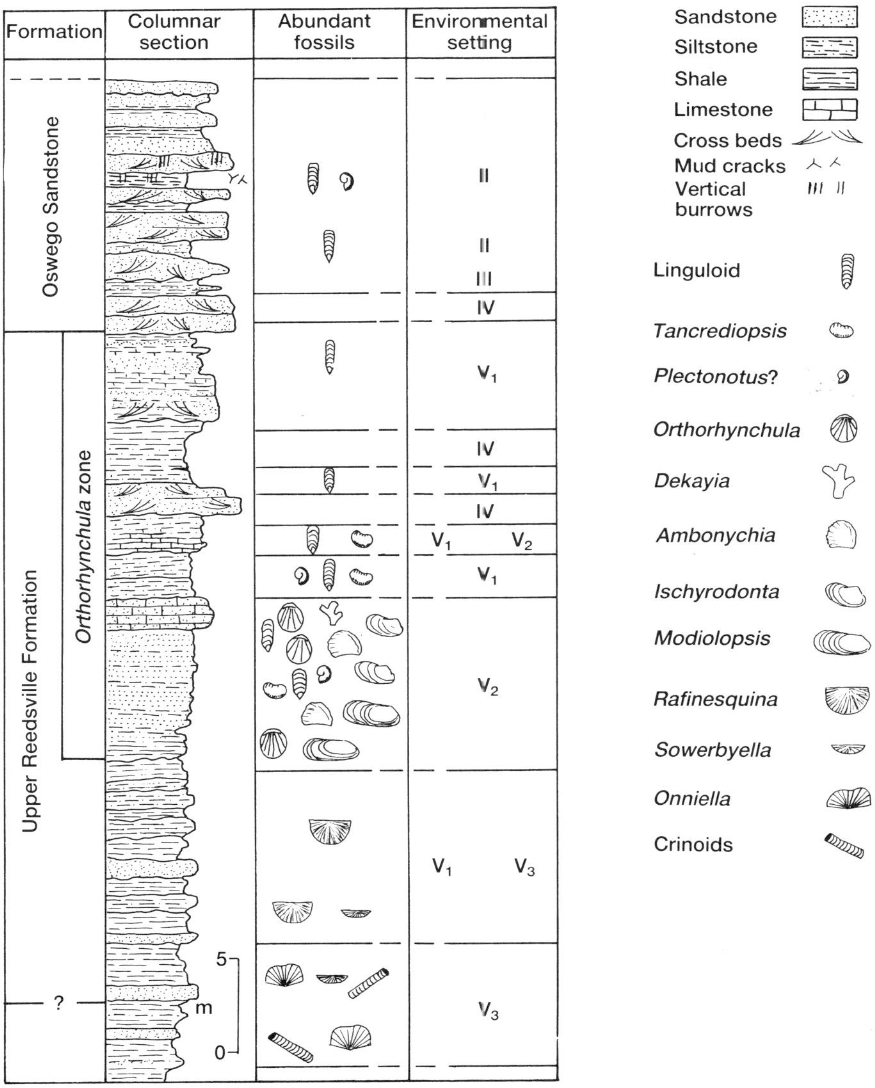

Fig. 11.9 A regionally composite stratigraphic section through the Middle Ordovician Reedsville and Oswego formations of the central Appalachians, showing distribution of the major fossils — including brachiopods, bryozoans, pelecypods, gastropods, and crinoids — and inferred environmental settings. Key to environments: I, beach, flood plain; II, silt-mud flats; III, lagoonal; IV, bars and barriers; V, shallow marine (V_1, intertidal; V_2, inner sublittoral; V_3, outer sublittoral). After Bretsky (1970).

How is the other ill adapted to soft substrates?

The species that was not adapted to soft substrates apparently was rooted in sediment by its pedicle. What attribute of its shell would have placed the posterior end against the substrate even if it were swept by water movement away from its original spot on the sea floor?

QUESTION 8

Examine the distributions of the two brachiopods in the columnar section in Fig. 11.9. Is their distribution within the lithologies and inferred environments consistent with or contradictory to your conclusions for Questions 6 and 7? Explain.

EXERCISE 12 Bryozoa

12.1 INTRODUCTION

Bryozoans are one of the most diverse and abundant groups of fossils in shallow marine carbonate and mixed carbonate/clastic rocks that accumulated during the Ordovician through Permian Periods. (Brachiopods and stalked echinoderms were on average about equally dominant with bryozoans during that time.) After a period of low diversity and uncommon occurrence during the early Mesozoic, bryozoans became an important constituent of the marine fossil record again during the Middle Jurassic. Following the Late Jurassic to Early Cretaceous origin and radiation of a new, rapidly evolving group (the Cheilostomata), bryozoans have been diverse and widespread in the seas up to the present, locally reaching the abundance that they had during the Paleozoic and even exceeding their previous diversity.

Cnidaria (Exercise 5) and Hemichordata (Exercise 16) include many colonial taxa, but with one or two debatable exceptions among living species, bryozoans are *exclusively colonial*. Their colonies have grown as simple to elaborate forms that may be reduced to three groups: encrusting, erect, and free-living (Fig. 12.1). Regardless of the superficial appearance of a great diversity of shapes, bryozoans have been constrained to a limited number of basic growth forms. Each of these has developed independently many times in different orders and different classes through **iterative evolution**. The limited number of basic growth forms and their repeated evolutionary development suggest that their morphologies are under a strong functional control.

The purposes of this exercise are:
1 to examine carefully a single polypide and enclosing body wall of a feeding zooid in a Recent bryozoan colony;
2 to examine skeletal morphology of representatives of the higher taxa of bryozoans with calcified skeletons; and
3 to analyze constraints imposed on colony morphology by the feeding process.

12.2 INDIVIDUAL MODULES OF THE COLONIES

The basic unit or module of a bryozoan colony is termed a **zooid**. It consists of a **polypide**, which both contains and is situated within a coelom, a few organs connected to the polypide, and an enclosing body wall. The polypide is organized into two parts, the **lophophore** (a set of ciliated tentacles, each of which contains a central coelom) and a U-shaped gut (Fig. 12.2). The body wall may be entirely organic or may be reinforced by calcite, mixed calcite and aragonite, or aragonite skeleton. The skeletal parts of a zooid are called the **zooecium**. The skeletal part of a colony is termed the **zoarium**, and it consists of the collective zooecia plus any skeleton deposited by the colony between the zooecia. A few non-mineralized bryozoans have been preserved, either as hollow spaces excavated into calcareous

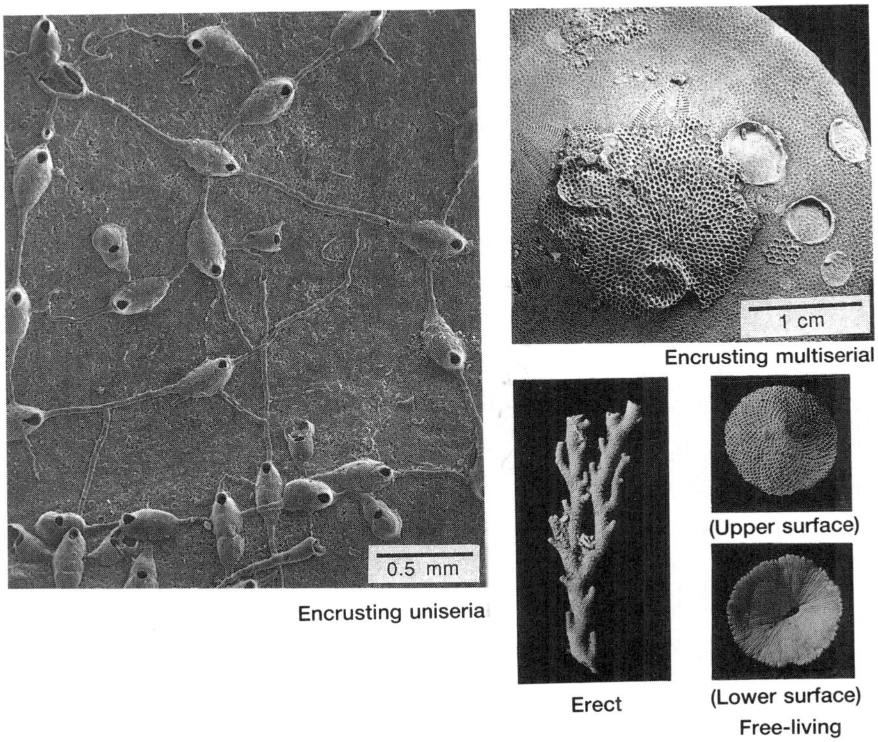

Fig. 12.1 Basic groups of bryozoan growth forms. Free-living bryozoans are represented by a cap-shaped form that has been locally common from Late Cretaceous to the Recent. The scale for encrusting forms also applies to illustrations of erect and free-living forms.

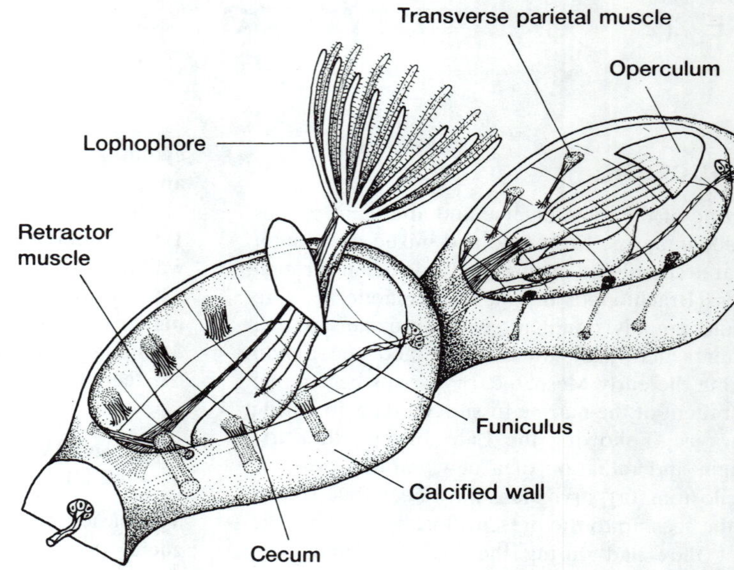

Fig. 12.2 Autozooids of bryozoans, with polypide retracted on the right and with polpide protruded to the feeding position on the left. Modified from McKinney & Jackson (1989).

shells or as impressions on the undersides of organisms that grew over them.
(Question 1)

12.3 CLASSIFICATION

Marine bryozoans are distributed among the classes Stenolaemata and Gymnolaemata. Stenolaemates are usually divided into four or five orders depending on whether the Fenestrata are recognized separately from the Cryptostomata, as is done in this laboratory manual, and gymnolaemates are divided into two orders (Table 12.1). The gymnolaemates are studied almost exclusively from their surface morphology, but study of stenolaemates usually involves oriented sections cut through the colony (Fig. 12.3). Representatives of each of the orders of mineralized bryozoans are illustrated in Figs 12.4–12.9.

Morphology of zooids is on average very different between

Table 12.1 Characteristics of the marine Bryozoa. From McKinney & Jackson (1989).

Class Stenolaemata	Zooids are elongate cylindrical and continue to lengthen through ontogeny, with long axis at angle to local colony growth direction Basal and vertical walls rigidly calcified Interzooidal communication through space between outer ends of vertical walls in most, but through pores in vertical walls in some Membranous sac encloses polypide and is deformed to protrude lophophore through orifice at outer end of skeletal tube Lower Ordovician–Recent
Order Trepostomata	Colonies encrusting or erect Elongate autozooids generally containing basal diaphragms and commonly other lateral structures Kenozooids common, extrazooidal skeleton in some Skeletons without communication pores, typically laminated Lower Ordovician–Upper Triassic or ?Recent
Order Cystoporata	Colonies encrusting or erect Autozooids short and without, or long and with, basal diaphragms In some, autozooids have a thickened strip (**lunarium**) of different microstructure along one side in thick-walled outer zone Skeletal structure laminated, granular or granular-prismatic, some with communication pores Kenozooids in some, maternal zooids uncommon Vesicular extrazooidal skeleton common and abundant Lower Ordovician–Upper Permian or ?Cretaceous
Order Cryptostomata	Colonies erect arborescent, or bilaminate sheets Generally short autozooids, some with basal diaphragms or hemisepta Skeletons without communication pores, typically laminated Kenozooids or extrazooidal skeleton may be present Lower Ordovician–Upper Permian or ?Triassic

Table 12.1 *(Continued)*

Order Fenestrata	Colonies erect, composed of narrow unilaminate branches Short autozooids commonly with hemisepta, but basal diaphragms in very few Primary zooidal skeleton without communication pores, typically granular Polymorphs in some Extrazooidal skeleton extensive, several times autozooidal skeleton in volume, consisting of laminae penetrated by small granular rods Lower Ordovician–Upper Permian or ?Triassic
Order Cyclostomata	Colonies encrusting or erect Autozooids commonly long, some with basal diaphragms and other structures Skeletal structure typically laminated, many with communication pores Maternal zooids common, kenozooids or **nanozooids** (small zooids with skeletal aperture restricted to diameter of single tentacle) in some Extrazooidal skeleton may be present Possibly a polyphyletic mixture (Boardman, 1984) Lower Ordovician–Recent
Class Gymnolaemata	Zooids generally box- or sac-shapes to short cylinders, with long axis roughly parallel to local colony growth direction Zooidal size is fixed early in ontogeny Zooidal body walls entirely organic to rigidly calcified Interzooidal communication by *funicular* network (network of tissue strands perhaps homologous to blood vessels) through tissue-plugged pores in vertical walls Vertical or frontal walls deformed to protrude lophophore Upper Ordovician–Recent
Order Ctenostomata	Zooidal walls membranous or gelatinous Orifice terminal and closed by pleated collar in most Polymorphs absent or only kenozooids Upper Ordovician–Recent
Order Cheilostomata	Zooidal walls calcified, flexible or rigid Orifice frontal and closed by proximally hinged operculum Polymorphs, especially avicularia, usually present and commonly diverse Suborders based on frontal calcification and mechanism of lophophore protrusion Upper Jurassic–Recent

stenolaemates and the calcified gymnolaemates, which constitute the Order Cheilostomata. Stenolaemate zooids are generally highly elongate, oriented at a high angle to the colony surface, and quite simple on the exterior; but they may contain one or more types of internal skeletal structures. These include complete transverse plates (**diaphragms**), plates that only extend about half-way across the interior space (**hemisepta**), strongly curved and commonly overlapped marginal plates (**cystiphragms**), **spines**, and other structures (Figs 12.4–12.8). **Acanthostyles** occur in many stenolaemates. These are rods within zooidal and colonial (**extrazooidal**) skeleton that are oriented at a high angle to the skeletal surface and generally consist of a core of hyaline skeleton surrounded by conical laminae.

In contrast with zooids of stenolaemates, cheilostome zooids generally are short relative to their width, have their long axis nearly parallel with the colony surface, and commonly have a complex skeletal exterior (Fig. 12.9). The exposed surface may bear abundant perforations, deep indentations that do not completely perforate the skeleton, vertical spines around the aperture, arched spines that meet along the plane of zooidal symmetry, or horizontal spines projecting into a centrally placed large uncalcified area.

Three fundamental grades of evolution are recognized in cheilostomes, based on the relationship between calcification of the frontal surface and the frontal membrane that is pulled inward toward the zooid's body cavity in order to cause the lophophore to be pushed out to the feeding position. The simplest grade comprises the **anascan** cheilostomes and is characterized by a flexible frontal membrane that is not defended by a calcium carbonate skeletal shield or overarching spines. The **cribrimorphs** are characterized by a deformable frontal membrane that is defended by a dense set of overarching spines with spaces between them through which water may flow. The **ascophorans** are characterized by calcification of the frontal region, with a locally inwardly ballooned region of deformable membrane (the **ascus**), which may be expanded within the zooid in order to push the lophophore out. The

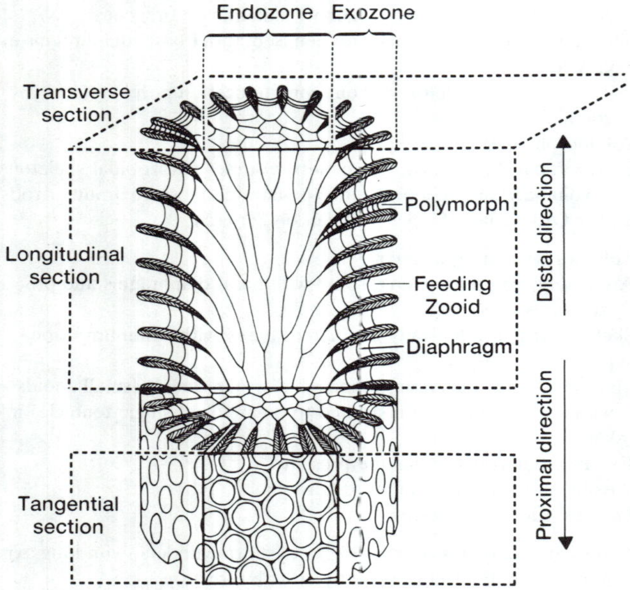

Fig. 12.3 Stenolaemate branch showing endozone, exozone, and other features in transverse, longitudinal, and tangential sections. From Boardman & Cheetham in Boardman *et al.* (1987).

ascus opens through a more-or-less well-defined pore that in many ascophorans forms a sinus in the proximal edge of the orifice.

Polymorphism (the presence of one or more distinctly different types of modules within a colony) is much more highly developed in most cheilostomes than it is in stenolaemates. In addition to normal feeding zooids (**autozooids**), there are **maternal zooids** that brood embryos, **kenozooids** that lack feeding structures but that serve to fill areas where other zooids are absent or areas of awkward shapes and sizes among other zooids, and zooids termed **avicularia** with hypertrophied (vastly overgrown) **mandibles** (Fig. 12.9). Mandibles are derived from **opercula** that in autozooids cover the orifice where the lophophore is protruded to its feeding position. Avicularia are diverse, ranging from tiny shoeprint- or beak-shaped structures on the surface of other zooids (**adventitious avicularia**), to structures situated among autozooids and almost entirely covered by the operculum, to structures resembling vulture's heads elevated on short or long stalks, to tiny polymorphs (**vibracula**) with long needle-like opercula that may even serve as walking appendages for the colony.

In this exercise, you will examine a few genera from each of the calcified orders given in Table 12.1, in order to study some of the skeletal characteristics that are useful for assignment to class and order. To see the features adequately, you must first very carefully examine the exterior and any broken surfaces of the zoarium or zoarial fragment, and then examine the zooecia in detail. This will require examination of the surface of the specimen with a binocular microscope and, for stenolaemates, study through a transmitted-light microscope of at least two differently oriented thin-sections that accurately correspond with cuts as shown in Fig. 12.3. The characters and character states to look for are listed below.

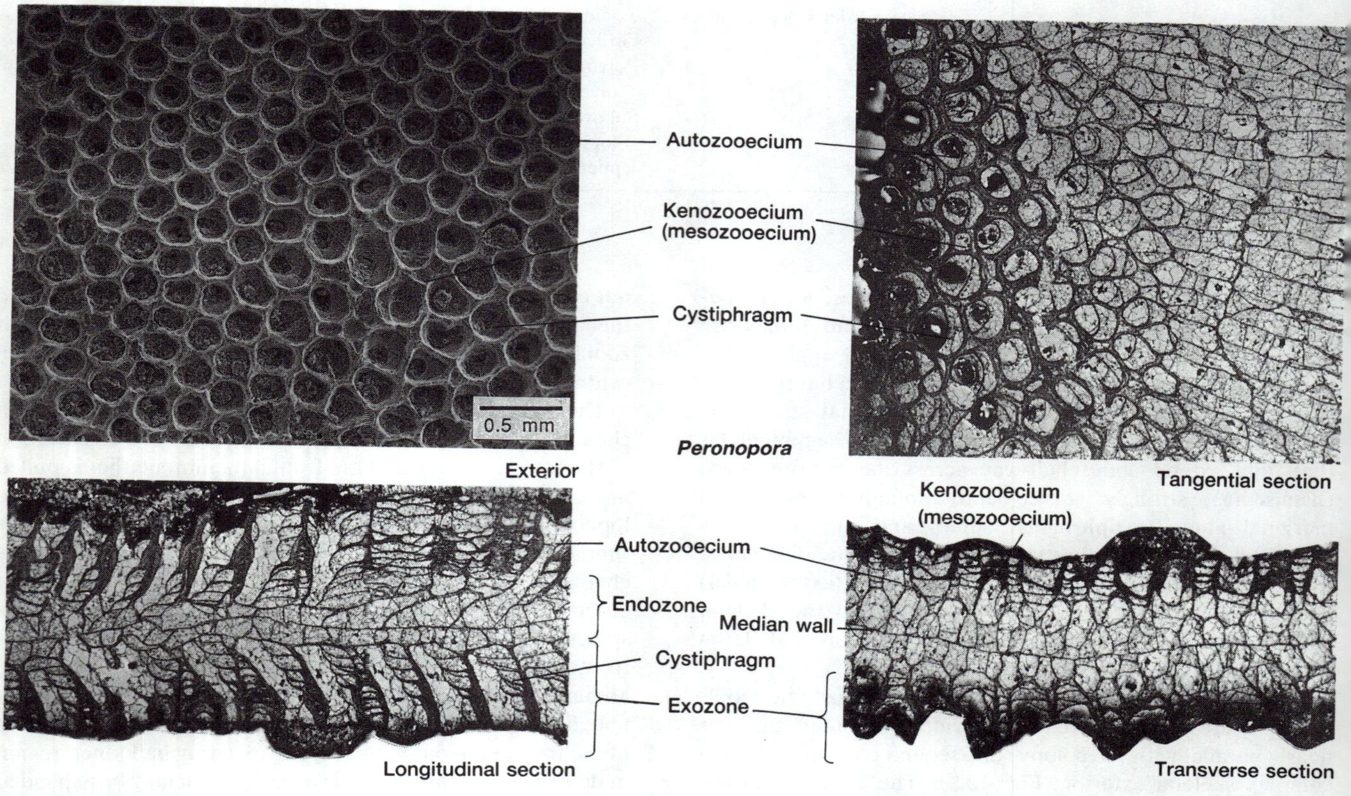

Fig. 12.4 Some skeletal features of the stenolaemate order Trepostomata. All sections have the same magnification.

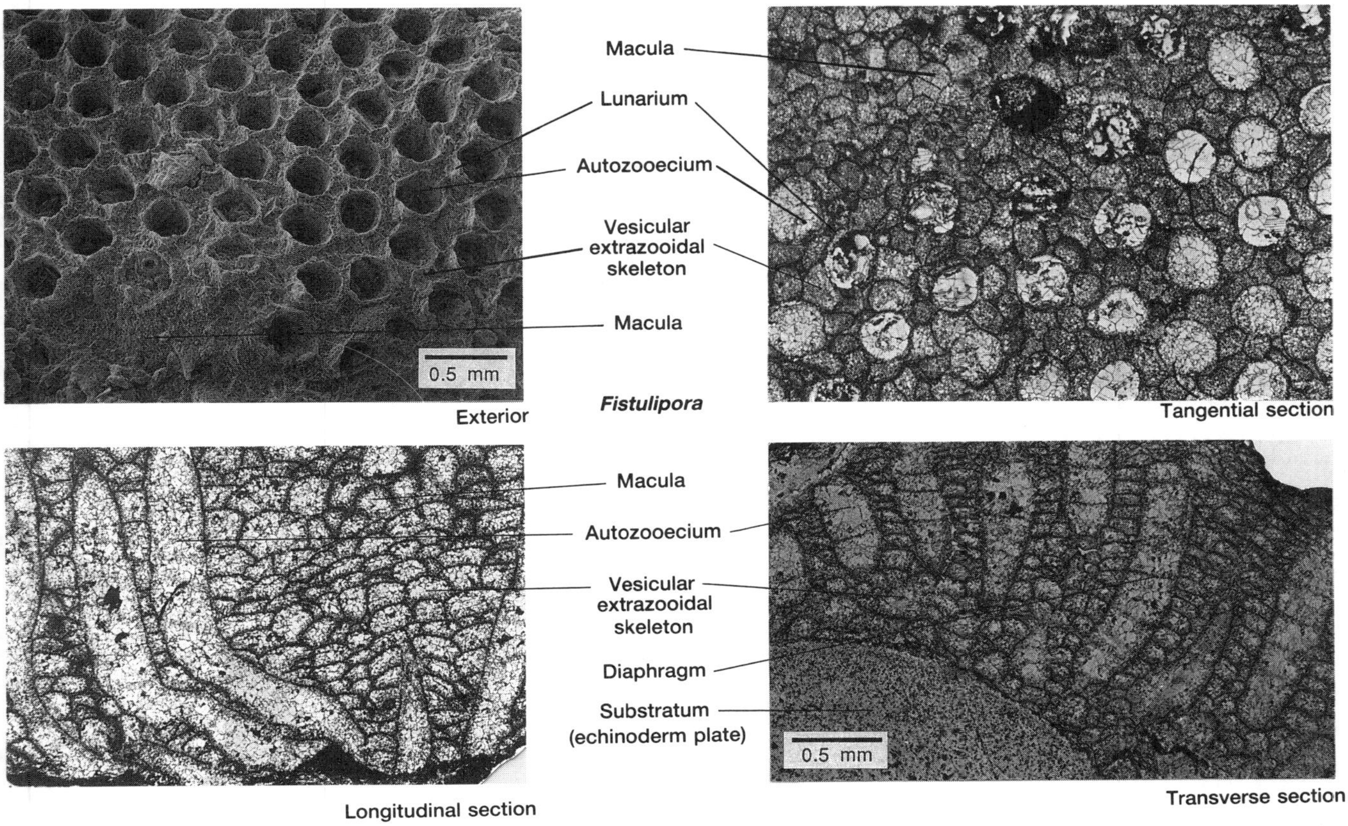

Fig. 12.5 Some skeletal features of the stenolaemate order Cystoporata. All sections have the same magnification.

1 Zooidal shape and orientation.
(a) Shape of the feeding zooids: (i) box- or sac-shaped; (ii) conical (distinctly tapered; length < 4× width); (iii) tubular (length > 4× width).
(b) Zonation of the feeding zooids: (i) divided into **endozone** (inner or proximal thin-walled portion) and **exozone** (outer or distal thicker-walled portion, usually with curve or bend between it and endozone) (Figs 12.4–12.8); (ii) not divided into endozone and exozone.
(c) Orientation of feeding zooids: (i) long axis oriented essentially parallel with local colony surface (Figs 12.4–12.6 and 12.8); (ii) long axis oriented at high angle to local colony surface (Figs 12.7 and 12.9).

2 Skeletal structures within zooecia.
(a) Diaphragms — skeletal plates crossing and partitioning space within feeding zooids (Figs 12.4 and 12.5): (i) present; (ii) absent.
(b) Hemisepta — partial plates often at the boundary between endozone and exozone (Fig. 12.6): (i) present; (ii) absent.

3 Skeletal texture and distribution.
(a) Wall texture: (i) **lamellar** — skeleton composed of many parallel or subparallel thin layers, at either a high or low angle to the local skeletal surface; (ii) **granular** — translucent skeleton with no apparent orientation of the small, equidimensional crystals of which it is composed; (iii) **hyaline/lamellar** — glassy-appearing skeleton partially around feeding zooids, with lamellar skeleton deposited on or around it.
(b) **Primary extrazooidal skeleton** — skeletal deposits between zooids that cannot be assigned to any particular feeding zooid or polymorph and that was deposited among zooids as the colony grew; may be solid skeleton or walls of blister-like vesicles (Figs 12.5 and 12.7). This contrasts with **secondary extrazooidal skeleton**, which forms in some bryozoans as a continuous skeletal deposit over pre-existing zooids, generally as a means to strengthen the colony at a point of structural hazard: (i) present; (ii) absent.

4 Polymorphs.
(a) Kenozooids — any polymorphic zooid lacking polypide and orifice or orificial wall, or any polymorph in extinct bryozoans with body cavity too small to have contained a polypide (Figs 12.4, 12.8 and 12.9): (i) present; (ii) absent.
(b) Avicularia — polymorphs with equivalent of orificial wall relatively larger and more intricately reinforced than those of ordinary feeding zooids, to form a mandible (Fig. 12.9): (i) present; (ii) absent.

This is by no means an exhaustive listing of polymorphs, characteristics of feeding zooids, or colony features. Rather, it contains some of the more widespread or commonly accessible characteristics of bryozoans that are useful in classifying many taxa.
(Question 2)

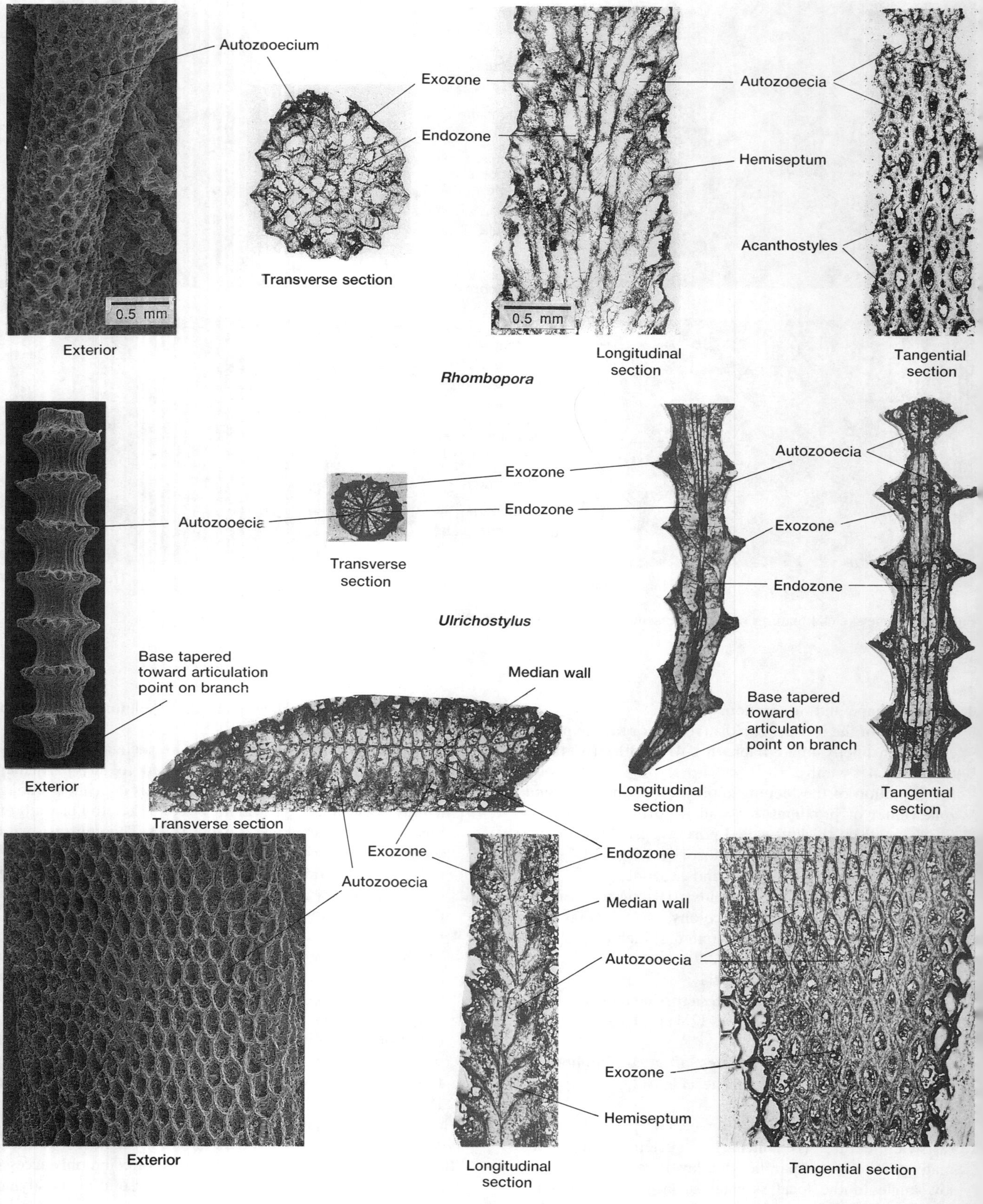

Fig. 12.6 Some skeletal features of the stenolaemate order Cryptostomata. All sections have the same magnification.

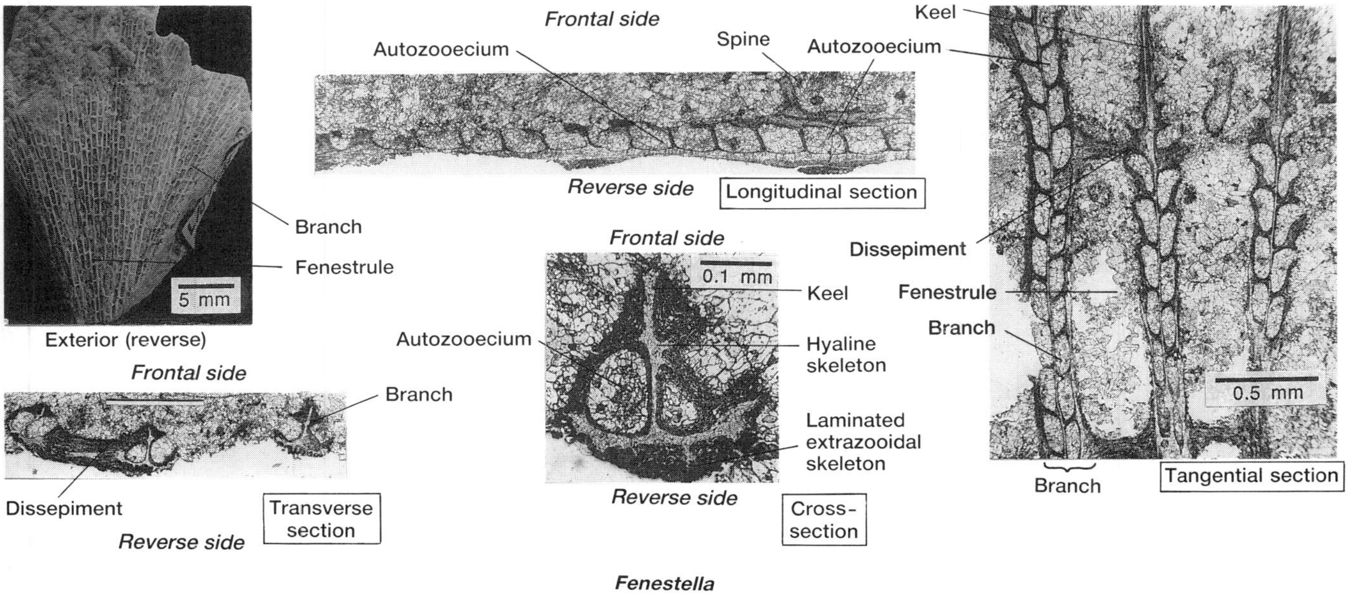

Fig. 12.7 Some skeletal features of the stenolaemate order Fenestrata. All sections except the more highly magnified single branch cross section have the same magnification.

12.4 FEEDING CURRENTS AND COLONY MORPHOLOGY

Bryozoans feed on food particles that are contained in the water that zooids actively pump toward the colony. Movement of the water is driven by cilia that line the tentacles of the protruded lophophores. Few records or estimates of the rate of cilia-generated flow have been published, but all are a few millimeters per second or less. In order for bryozoans to continue feeding, the filtered water that has passed through the lophophores must move on. However, most bryozoans form continuous surfaces that are more than a few zooids wide, which is a condition that prevents the continued flow of water in the same direction that it approached the colony. On relatively flat surfaces, the outflow is directed laterally, as in the diagram of a flat, circular encrusting colony in Fig. 12.10. In this case, the water flows between the lophophores and colony surface to the margins of the colony. Obviously, some problems will arise as a flat colony grows: if all the water flows to the margins of the colony, there will be more and more water rushing toward the colony's edge, all confined in the narrow space between the protruded lophophores and the colony surface. Colonies can grow to large sizes only by withstanding or circumventing the physical stresses that such vigorous flow would generate. The final questions in this exercise are intended to demonstrate the problem more clearly and to examine the morphological consequences that it generates.
(Questions 3–6)

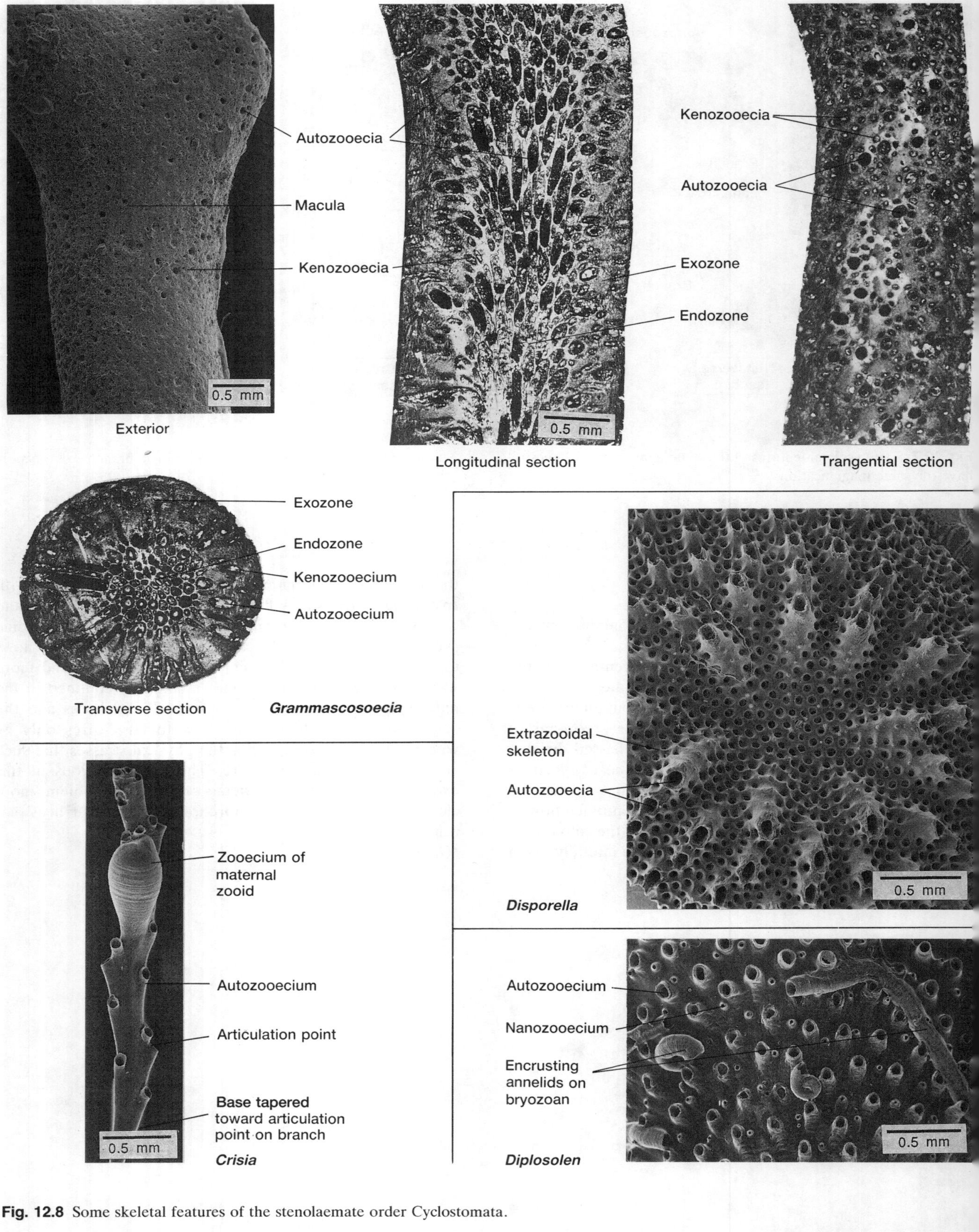

Fig. 12.8 Some skeletal features of the stenolaemate order Cyclostomata.

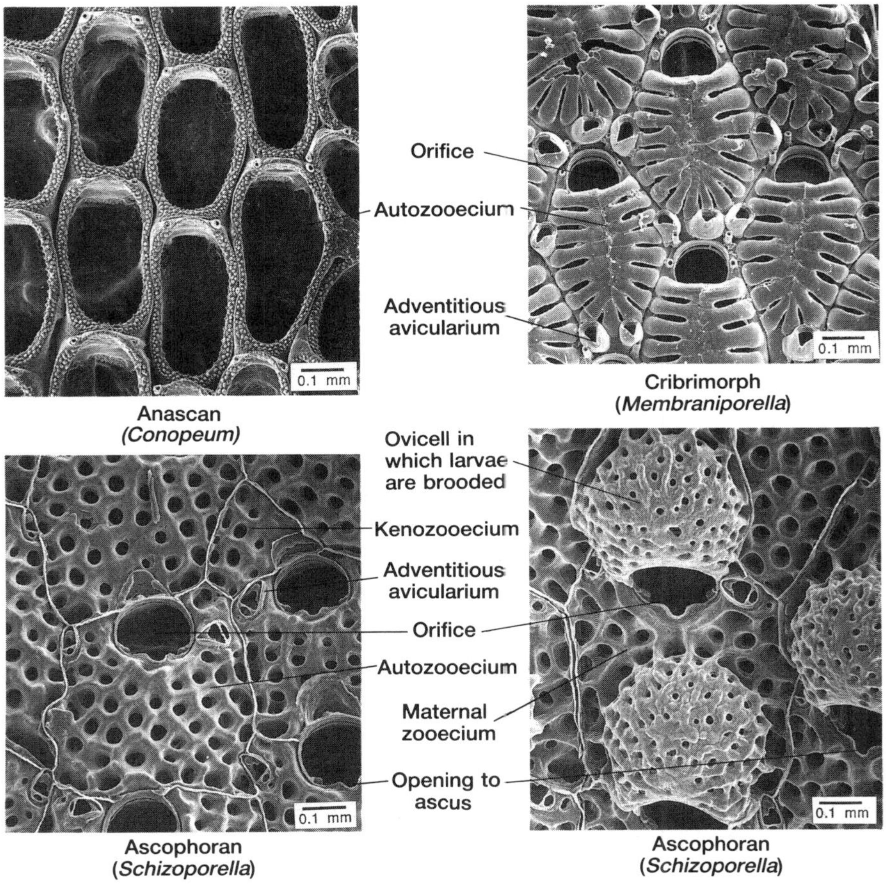

Fig. 12.9 Some skeletal features of the gymnolaemate order Cheilostomata.

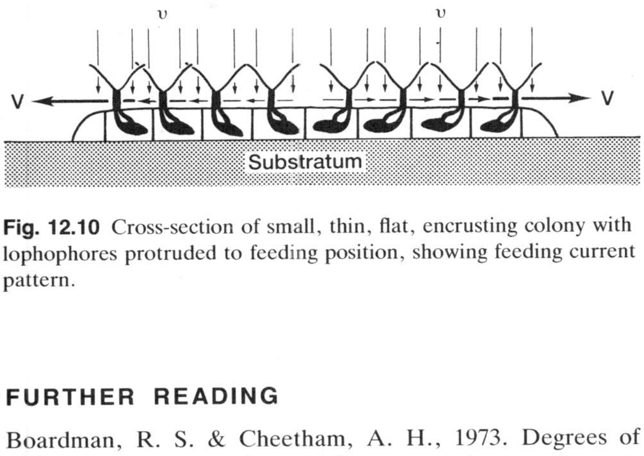

Fig. 12.10 Cross-section of small, thin, flat, encrusting colony with lophophores protruded to feeding position, showing feeding current pattern.

FURTHER READING

Boardman, R. S. & Cheetham, A. H., 1973. Degrees of colony dominance in stenolaemate and gymnolaemate Bryozoa, in R. S. A. Boardman, A. H. Cheetham & W. A. Oliver (eds), *Animal colonies*, pp. 121–220. Dowden, Hutchinson & Ross: Stroudburg, PA.

Boardman, R. S. *et al.*, 1983. Bryozoa, Vol. 1, in R. A. Robison (ed.), *Treatise on invertebrate paleontology*, Part G (Revised). Geological Society of America and University of Kansas Press: Lawrence, KS.

Dutro, J. T. Jr. & Boardman, R. S. (eds)., 1981. *Lophophorates, notes for a short course*. University of Tennessee Department of Geological Sciences Studies in Geology 5.

McKinney, F. K., 1989. Feeding and associated colonial morphology in marine bryozoans. *Reviews in Aquatic Sciences* **2**: 255–280.

McKinney, F. K. & Jackson, J. B. C., 1989. *Bryozoan evolution*. Unwin Hyman: Boston, MA.

Ryland, J. S., 1970. *Bryozoans*. Hutchinson University Press: London.

EXERCISE 12

Worksheet

Name

QUESTION 1

The organization of polypides is quite similar throughout the phylum, so study in detail of polypides in almost any species will represent the entire phylum fairly well. A specimen of a Recent bryozoan with at least a locally clear body wall is provided so that the polypides and their positions relative to the enclosing body wall can be studied. Choose one of the clearest zooids, and draw and label it. Draw an outline of the zooid at 8–10 cm length, which will be roughly 100× magnification. Draw the polypide and any membranes, muscles and other tissues that can be seen. Label all the features from the following list that are visible.

Body wall
Orifice or operculum
Tentacles
Tentacle sheath
Esophagus (between mouth and cecum)
Cecum (main chamber of digestive tract)
Retractor muscle
Brown body (residual mass from former polypide)

QUESTION 2

Examine whole specimens and thin-sections or acetate peels of the genera provided. Each is identified to order and genus. For each, determine the state for each character as given in the list in Section 12.3 and enter the number assigned to that state in the list.

	Feeding zooids								
Genus (and order)	*Shape*	*Zonation*	*Orientation*	*Diaphragms*	*Hemisepta*	*Wall texture*	*Kenozooids*	*Avicularia*	*Extrazooidal skeleton*
1 ______	___	___	___	___	___	___	___	___	___
2 ______	___	___	___	___	___	___	___	___	___
3 ______	___	___	___	___	___	___	___	___	___
4 ______	___	___	___	___	___	___	___	___	___
5 ______	___	___	___	___	___	___	___	___	___
6 ______	___	___	___	___	___	___	___	___	___
7 ______	___	___	___	___	___	___	___	___	___
8 ______	___	___	___	___	___	___	___	___	___
9 ______	___	___	___	___	___	___	___	___	___

10 ______ ______ ______ ______ ______ ______ ______ ______ ______ ______

11 ______ ______ ______ ______ ______ ______ ______ ______ ______ ______

12 ______ ______ ______ ______ ______ ______ ______ ______ ______ ______

13 ______ ______ ______ ______ ______ ______ ______ ______ ______ ______

14 ______ ______ ______ ______ ______ ______ ______ ______ ______ ______

15 ______ ______ ______ ______ ______ ______ ______ ______ ______ ______

QUESTION 3

Assuming that the tentacle sheath projects 0.02 cm above a colony similar to the one in Fig. 12.10 and that pumping velocity (V) is 0.4 cm/s (a "typical" figure as far as is known), what would be the volume of water (Q) that must escape from the margin of a 0.5, 1, 2, 4, 8, 12, 16, and 20 cm colony, assuming that all zooids feed almost continuously and that all water flows to the colony margin? [Q (in cm^3/s) = $V \times \pi \times r^2$]. Plot the answers as points on the graph provided.

$Q = V \times \pi \times r^2$

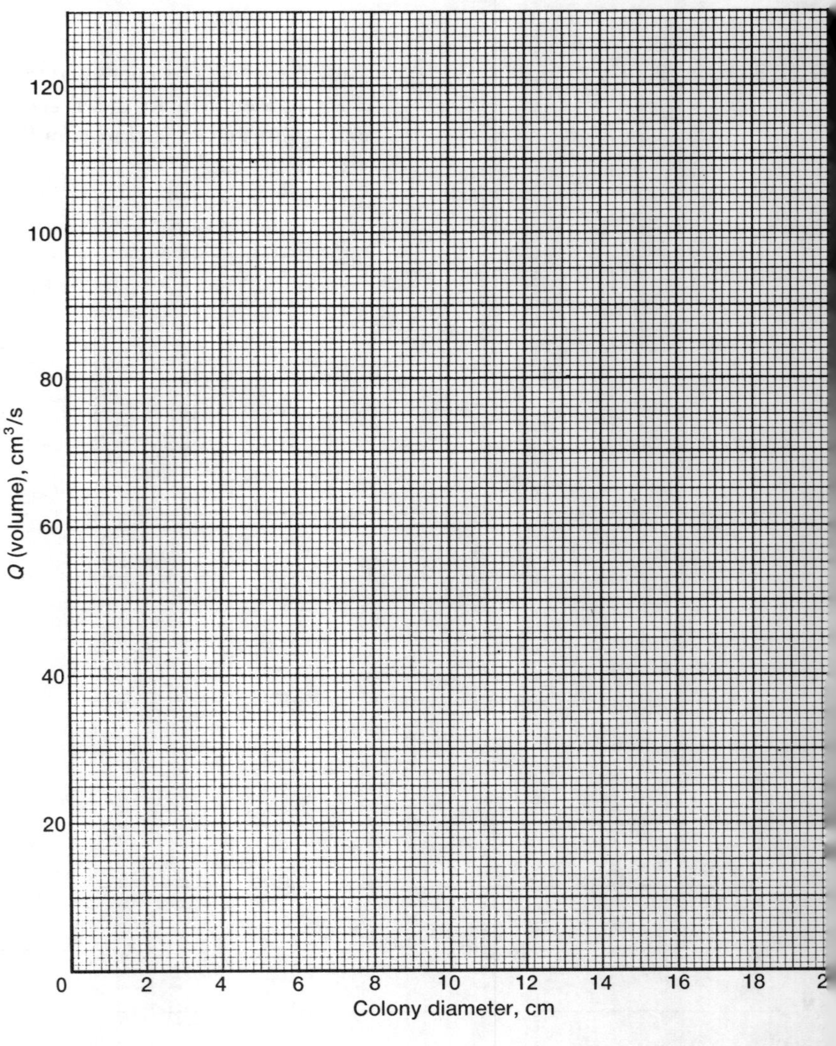

Colony diameter	*Colony radius (r)*	*Q*
0.5 cm	______ cm	______ cm^3/s
1	______	______
2	______	______
4	______	______
8	______	______
12	______	______
16	______	______
20	______	______

QUESTION 4

What would velocity (V) of escape for the filtered water be around a colony at each size and conditions given for Question 3? [V (in cm/s) $= Q/A = Q/(\pi \times d \times h)$, where d is the diameter of colony and h is the height that tentacle sheath projects above the colony.] Plot the answers as points on the graph provided on p. 182. With the conditions given, does V remain constant as colony radius increases? If so, why? If not, is the rate of change linear or exponential? Under such conditions as stipulated in Question 3, are large surface areas of colonies feasible? As a reminder, here are the conditions, which at this point will be treated as assumptions:

1 velocity of incoming water = 0.4 cm/s;
2 everted tentacle sheaths place lophophores 0.02 cm above colony;
3 virtually all feeding zooids across the entire surface feed simultaneously; and
4 all filtered water leaves the colony at its margin.

Are all the assumptions reasonable? What would be the effect of varying each of them? How could velocity of escaping water most effectively be kept below 10 × velocity of incoming water?

$V = Q/(\pi \times d \times h)$

Colony diameter	*Q (from Question 3)*	*V*
0.5 cm	_____ cm^3/s	_____ cm/s
1	_____	_____
2	_____	_____
4	_____	_____
8	_____	_____
12	_____	_____
16	_____	_____
20	_____	_____

Does V remain constant with increase in d?________________________. Explain.

____

____

Are large surface areas fèasible?______________________________

Are assumptions reasonable?______________________________
Effect of varying each:

1 ______________________________

2 ______________________________

3 ______________________________

4 ______________________________

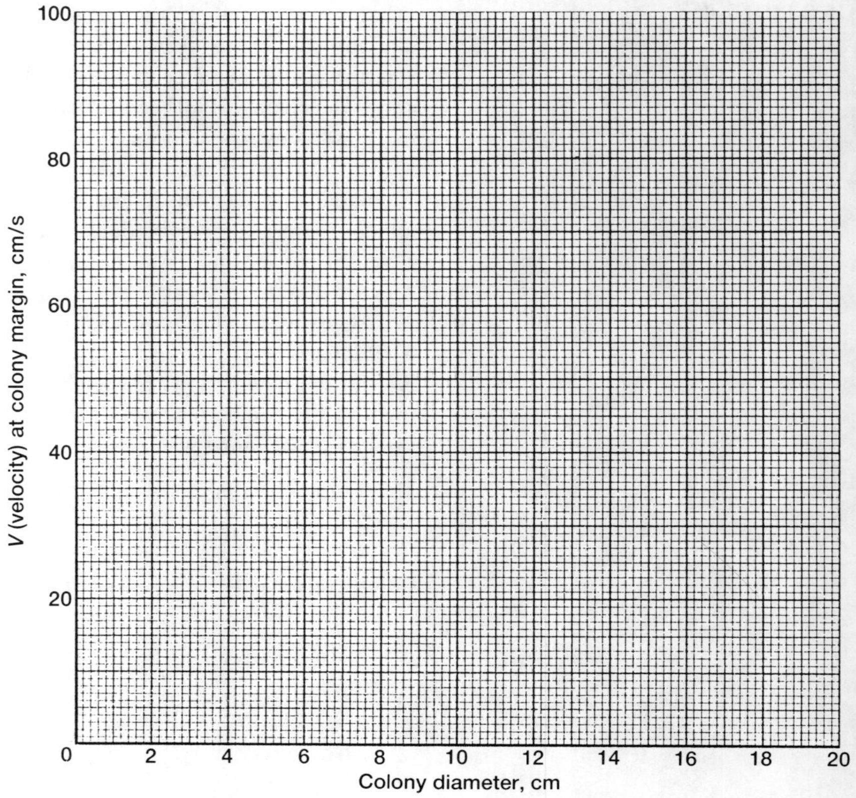

How best to keep $V < 10\times v$?____________________

____

QUESTION 5

Figure 12.11 shows an actively feeding colony of *Membranipora membranacea* in which velocity of incoming water is about 0.4 cm/s, height of everted tentacle sheath is about 0.02 cm, and most zooids across the entire surface are feeding, yet velocity of escaping water along the colony margin is less than would be calculated in Question 4. Why?

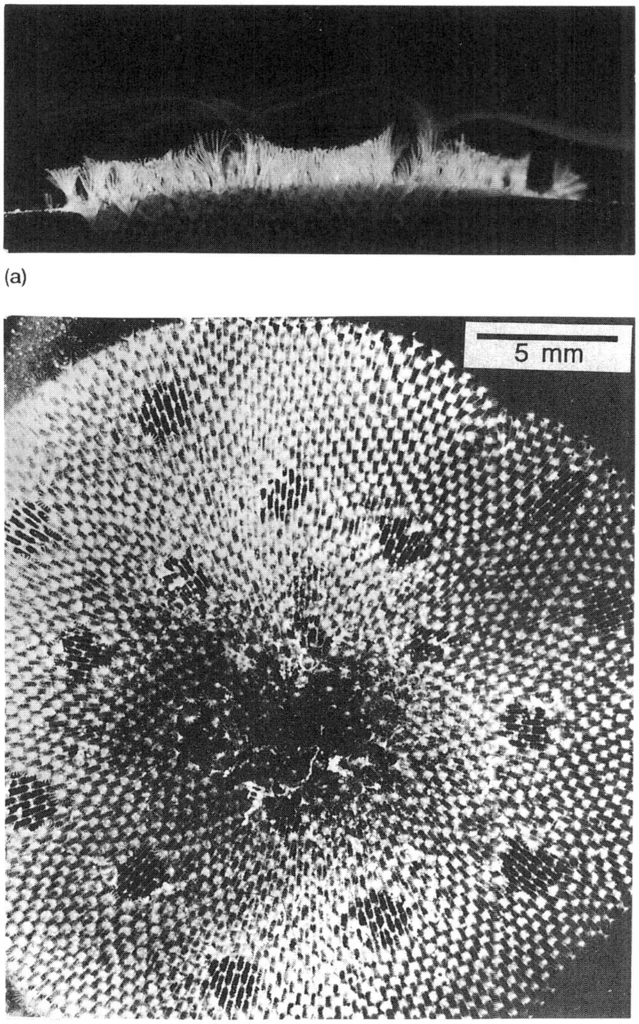

(a)

(b)

Fig. 12.11 Feeding colonies of the sheet-like encrusting cheilostome *Membranipora membranacea*, (a) in profile and (b) surface view. Part (a) is magnified about 2× as much as part (b). From Lidgard (1981).

QUESTION 6

Almost all large surface areas in encrusting stenolaemate bryozoans and many in cheilostomes have regularly spaced *maculae* (spots, either raised or flush) as seen in Fig. 12.12; these are conspicuous on many specimens examined for Question 2. What is one probable function of the maculae?

____

____

____

____

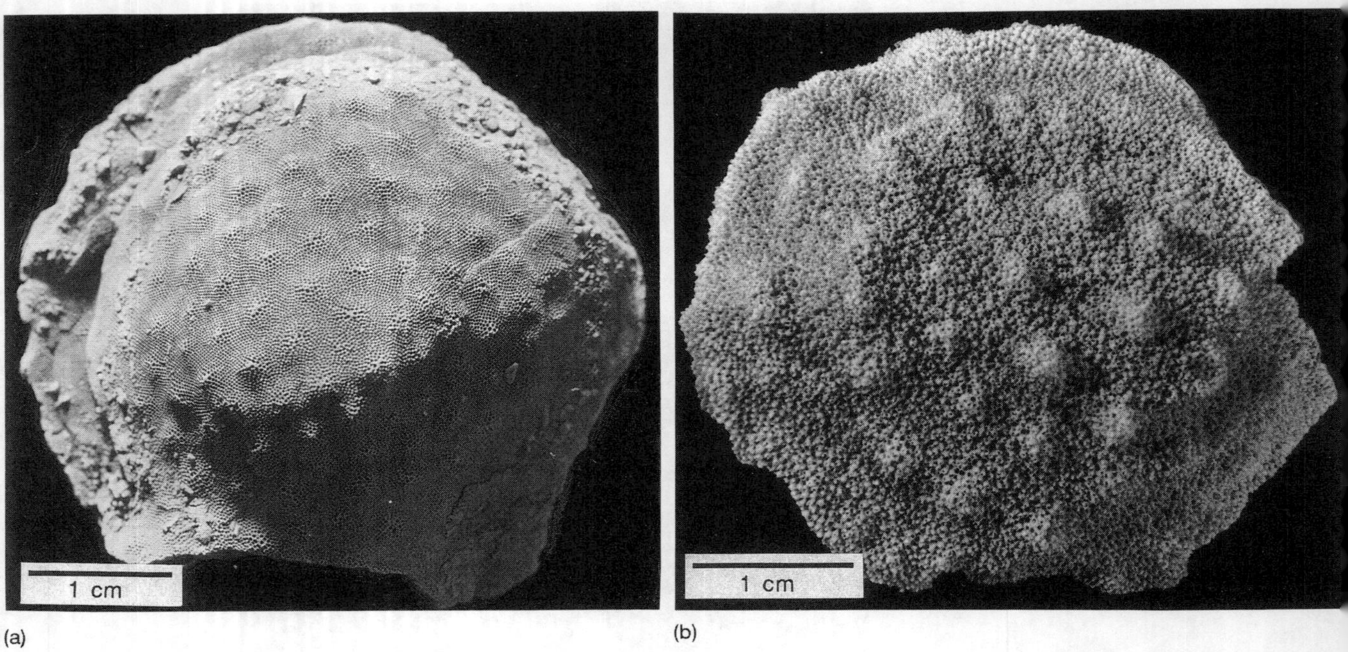

Fig. 12.12 Maculate skeletons of (a) a Lower Carboniferous trepostome stenolaemate and (b) a Recent cheilostome gymnolaemate.

EXERCISE 13 Echinodermata I: General features and non-stalked classes

13.1 INTRODUCTION

Many echinoderms are among the most esthetically pleasing of the living organisms, and fossil echinoderms commonly are also quite beautiful and highly sought by amateur collectors as well as by museums. Perhaps the most geologically important feature of echinoderms, however, is the volume of sedimentary rock that they constitute. Their skeletons commonly dissociate into individual calcitic structures that have various shapes. Some are tabular structures termed **plates**, which are most commonly a few millimeters in diameter. Other skeletal structures range from microscopic **spicules** to **spines** tens of centimeters long. The dissociated skeletal structures may be concentrated into huge accumulations or may be widely dispersed, occurring as isolated grains in almost any marine deposit. Some sedimentary units, including entire formations, are composed predominantly of echinoderm plates piled one on top of another.

Echinoderm morphology is closely determined by functional requirements of the animal and its relation to the environment. In addition, physiology restricts echinoderms to a narrower range of salinities than most other readily fossilizable phyla. Consequently, echinoderms are quite useful for paleoecological as well as evolutionary studies.

The purposes of this exercise are:

1 to introduce the fundamental characteristics of echinoderms;
2 to briefly survey the major classes of non-stalked members of the phylum; and
3 to examine contrasting morphologies and life modes found in the class Echinoidea.

There is an emphasis on the Echinoidea in this exercise because they are easily obtained as living or preserved specimens and also as fossils, so are a good focal point for understanding the essential features of echinoderms. Study of stalked echinoderms is reserved for Exercise 14.

13.2 CHARACTERISTICS OF THE PHYLUM

Echinoderms are complex organisms characterized by several features that are unique to the phylum. The most conspicuous of the unique features is the generalized pentameral symmetry (Fig. 13.1), which is found in most members of the phylum excepting some early, bizarre classes and occasional aberrant members of other normally pentameral classes. However, in even the apparently perfectly pentameral forms, the location of some feature, such as the rectum and anus or the stone canal and madreporite, almost always superimposes some degree of bilateral symmetry.

Perhaps the most singular feature of echinoderms is the **water vascular system** (Fig. 13.2). This is a set of intercommunicating coelomic tubes, filled with fluid and open only at the end of one tube (the **stone canal**), where the volume of water in the water vascular system may be adjusted by movement of water into or out of the system through an opening termed the **hydropore** (if a single opening) or **madreporite** (a modified genital plate that characteristically has multiple openings). The stone canal is continuous with the **ring canal** (also called the circumoral ring) that encircles the digestive tract. From the ring canal, five **radial canals** extend out along the rays of pentameral symmetry. In almost all cases, short lateral extensions from both sides of the radial canals connect to **ampullae** (singular, ampulla) and **tube feet**, which are terms for the two ends of one continuous unit. The ampullae resemble and function as squeeze-bulbs. They are seated within the body of the organisms whereas the tube feet extend outside the body. The tube feet are capable of being highly deformed: they may be extended to great lengths by hydrostatic pressure as the ampullae contract, and longitudinal muscles may cause them to wave around in any direction or may contract and collapse the tube feet into very short extensions. A muscular valve at the junction of the ampulla and the lateral extension from the radial canal can isolate the ampulla/tube-foot system for its hydrostatic operation. Each radial canal, its lateral extensions, and the associated ampullae and tube feet are part of a specially plated, radiating region termed an **ambulacrum**.

The ambulacra are the most easily recognizable and obvious structures in the pentameral symmetry of echinoderms. The ambulacra and the tube feet that characterize them are basically food-gathering structures, although the tube feet may also serve in respiration, locomotion, and sensory perception.

Echinoderms have interior calcitic skeletons (Fig. 13.3) that consist of plates, plates and spines, or diminutive spicules, in all cases covered by a skin. The plates may be firmly joined and constitute a rigid skeleton as in most sea urchins, or they may be loosely joined and constitute a firm but flexible skeleton as in starfish. Where mineralized skeletons are reduced to spicules, the body is highly flexible.

Individual plates and spines of echinoderms are characterized by a microstructure that is unique to the phylum. Each is permeated at a very fine scale by tissue, so that a plate or spine normally consists of a microscopic set of rods that are continuous as a three-dimensional lattice (Fig. 13.4) collectively termed **stereom**. The tissue-filled interconnected spaces between the elements of the stereom are referred to as **stroma**. With a few specific exceptions, all elements of the stereom are in crystallographic continuity, like a highly perforated crystal of calcite. If a petrographic thin section is cut from a plate of a modern echinoderm, it appears under cross-polarized light in a microscope as tiny black spots (the non-mineralized stromal spaces) within a fine lattice of skeleton that exhibits unit extinction as the specimen is rotated.

With very few exceptions, modern adult echinoderms are benthic. Some are **sedentary**, permanently attached to a single site. Most, however, are mobile (**vagrant**). Their speed of movement is small compared to their body size (again, with

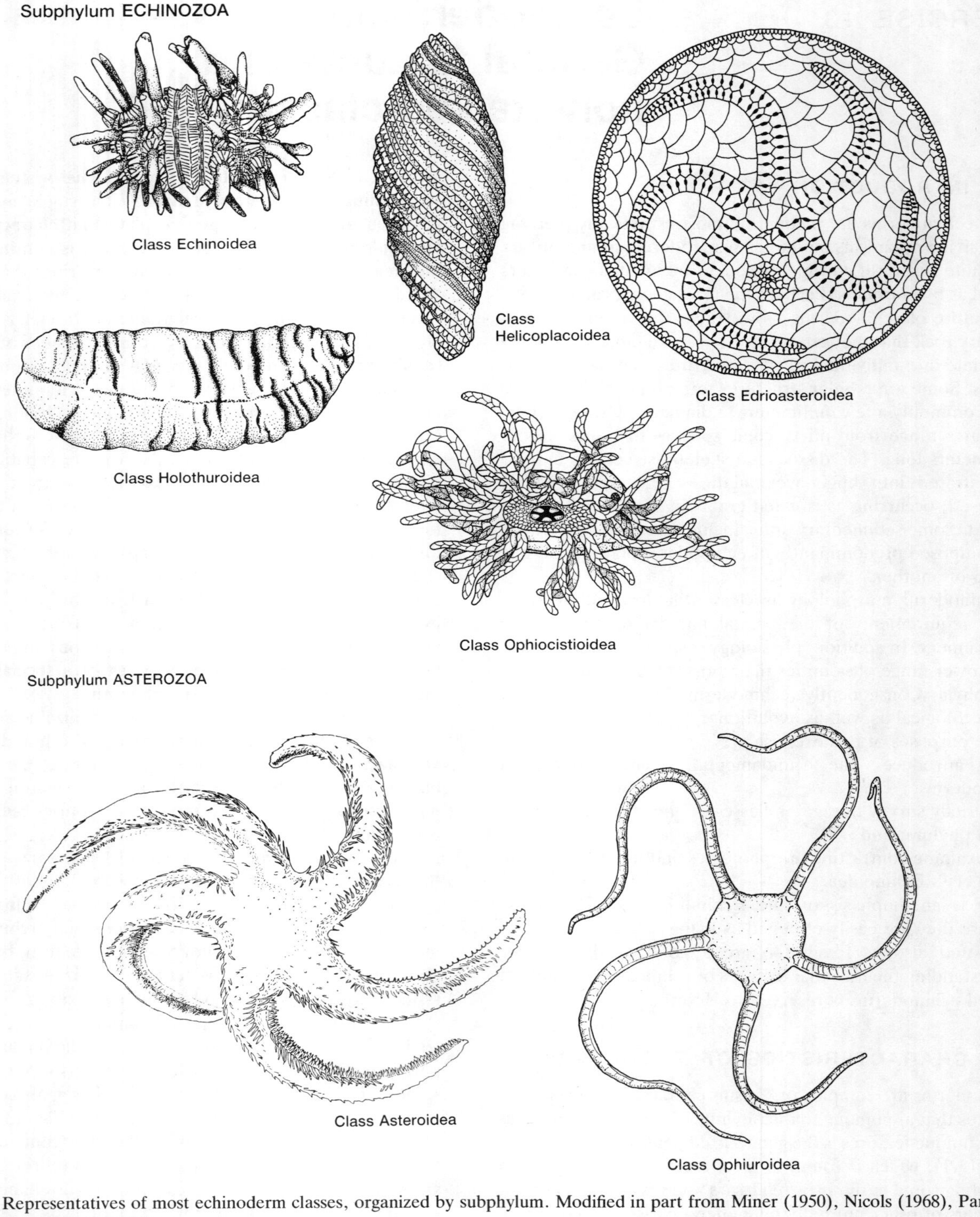

Fig. 13.1 Representatives of most echinoderm classes, organized by subphylum. Modified in part from Miner (1950), Nicols (1968), Parsley & Mintz (1975), and Bell (1976).

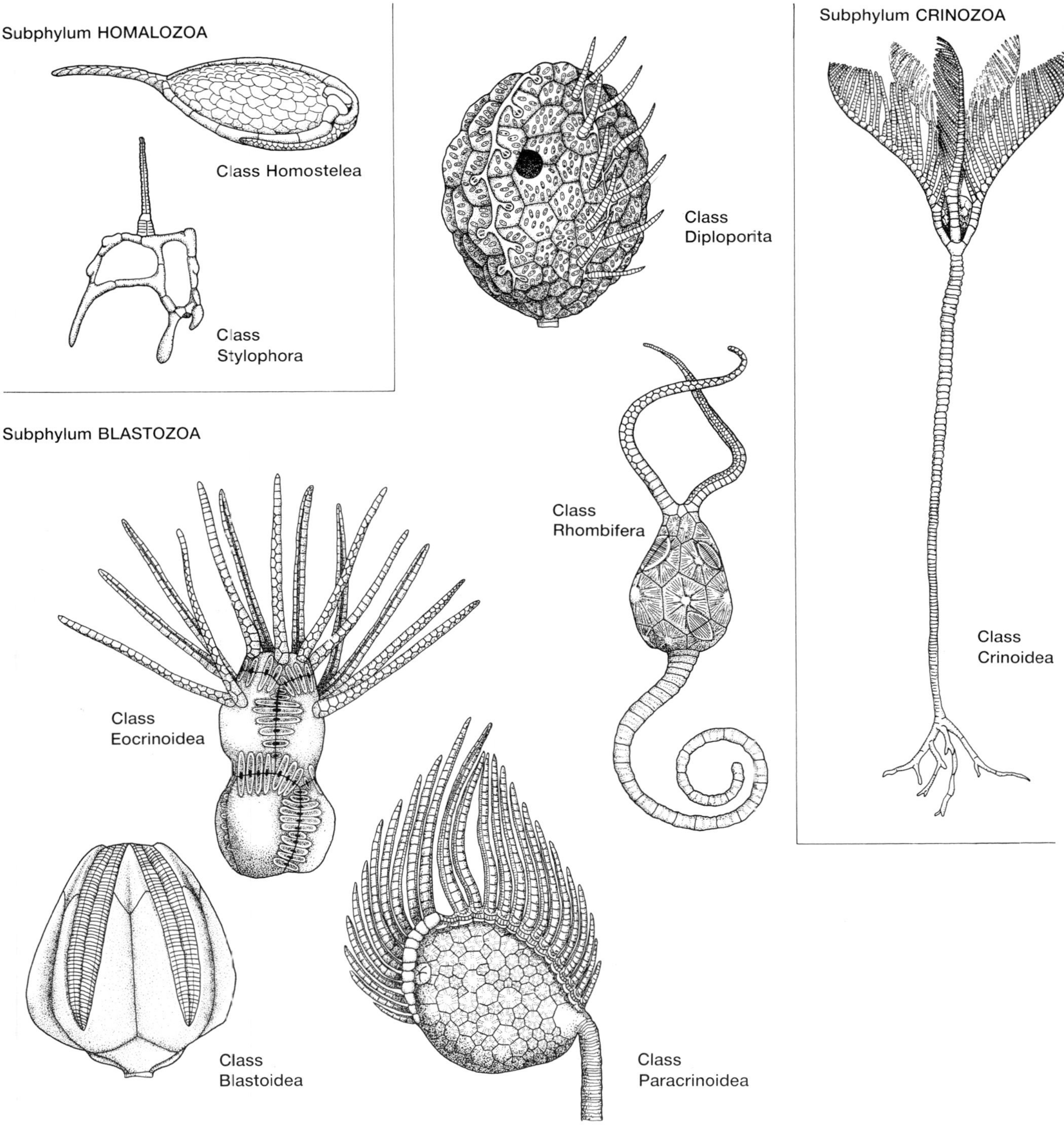

Fig. 13.1 *(Continued)*

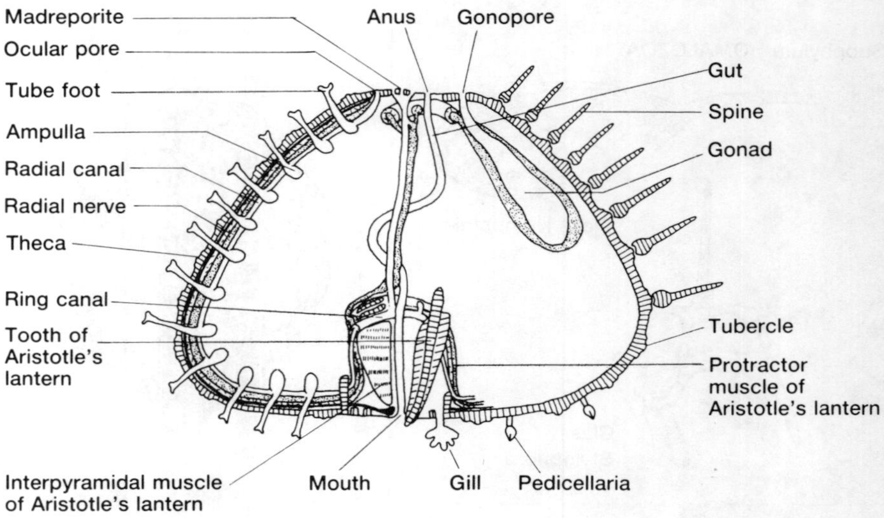

Fig. 13.2 Cross-sectional cut through an echinoid, showing many of the fundamental echinoderm features. Modified from Nichols (1969), Smith (1984) and Kier in Boardman *et al.* (1987).

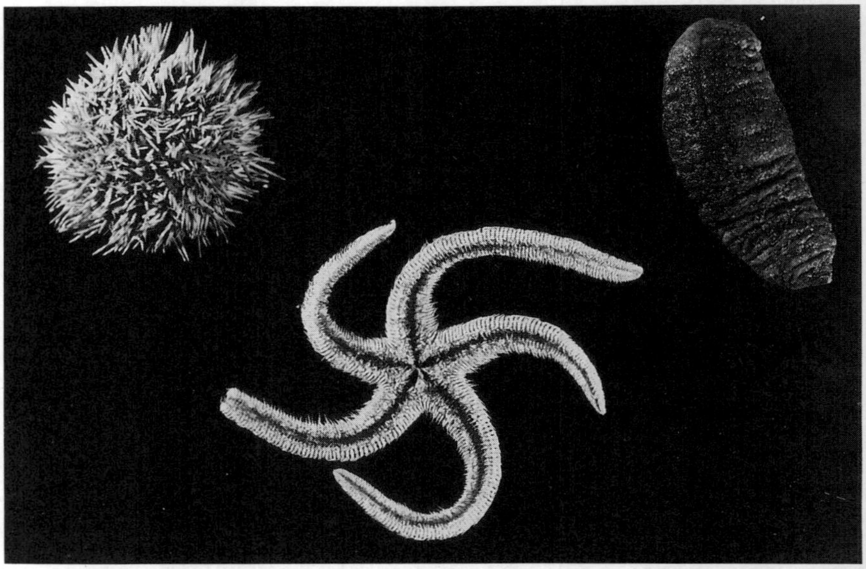

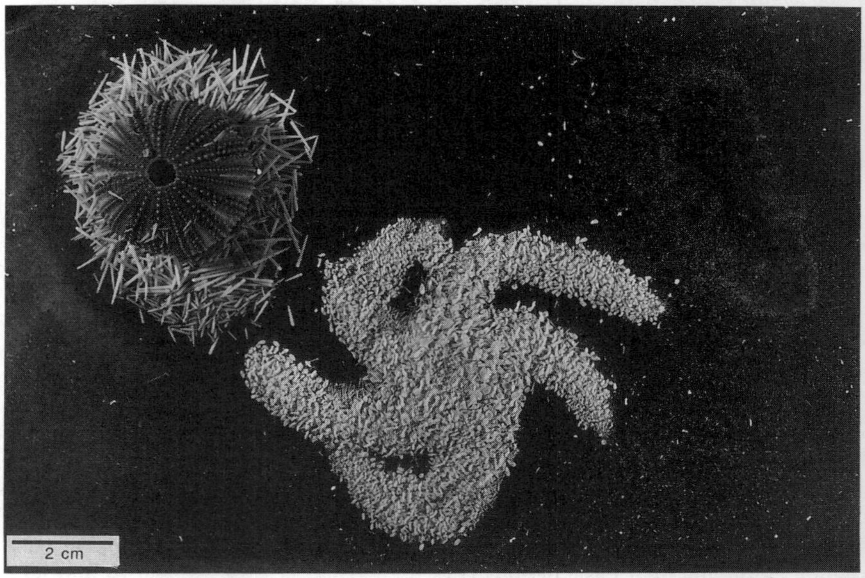

Fig. 13.3 Skeletal products of a regular echinoid (sea urchin; left), asteroid (starfish; *middle*), and holothurian (sea cucumber; *right*). Upper figure shows specimens with soft tissues preserved, but the specimens were treated with sodium hypochlorite so that only their skeleton remains in the lower figure.

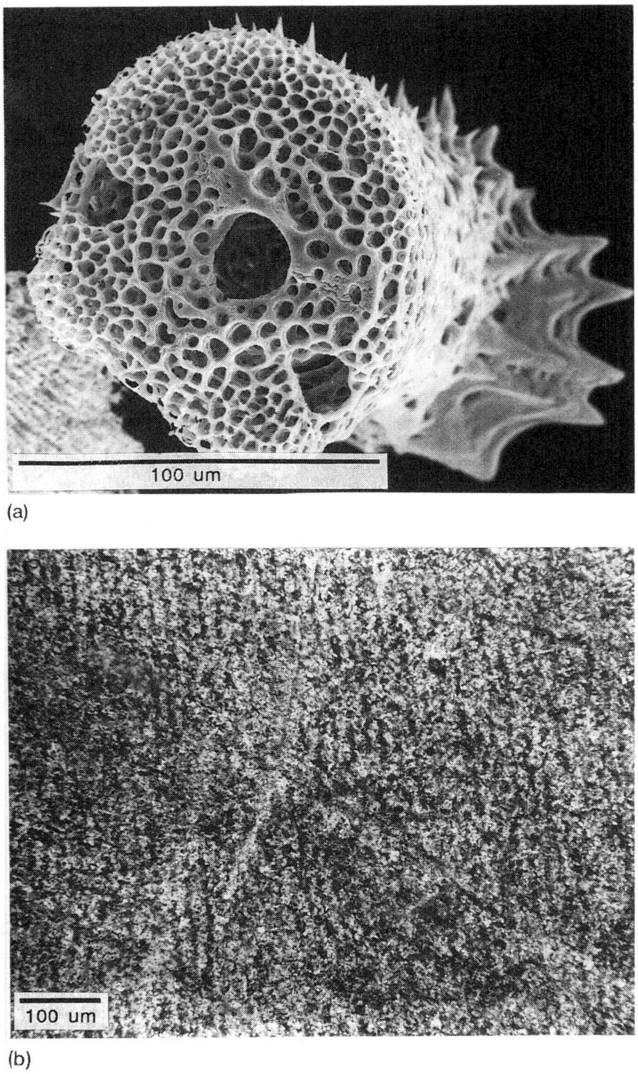

Fig. 13.4 Structure of echinoderm stereom as seen in (a) a scanning electron micrograph of a plate from an arm of the crinoid *Antedon mediterranea* and (b) a thin section through a calyx plate of the carboniferous stalkless crinoid *Agassizocrinus*.

very few exceptions), so that even the speediest of them can move only a few centimeters per minute. More commonly they appear at a glance to be sitting perfectly still when they are actually actively grazing or plowing ahead at full throttle.

13.3 ORGANIZATION OF THE PHYLUM

Various terms are used to refer to the body in different echinoderm groups: theca, test, calyx, central disk, etc. Throughout Exercises 13 and 14, I use **theca** to designate the visecra-enclosing skeletal cavity, although this extends the term into groups where echinoderm specialists normally would use a different term.

Echinoderms have evolved a large number of body forms that exploit various ecological roles, without violating or by slightly modifying their basic biology (Fig. 13.5; Table 13.1). The body theca may be variously shaped, the ambulacra may be recumbent or extend as separated rays or arms, a stalk of stacked plates may be present or absent, the ambulacra may be oriented away from or toward the substrate, and so on. These are simply variations on a theme, but the various combinations have resulted in unique, distinctly different subgroups of the phylum. Echinoderm specialists disagree to some extent on the hierarchical level (subphylum, class, or order) to which some groups should be assigned and how many fundamentally different subgroups actually have existed. In this manual, I most closely follow a classification that subdivides the phylum into five subphyla and 20 classes (Fig. 13.1; Table 13.2).

The subphyla Crinozoa and Blastozoa are characterized by the presence of a uniserial stalk and are the subject of Exercise 14. The Asterozoa, Homalozoa, and Echinozoa are treated in this exercise.
(Question 1)

13.4 ECHINOZOA: CLASS ECHINOIDEA

The fossil record of echinoids is generally poor in Paleozoic rocks, with local exceptions. In part the record is poor because the thecae were not rigidly sutured, but in addition they seem to have been neither abundant nor diverse. Later, after rigid thecae became common during the early Mesozoic, echinoids underwent and adaptive radiation — including burrowing forms — and apparently became much more abundant so that they are characteristic components of post-Paleozoic marine faunas.

13.4.1 Morphology of echinoids

Characteristics of echinoids that distinguish them from other echinoderms are given in Table 13.2.

The earliest echinoids, as well as many species up to the present time, had pentaradial ("regular") thecae (Fig. 13.6). These are the **regular echinoids**, which are characterized by spherical to subspherical, globular thecae with the mouth on the lower side, the anus centered on the upper side (opposite the mouth), and the ambulacra extending the entire distance from the mouth to the group of 10 plates (the **apical system**) that encircle the opening (**periproct**) containing the anus and covered by a set of movable plates. The apical plates that are aligned with the interambulacral areas are the **genital plates** through which the sperm and eggs are extruded; one also serves as the madreporite. Apical plates that are aligned with the ambulacra are the so-called **ocular plates**, which actually have nothing to do with light perception.

All regular echinoids have a complex masticatory apparatus termed the **Aristotle's lantern**, which contains 40 pieces of skeleton (an exact count, not a biblical number meaning "many"). Five of the pieces are teeth that can be protruded through the mouth to scrape or gouge various foods below the theca. The other pieces are important for translating forces from a complex of muscles for operation of the teeth. The complex of muscles ultimately ties the Aristotle's lantern to inner extensions (**apophyses**) of the thecal plates that surround the mouth (**perignathic girdle**). Evolution of the Aristotle's lantern in regular echinoids resulted in progressively greater thrust and scraping forces translated through the teeth.

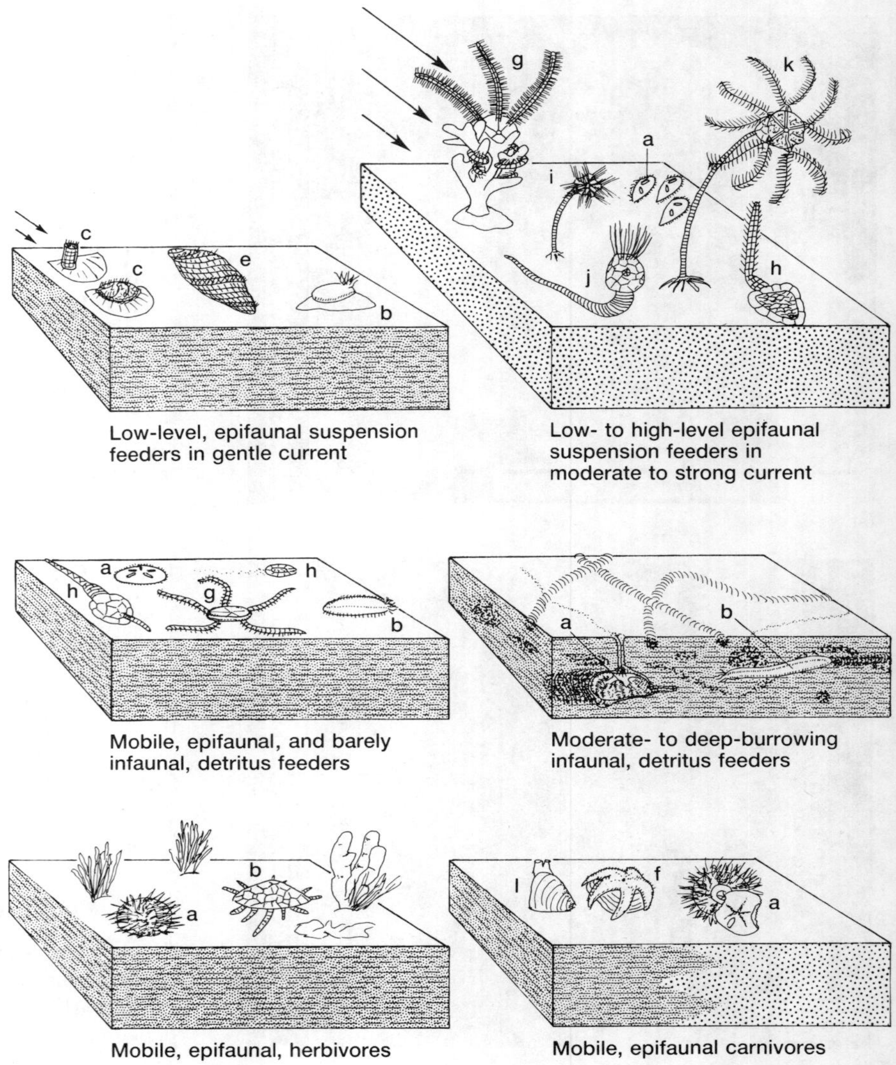

Fig. 13.5 Inferred life habits of fossil echinoderms. (a) Echinoid, (b) holothurian, (c) edrioasteroid, (d) ophiocistioid, (e) helicoplacoid, (f) asteroid, (g) ophiuroid, (h) homalozoan, (i) blastoid, (j) rhombiferan, (k) crinoid, (l) pelecypod prey of asteroid. After Sprinkle (1980).

One or more movable exterior spines articulate on each echinoid plate. The spines vary greatly in size, shape, and degree of differentiation across the theca. Their basal surface is concave, resting on the central portion of a distinctive articulating structure, the **tubercle**, on the plate. Size of the tubercle corresponds with massiveness of its associated spine. Tubercles in regular and in most irregular echinoids are radially symmetrical, so the spines can be rotated equal distances and with equal force in a 360° arc.

Pedicellariae are minute, articulated, pincer-like claws elevated on flexible stalks above plate surfaces. Their apparent role is to guard the echinoid against small organisms and settling larvae. They are almost invariably lost from fossil echinoids, even when spines are still preserved in place. *(Questions 2–4)*

During the Mesozoic, at least two lineages of echinoids developed bilateral symmetrical ("irregular") thecae associated with evolution of an actively burrowing habit. These are the **irregular echinoids**, which are characterized by slightly compressed and elongate globular thecae to thinly flattened, elongate or circular thecae. The mouth is centered on one side or displaced towards the edge that leads as the animal burrows; this leading edge is centered on an ambulacrum designated the **anterior ambulacrum**. The anus is not situated within the apical system but is found somewhere within the interambulacral area (**posterior interambulacrum**) opposite the anterior ambulacrum, and may be on the aboral side, at the posterior edge, or on the oral side.

Some irregular echinoids have an Aristotle's lantern, but it is absent in others. Additional major functional features found in some irregular echinoids include: **phyllodes**, which are abundant and crowded ambulacral pores for passage of numerous tube-feet near the peristome (may also occur in regular echinoids); **bourrelets**, which are inflations of the ambulacra bordering the peristome in some echinoids, where abundant special spines for scraping may occur; **fascioles**, which are narrow, ciliated bands of small spines that encircle the test of some heart-shaped echinoids and generate a flow of water over the surface; and a **plastron**, a flattened portion of the posterior interambulacrum on the oral surface where special spines for locomotion are located in some echinoids.

Table 13.1 Life habits of the living echinoderms. From Meyer (1980).

Food location:	Benthic				Pelagic	
Food type:	Organic detritus, sediment	Plants	Animals	Carrion	Organic detritus, plankton	Dissolved organic substances
PLANKTONIC	—	—	—	—	Holothuroidea	—
EPIFAUNAL	Sessile (Crinoidea?) Mobile *Ophiuroidea* Asteroidea *Holothuroidea*	— *Regular* *Echinoidea*	— *Asteroidea* Ophiuroidea (Echinoidea?)	— Ophiuroidea Asteroidea	Sessile *Crinoidea* Ophiuroidea Asteroidea Holothuroidea	Sessile/mobile Crinoidea Ophiuroidea Asteroidea Echinoidea Holothuroidea
INFAUNAL	Sessile Ophiuroidea Holothuroidea Mobile *Irregular* *Echinoidea* Holothuroidea Asteroidea	— —	— —	— —	Sessile Ophiuroidea Echinoidea Holothuroidea —	— —

Italic entries indicate the major feeding habit for the group.

Spines are numerous and relatively small in irregular echinoids. One of their most important functions is to move sediment as the animal burrows. Consequently, they must be able to exert more force in the direction in which they push sediment than in other directions. Accomodation of greater muscles on the side toward which sediment is pushed commonly causes eccentricity of the tubercle so that there is greater space between the tubercular rim and the domed surface (mammelon) on which the spine articulates.

Characteristics of the major subdivisions of the echinoids are given in Table 13.3. The classification is somewhat complex, and the table includes only the various formal and informal groups that are the most commonly used or readily recognizable.
(Question 5)

13.4.2 Life habits of echinoids

Regular echinoids are surface grazers of algae, encrusting organisms, or detritus, and may move in any direction in search of food. A few are carnivorous on even more slowly moving sand dollars (clypeasteroid echinoids). Some live on rocky substrata, where they may range widely or may be confined to a small indentation that their own scraping activity has generated or enhanced. Others, including the carnivores that eat clypeasteroids, crawl across loose sediment in search of food.

The burrowing habit of irregular echinoids evolved rapidly and was accompanied by a distinctive set of morphological changes (Fig. 13.7). The rather thick irregular echinoids are quite sluggish semi-infaunal or infaunal burrowers, but the flat forms (sand dollars; Order Clypeasteroida) burrow more rapidly, generally just below the sediment surface in shifting sands.

Each life habit (non-burrowing versus burrowing, sluggish burrowing in relatively quiet water versus more rapid burrowing in sands shifted by agitated water) is characterized by a range of distinctive morphologies (Figs 13.7 and 13.8).
(Question 6)

13.5 ASTEROZOA

Plates of asterozoans are not firmly sutured together, so upon death and deterioration of soft tissues, their plates are readily moved by scavengers, currents or passing organisms. Therefore, well-preserved asterozoan skeletons are not common, although their isolated plates are widespread even if generally unrecognized as asterozoan fossils.

The two asterozoan classes (Asteroida and Ophiuroida) have several features in common, including a central disk from which ambulacra-bearing arms radiate, arms that consist of multiple articulating rows of plates, and orientation of the body such that the ambulacra are against the substratum. They are easily distinguished from one another, however, in degree of distinctiveness of arms from the central disk, whether the stomach is confined to the central disk or extends into the arms, robustness of the tube-feet, and organization of plates that constitute the arms.

Asteroids (starfish) are predominantly carnivores, most of which extrude their stomach to digest the prey. Basic organization of asteroids is shown in Fig. 13.9. Plates in the arms (Fig. 13.10) include two columns of ambulacral plates that are situated in a deep groove and have a series of pores along the suture between them for passage of fluid from the ampullae to the robust tube-feet. Along each side of the double row of ambulacrals are a single row of adambulacral plates, which bear robust spines in many starfish. The spines from the adambulacral plates may be folded axially and overlapped to

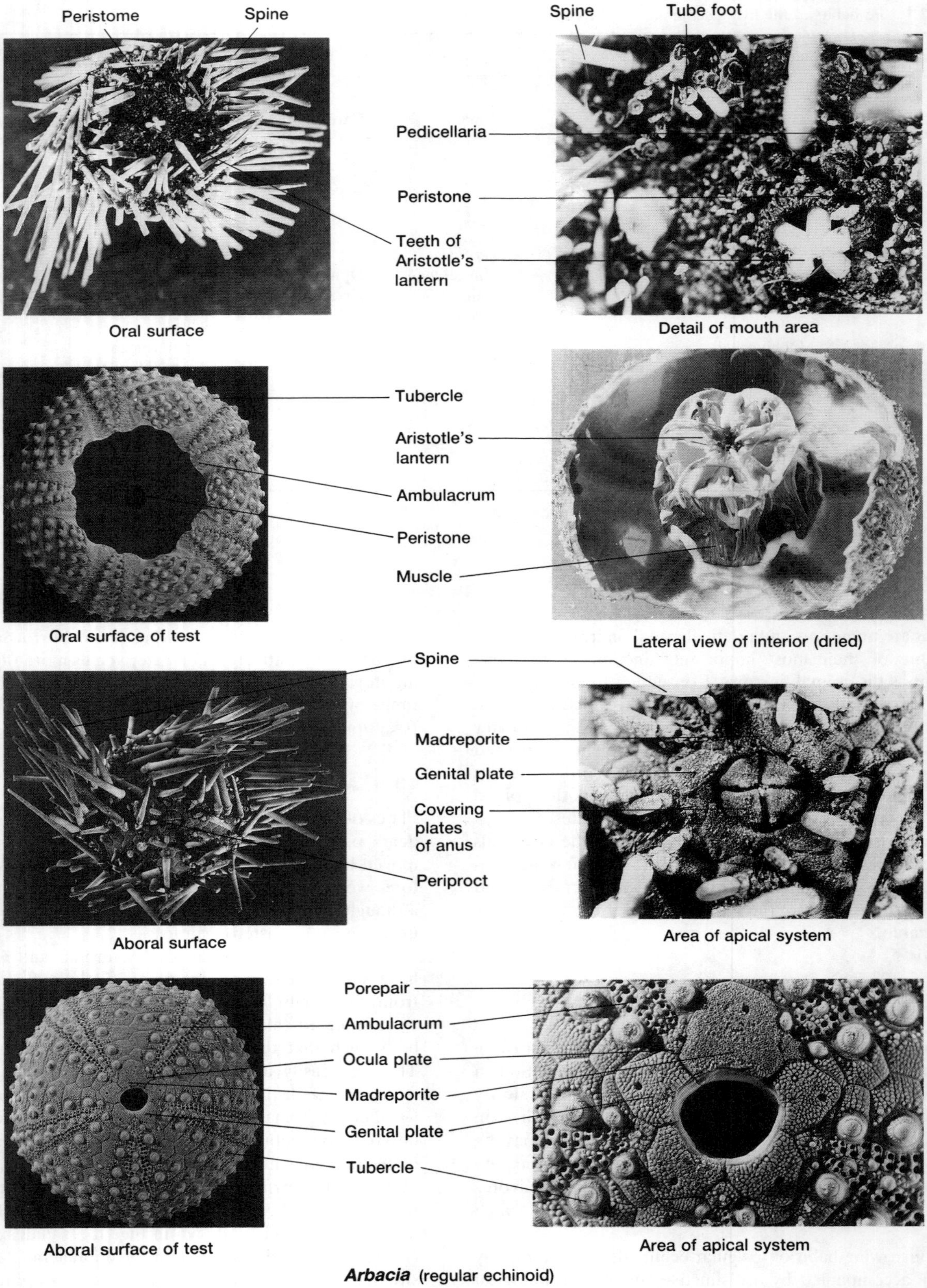

Fig. 13.6 Features of a regular echinoid apical system and of the peristome-lower ambulacra, and features of irregular echinoids. In part from Kier (1975, 1980).

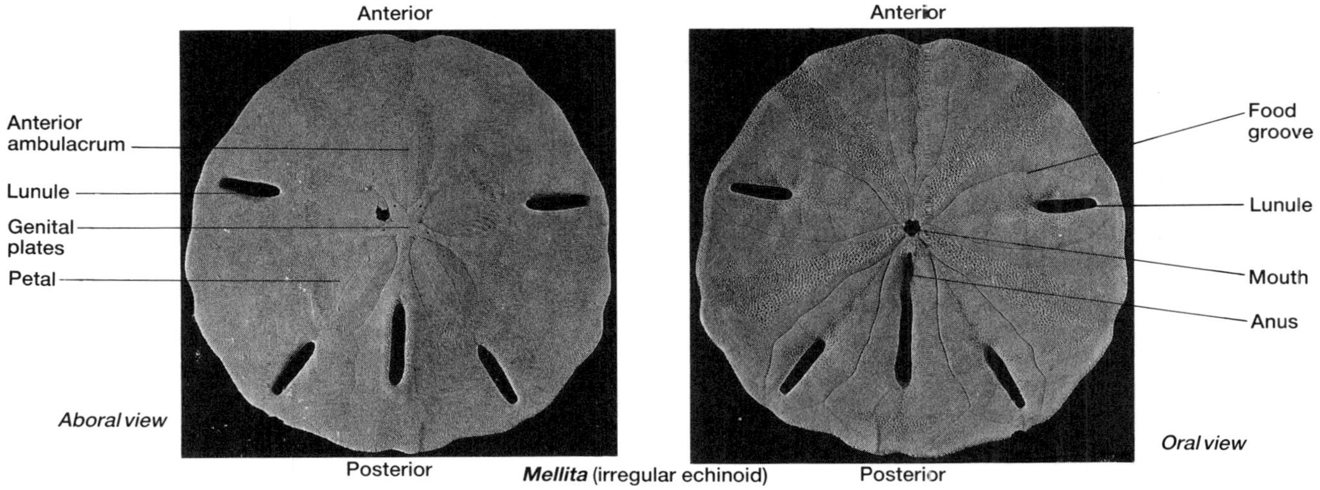

Mellita (irregular echinoid)

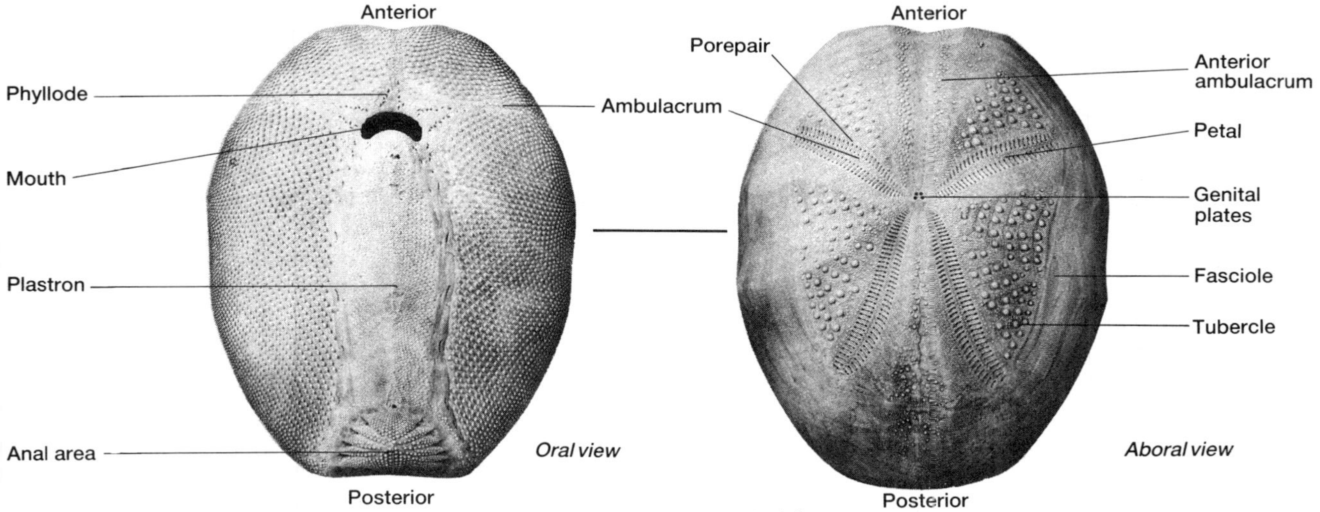

Plagiobrissus (irregular echinoid)

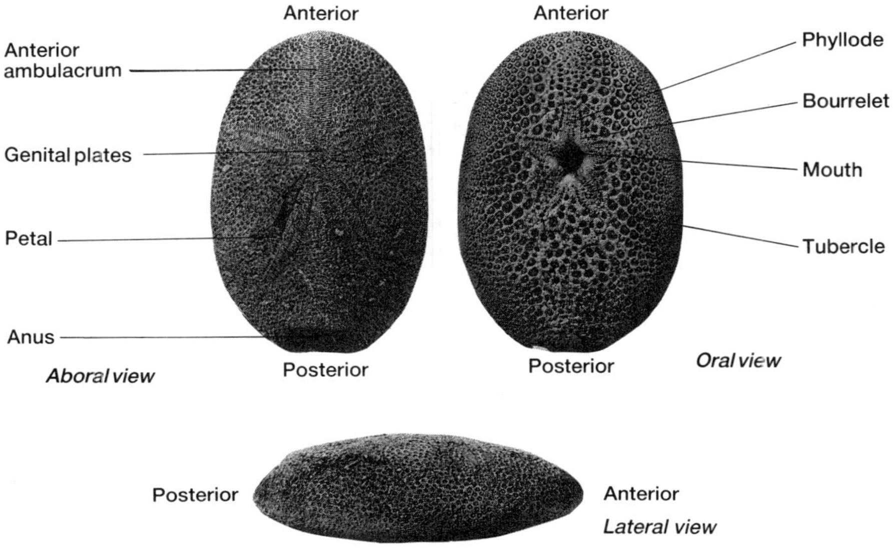

Eurhodsa (irregular echinoid)

Table 13.2 Classification and characteristics of the subphyla and selected classes of echinoderms. Summarized variously from Nichols (1969), Broadhead (1980), Kier in Boardman *et al.* (1987), and Sprinkle in Boardman *et al.* (1987).

Subphylum Echinozoa	Generally stemless and armless, with primitively globular thecae that may be secondarily flattened or elongated Skeleton varying from rigid with interlocked stereom along plate boundaries to isolated spicules, most having numerous plates Ambulacra with pores through which tube-feet protrude Mouth and anus generally on opposite sides of theca Lower Cambrian–Recent
Class Echinoidea	Globular or flattened Thecal skeleton of hundreds of interlocked, spine-bearing plates Twenty series of plates organized into two series each in the five ambulacra and five interambulacral areas in post-Paleozoic representatives, but commonly less regular in Paleozoic representatives Mouth on lower surface, anus at convergence of ambulacral tips on upper surface or displaced eccentrically to a location within an interambulacral area (the posterior interambulacrum) Radial canals are internal to the skeleton Upper Ordovician–Recent
Class Holothuroidea	Elongate, sac-shaped body of leathery skin composed of epidermis and thick, muscular mesoderm, containing small spicules, with mouth and anus at opposite ends Mouth surrounded by ring of tube-feet modified as tentacles Recumbent on one side on substratum, resulting in differentiation of ambulacra and conspicuous bilateral symmetry Unique, elaborate respiratory "tree" within body developed from posterior end of alimentary canal Middle Cambrian?, Middle Ordovician–Recent
Class Edrioasteroidea	Thecae discoidal in most but columnar in some, with five straight or curved ambulacra with small cover plates Mouth central Anus interambulacral Interambulacral plates imbricated, commonly merging into very small rimming plates around periphery Attached to solid substratum but possibly capable of some movement Lower Cambrian–Upper Carboniferous
Class Ophiocistioidea	Mouth central on lower surface of domal thecae Five well-defined radiating ambulacra and intervening interambulacral areas on lower side of thecae, but plates of upper surface irregular in shape Anus aboral, eccentrically placed in same interradial region as madreporite Tube-feet covered by imbricated plates Five large teeth in mouth Lower Ordovician–Lower Carboniferous
Class Helicoplacoidea	Spirally pleated, flexible, free-living, fusiform (elongate with tapering ends) thecae Two or three spiraling ambulacra of different lengths Lower Cambrian
Class Cyclocystoidea	Thecae consisting of a ring of heavily ornamented plates encircling two sheets of thinner plates, with peripheral imbricated plates forming a marginal skirt Rare and poorly understood Middle Ordovician–Upper Devonian
Subphylum Asterozoa	Mouth central on lower surface of theca Anus, where present, on upper surface Five or more large, radiating arms bear ambulacra on their lower surfaces Pentameral symmetry well developed Arms merging with or distinct from the central disk Radial canals of water vascular system not enclosed by ambulacral skeleton and bear numerous tube-feet Lower Ordovician–Recent.

Table 13.2 *(Continued)*

Class Asteroidea	Five to 25 large arms containing extensions of the coelomic cavity and merging with indistinct central disk Tube-feet large Theca with numerous small plates Gut short and straight Anus usually present, on upper surface adjacent to madreporite Lower Ordovician–Recent
Class Ophiuroidea	Distinct, large central disk of numerous plates from which five elongate, thin arms extend Tube-feet small Gut blind (no anus) Madreporite on oral surface or absent Flexible arms each having a single series of large, articulating plates (vertebrae) encased by smaller plates Lower Ordovician–Recent
Subphylum Homalozoa	Thecae flattened, asymmetrical to bilaterally symmetrical Plates tightly sutured or loosely joined One or more arm- or tall-like appendages common Cover plates on arm-like structures Mouth at base of arm-like feeding appendage or in homologous position in representatives with no such appendage Anus lateral to or on opposite side of body from mouth (Classes of this subphylum are differentiated by presence or absence of an arm-like appendage that bears a single ambulacrum, presence or absence of a tail-like appendage on the side of the theca opposite to where the arm-like appendage might be, and degree of differentiation of plates on the theca. There are only about 50 known genera in the entire subphylum, for which there seem to have been too many classes erected.) Middle Cambrian–Upper Carboniferous
Subphylum Blastozoa	Stalked, with globular, multiplated, tightly sutured thecae enclosing viscera Gut is U-shaped, with mouth opposite the stem attachment, and the anus lateral to and beside the mouth Ambulacra erect or radiate across the surface of the theca from the mouth and bear small, erect, biserially plated, unbranched appendages (**brachioles**), with an extension of the ambulacrum on one side but no evidence of tube-feet Most with hydropore and special respiratory thecal structures Lower Cambrian–Upper Permian
Class Blastoidea	Theca bud-shaped, with four circlets of plates Pentameral symmetry well developed Ambulacra each with single central, elongate lancet plate with rows of two types of side plates Three (rarely four) basal plates Internal, thin, calcified respiratory folds open as slits or pores along ambulacral areas Middle Ordovician–Upper Permian
Class Rhombifera	Theca globular, with four or five circlets of plates Two to five ambulacra extending from mouth to edge of upper surface or down over side of theca Ambulacra of two types of flooring plates, in some reduced to one or a few large, erect, arm-like brachioles near mouth Anus on side of theca Respiratory structures are rhomboid sets of folds or canals that cross between adjacent thecal plates Lower Ordovician–Upper Devonian
Class Diploporita	Theca globular, covered with numerous irregularly shaped and arranged plates or plates arranged symmetrically in pentameral fashion Three to five brachiole-bearing food grooves extending across theca, or small brachioles only in region of mouth Anus at edge of upper surface or side of theca

Table 13.2 *(Continued)*

	Respiratory structures are paired pores (**diplopores**) extending through thecal plates Stalk commonly absent, or attached to four basal plates Lower Ordovician–Lower Devonian.
Class Eocrinoidea	Theca globular or flattened Symmetry pentameral or bilateral, poor or well developed Two to five short ambulacra support a few long, erect brachioles Anus lateral Pores along sutures in early forms, generally absent in later forms that have thin plates Lower Cambrian–Upper Silurian
Class Parablastoidea	Theca bud-shaped, with well-developed pentameral symmetry, composed of a few circles of plates Two rows of single type flooring plates in each ambulacrum Respiratory structures are deeply incised longitudinal folds (**cataspires**) in large deltoid plates between ambulacra Middle Ordovician
Class Paracrinoidea	Theca lenticular or globular, plates irregularly arranged Two to five ambulacra, recumbent and asymmetrical with erect pinnules or on erect, uniserial, pinnule-bearing arms of unequal length Mouth offset from center of summit, anus nearby and nearly opposite stem attachment Hydropore beside mouth Slit-like respiratory structures present in some Stem attached to the three basal plates Lower Ordovician–Lower Silurian
Subphylum Crinozoa	Usually stalked, with globular theca (usually termed **calyx** in this subphylum) that is partially or entirely tightly sutured Radial canals of water vascular system not embedded within skeleton, generally extending along long, erect arms that bear numerous tube-feet Arms may bear small lateral branches (pinnules) Two or more circlets of plates in calyx pentamerally arranged Gut is U-shaped Mouth nearly centered opposite stem on upper surface of calyx, with anus lateral but on same surface Most lacking hydropores and accessory respiratory structures Middle Cambrian–Recent
Class Crinoiodea	Calyx divided into conical to bowl-shaped lower **cup** and upper flat to domal **tegmen** Cups with well-developed pentameral symmetry expressed in two or more circlets of plates Tegmen covered by generally irregularly shaped, small plates Arms branched or unbranched, short to long, uniserial or biserial, with or without lateral uniserial pinnules Calyx permanently or temporarily attached to substratum by uniserial stem or by numerous short, thin uniserial appendages (**cirri**) extending from lower plates of cup Middle Cambrian–Recent

Table 13.3 Characteristics of the major subdivisions of the Echinoidea. Generally after Kier in Boardman *et al.* (1987).

Order Echinocystitoida	Theca regular, flexible Ambulacra and interambulacra each with two or more columns of thin, imbricated plates One or five genital plates Upper Ordovician–Permian
Order Palaechinoida	Theca regular, plates thick Ambulacra each with two or more columns of plates Interambulacra with one or more columns of plates Five genital plates Upper Silurian–Permian
Order Cidaroida	Theca regular Plates thin to moderately thick, rigidly joined or imbricated Ambulacra each with two columns of plates In Paleozoic species, interambulacra each with one or more columns but only two columns in post-Paleozoic species Most commonly single large spine articulating on each interambulacral plate Five genital plates Lower Carboniferous–Recent
Order Echinothurioida Order Diadematoida Superorder Stirodonta Superorder Camarodonta	Lower Jurassic–Recent Upper Triassic–Recent Lower Jurassic–Recent Lower Cretaceous–Recent { Within all these groups: Theca regular; Thin to moderately thick extend plates, generally not imbricated; Ambulacra and interambulacra each of two columns of plates; Generally more than one spine articulation on each interambulacral plate; Five genital plates
Order Cassiduloida	Theca irregular All ambulacra developed on upper surface as **petals**, with elongate slits for respiratory tube-feet Phyllodes and bourrelets present One to four genital plates No accessory pores for tube-feet, food grooves, plastron, fascioles, nor (in adults) Aristotle's lantern Lower Jurassic–Recent
Order Spatangoida	Theca irregular Generally only four petals (all but anterior ambulacrum) Plates very thin to moderately thick Anterior ambulacrum commonly indented, pores small Phyllodes, plastron, and commonly fascioles present One to four genital plates No accessory pores, food grooves, bourrelets, nor Aristotle's lantern Lower Cretaceous–Recent
Order Clypeasteroida	Theca irregular, usually flattened, commonly with internal reinforcement Five petals Accessory pores present One genital plate Aristotle's lantern with broad teeth **Lunules** (which are slots to allow shorter passage of food from aboral to oral surface) absent or up to five No phyllodes, bourrelets, plastron, nor fascioles Paleocene–Recent

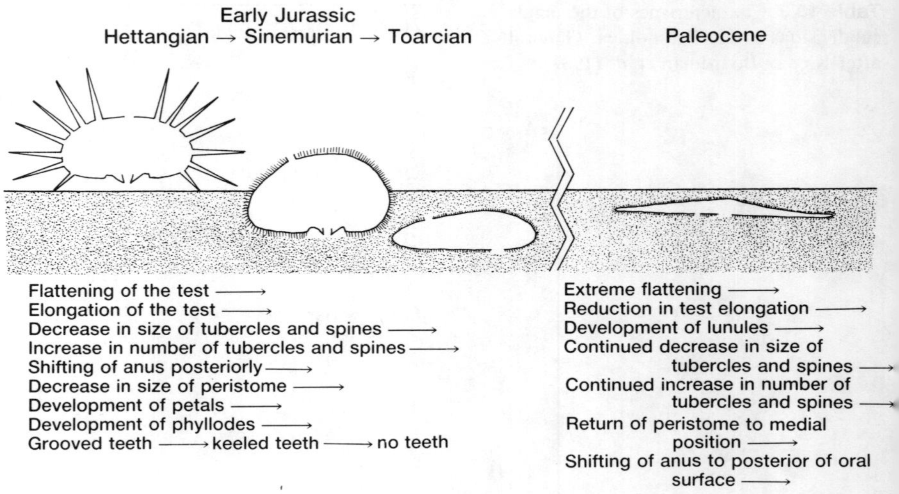

Fig. 13.7 Critical changes in the evolution of irregular echinoids from ancestral regular echinoids during the Early Jurassic, and a listing of the morphological changes that occurred. Modified from Kier (1982).

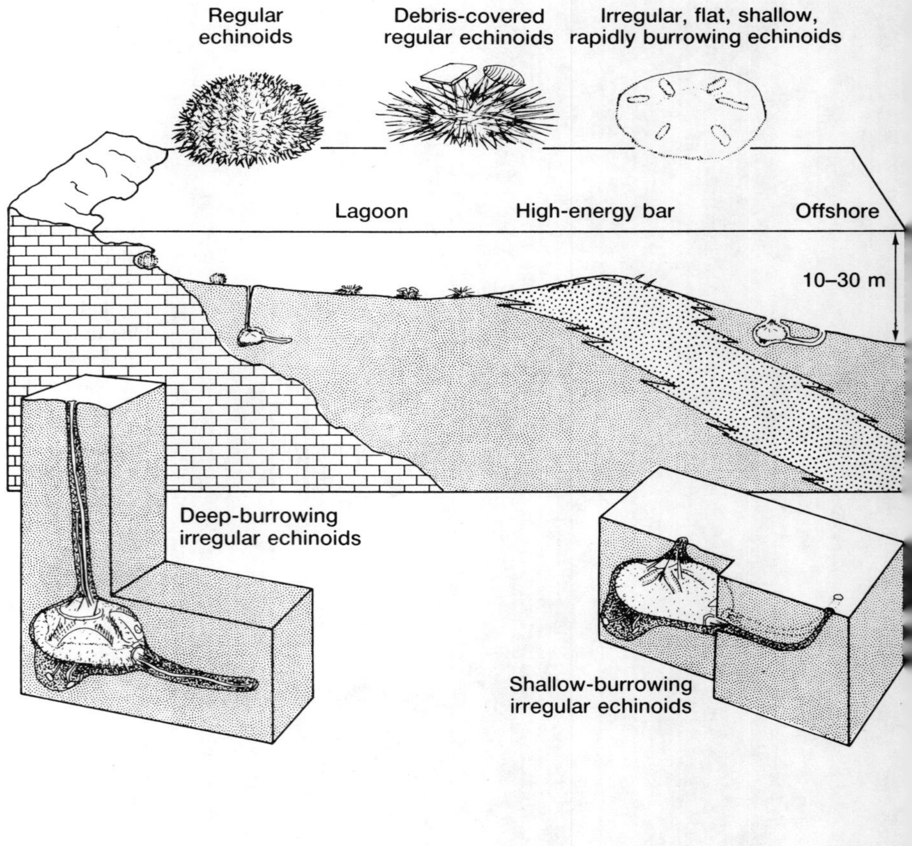

Fig. 13.8 Life habits of echinoids. Modified from Kier (1972).

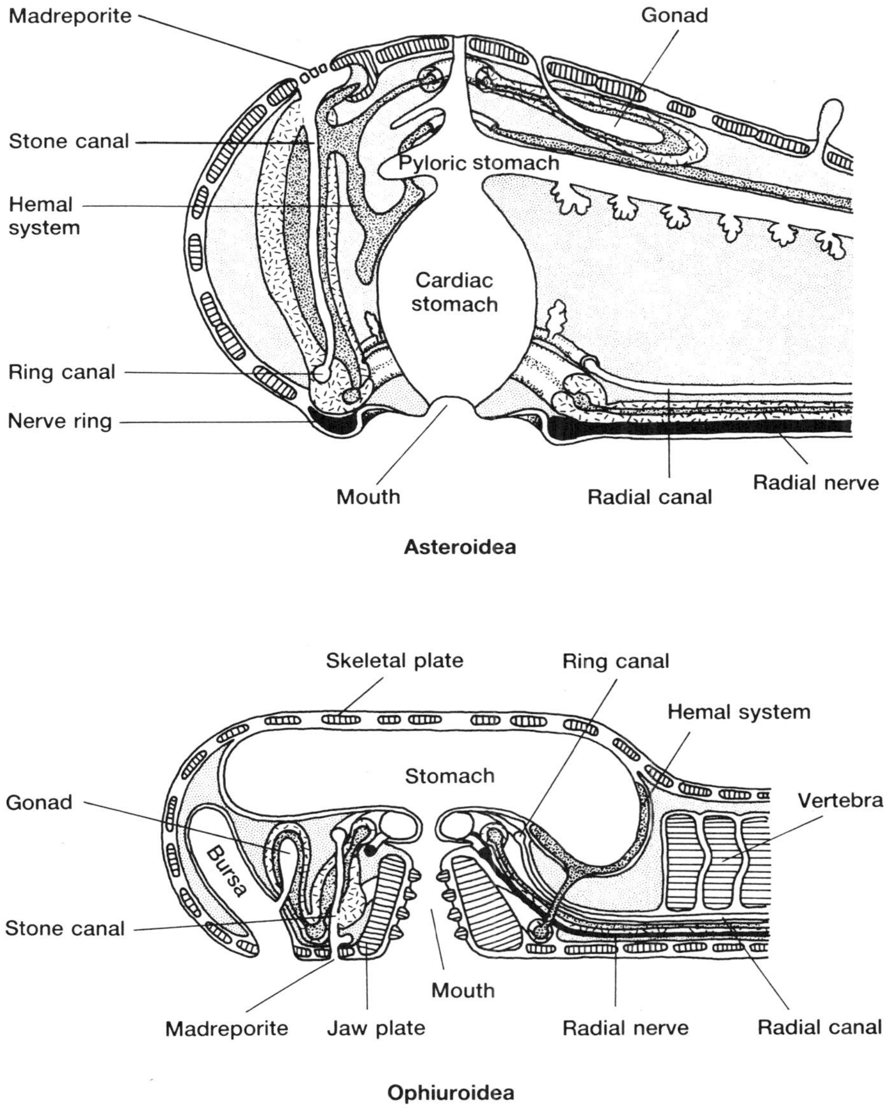

Fig. 13.9 Basic organization of skeleton and soft-tissue organs of asteroids and ophiuroids shown in sectional cuts passing from oral to aboral sides. The central disk is on the left, and the right portion of each diagram cuts along the midline of an arm. Modified from Nichols (1969).

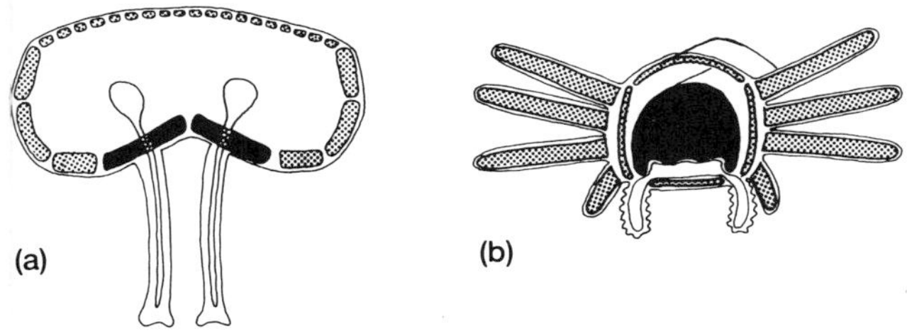

Fig. 13.10 Diagrammatic sketches of cross-sections showing arm plate systems typical of (a) an asteroid and (b) an ophiuroid. Ambulacral plates of asteroids and vertebrae of ophiuroids are thought to be homologous and are shaded differently from the other plates. Tube-feet and ampullae are on the lower, oral surface.

protect the tube-feet when needed. Beside the adambulacrals, on the oral or lateral surfaces of the arms, there typically are situated marginal plates, usually in two rows that are enlarged or show some other morphologic differences from the adjacent rows of plates. The aboral surface of the arm is covered by numerous rows of small dorso-lateral plates and in some a single medial row of carinal plates.

Ophiuroids (brittle stars) are predominantly detritus feeders or elevate two or three arms for suspension feeding. Basic organization of ophiuroids is shown in Fig. 13.9. Plates in the arms (Fig. 13.10) include a single column of massive plates termed the vertebrae, each of which is homologous with a pair of laterally fused ambulacral plates in asteroids. The vertebrae are surrounded by four series of thin plates: a

column of oral plates, a column of lateral plates on each side, and a column of aboral plates. The laterals are the only plates that bear spines.
(Questions 7 and 8)

FURTHER READING

Beaver, H. H. *et al.*, 1967. Echinodermata 1, in R. C. Moore (ed.), *Treatise on invertebrate paleontology*, Part S. Geological Society of America and University of Kansas Press: Lawrence, KS.

Broadhead, T. W. & Waters, J. A. (eds), 1980. *Echinoderms, notes for a short course*. University of Tennessee Department of Geological Sciences Studies in Geology 3.

Durham, J. W. *et al.*, 1966. Echinodermata 3, in R. C. Moore (ed.), *Treatise on invertebrate paleontology*, Part U. Geological Society of America and University of Kansas Press: Lawrence, KS

Nichols, D., 1969. *Echinoderms*, 4th edition. Hutchinson & Co.: London.

Smith, A., 1984. *Echinoid palaeobiology*. George Allen & Unwin: London.

EXERCISE 13

Worksheet

Name

QUESTION 1

Various specimens of Echinozoa, Asterozoa, and Homalozoa are available for study and comparison. For each, record the information requested in the table on p. 202.

QUESTION 2

Examine a preserved or dried modern regular echinoid with at least some spines still attached, along with a theca and dissociated spines and Aristotle's lantern, belonging to the same species, but from which all organic matter has been removed. Closely study the specimens with a microscope or good-quality handlens. By comparison with Figs 13.2 and 13.6, examine the following features, checking them off as seen.

Preserved or dried specimen (exterior)

_____ mouth

_____ tips of teeth of Aristotle's lantern

_____ peristome

_____ tube-feet (best seen on oral surface)

_____ pedicellariae (perhaps also best seen on oral surface)

_____ "typical" large spines

_____ specialized spines on oral surface?

_____ ambulacrum

_____ interambulacrum (how many spines per interambulacral plate? _____)

_____ apical system

_____ genital plates

_____ madreporite

_____ ocular plates

_____ periproct (how many plates within it? _______)

Preserved or dried specimen (interior)

_____ complex of plates in Aristotle's lantern

_____ apophyses of perignathic girdle

_____ complex of muscles associated with Aristotle's lantern and perignathic girdle

Cleaned specimen

_____ ambulacrum

Specimen number	*Classification*			*Body symmetry or shape*	*Type of appendages if present*	*Probable feeding method or food source*
	Subphylum	*Class*	*Genus*			

_____ interambulacrum

_____ pore or porepair in ambulacral plate

_____ tubercles on which spines articulate (how many different sizes? _____)

_____ apical system

_____ genital plates

_____ madreporite

_____ ocular plates

QUESTION 3

Again examine the large spines and their association with the tubercles of the interambulacral plates of the preserved or dried modern regular echinoid studied for Question 2. A radial sheet of muscles connects the shoulder of the spine (near its concave base) to the ring around the tubercle. Interior to the muscle is a radial collagenous sheet plus a central vertical thread of collagen. Based on this information plus your observations, draw a reconstruction of a medial section through the spine and tubercle, showing the soft tissues as well as the skeleton. Explain how the spine is moved or waved in multiple directions.

____

____

____

____

QUESTION 4

Based on your observations for Question 2, label the following features in the set of illustrations on p. 206.

Genital plates (two of them)
Madreporite
Ocular plates (two of them)
Porepair (a pair of laterally adjacent pores through which tube feet extend, located in ambulacra)
Tubercle
Ambulacrum (indicate both columns of plates)
Interambulacrum (indicate both columns of plates)
Peristome
Periproct

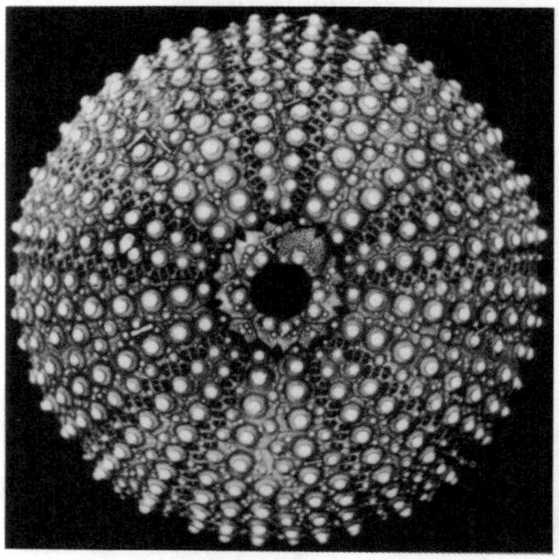

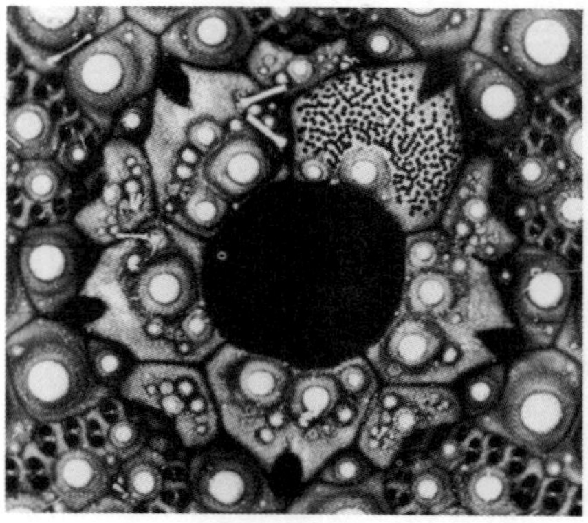

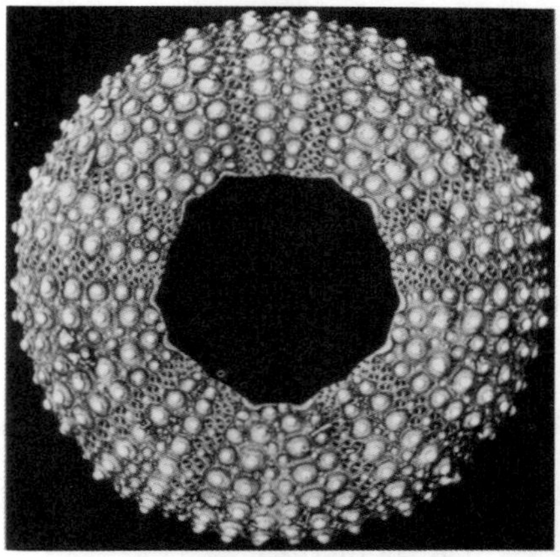

QUESTION 5

One or perhaps two different irregular echinoids are provided for this question. First, identify the echinoid(s) to order. Then, draw the outlines of each, as seen from the oral side and also as seen from the aboral side. Keep the anterior edge at the top of the outline. Label which is oral and which is aboral. Draw in and label the positions of the mouth, anus, apical system, and petal portions of the ambulacra. Label the anterior and posterior edges of each sketch.

QUESTION 6

For each species of echinoids from a single fauna, give the information required in the table opposite and, by comparison with Figs 13.7 and 13.8, give the probable life habit for each species.

QUESTION 7

Two preserved or dried asterozoan specimens are provided. Assign each to its correct class, and list the characteristics of the specimen on which the assignment was based.

Specimen 1. Class: ______________________________

____

____

____

Specimen 2. Class: ______________________________

____

____

____

QUESTION 8

Specimens of a regular echinoid, an irregular echinoid, an asteroid, and an ophiurold have been cleaned of organic material by soaking them in a dilute solution of sodium hypochlorite, then washing them. Presuming the residue to be typical of fossils of these groups:
Which would be most similar to the body form of the living organism after the parts became dissociated?

____

Which would be least similar?

____

For which group(s) would be isolated small skeletal elements be most easily placed to class?

____

For which group(s) would isolated small skeletal elements be most difficult to place correctly to class?

____

Species	*Measurements*			*Ratio of measurements*	*Position of:*		*Notable morphology (phyllodes, bourrelets, lunules, etc.)*	*Probable life habit*
	Length	*Width*	*Height*	*(let length = 1)*	*Mouth*	*Anus*		

EXERCISE 14

Echinodermata II: Blastozoa and Crinozoa

14.1 INTRODUCTION

Most species in the subphyla Blastozoa and Crinozoa are characterized by a two-part division: a stalk (**pelma**) and a crown. The pelma is secondarily absent in some, but it is typically present and consists of a stack of articulating plates attached by ligaments inside an organic skin (**integument**), thereby forming a strong but flexible support for the crown. The crown consists of a test (**calyx**) and either coelom- and tube-feet-bearing **arms** or smaller, non-coelomate **brachioles**, which are located on ambulacra that are recumbent on the test. Usual presence of the pelma caused these organisms to be recognized for many years as a single subphylum, the Pelmatozoa, a term that often is still used informally for them.

Members of the Blastozoa and Crinozoa typically have the mouth and ambulacra on the upper surface, oriented away from the substratum when the organism is standing erect. This is in contrast with the position of the mouth and ambulacrum in the majority of the classes in the subphyla covered in the previous exercise, most of which have the mouth and ambulacra on the lower surface, oriented toward the substratum. Therefore, most if not all blastozoans and crinozoans have been suspension feeders, whereas a diversity of feeding types (suspension feeders, detritus feeders, herbivores, carnivores, scavengers) characterizes echinozoans, asterozoans, and homalozoans.

The purposes of this exercise are:

1 to study attributes of some of the blastozoan and crinozoan classes;

2 to work through the procedure of classifying the Crinoidea based on plate arrangement in the calyx; and

3 to examine the remarkably uniform organization of tests in the Blastoidea.

In addition, there are questions on interpretation of paleoecology of some crinoids and recognition of plates from stalks in thin-sections or polished surfaces of sedimentary rocks.

14.2 BLASTOZOAN AND CRINOZOAN CLASSES

The pelmatozoans diversified rapidly during the early Paleozoic. However, only two classes survived into the Late Paleozoic, and only one of these two survived the great extinction at the end of the Paleozoic. The class Crinoidea, which survives to the present and has been typified by flower-like forms, is generally the most familiar, abundant, and diverse pelmatozoan class. However, the other classes were more diverse and abundant during the Cambrian through the Middle Ordovician, and even in some instances into the Early Carboniferous. Most of these less familiar groups had bilateral symmetry or even lacked symmetry altogether in the test. *(Question 1)*

14.3 CLASS CRINOIDEA

14.3.1 Classification and plate arrangement in calyx

Three subclasses of crinoids lived during the Paleozoic and died out near the end of the Permian, although members of the subclass Inadunata survived long enough to give rise to the single post-Paleozoic subclass, the Articulata. Fossils of each of the subclasses are usually easily assigned to the proper subclass, and often to order, because typical characteristics of each are quite distinctive (Fig. 14.1, Table 14.1). Distinctions among the major groups of crinoids include number of rows (**circlets**) of plates in the calyx below the arms; occurrence of accessory plates between the basic plates in the upper circlet; whether or not the lower arm plates are fused into the test; whether or not plates that make the arms are organized in a single row (**uniserial** arms) or in two rows (**biserial** arms); and presence or absence of small, linear side branches on the arms. The side branches are called **pinnules** if they occur on each arm plate and **ramules** if they are present on some, alternating with arm plates that have no side branches.

It is easiest to identify test plates for crinoid classification if the crown is entire. In such cases, the line of plates is traced downward along the arms toward the base of the calyx, following only the sequence of plates that are stacked in alignment with the arm that is being traced (Fig. 14.2). Finally, a level is reached where the next plates below are not in alignment, but instead are offset so that precisely below the middle of the final aligned plate there is a plate boundary. The final plate that is in precise alignment at the base of the arm belongs to the cup-shaped lower part of the calyx and is a **radial plate**, named for its alignment with the overlying arm or set of arms.

There are five radial plates around the cup in the vast majority of crinoids, with usually extra plates between the two radial plates nearest the position of the anus. The anus typically is located off-center on the **tegmen**, which is the covering of irregular plates over the top of the calyx, extending between the bases of the arms. In the subclass Camerata there may be **interradial plates** between each of the radial plates, with a precisely or approximately equal number in four of the spaces but a larger number in the anal interray.

The circlet of plates below the radial plates is offset so that the individual plates are centered below the boundary sutures between the overlying radials or below the interradials where they occur. The plates in this row are the **basal plates**, and their offset placement makes them interradial in position. There are five basal around the cup, except where fusion has reduced the number.

Another circlet of plates *may or may not be present* below the basals. If present, plates in the circlet are offset from the basals, so that their position is radial. These are called the **infrabasal plates** (Fig. 14.2b). They may be about the same size as basals or greatly reduced in size, and there may be five or fewer (including zero) of them. If only one row of

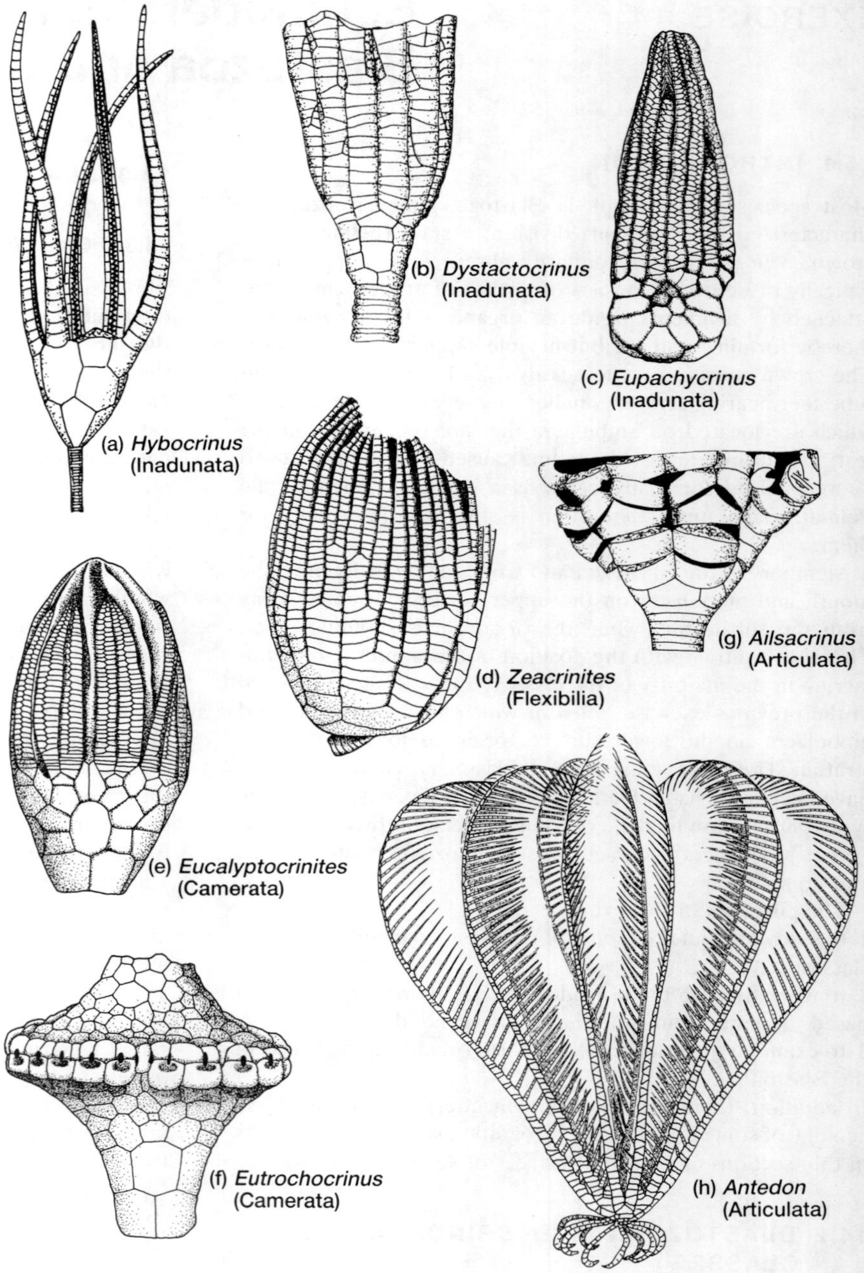

Fig. 14.1 Representatives of the subclasses of the class Crinoidea. (a)–(c) Inadunata; (d) Flexibilia; (e) and (f) Camerata; (g) and (h) Articulata. Modified from Chestnut & Ettensohn (1988), Taylor (1983), and Ubaghs *et al.* (1978).

plates, the basals, is present below the radials, the crinoid is **monocyclic** (Fig. 14.2a). If two rows of plates, basals and infrabasals, are present below the radials, the crinoid is **dicyclic** (Figure 14.2b). The stalk is attached directly to the base of the cup: to the basals in a monocyclic cup, or to the infrabasals in a dicyclic cup.

If arms are not preserved, the circlets of plates may still be identified. The uppermost regular circlet is composed of radial plates, which often contain facets on the upper surfaces where the arms articulated. There may be interradial plates, particularly anal plates, between the radial plates in the upper circlet. Basal plates and infrabasal plates are identified as in the procedure given above for more complete crowns. Note, however, that partially complete crowns that contain four or more circlets of plates *must* have some of the lower plates of the arms or tegmen preserved, because there are never more than three circlets of plates in the cup below the arms. *(Questions 2 and 3)*

14.3.2 Enviroments inhabited by crinoids

Modern crinoids are distributed from barely subtidal in tropical reefs to polar shallows, to abyssal floors of the oceans. The crinoids dwelling in the shallowest regions are stalkless as adults, using short clasping appendages termed **cirri** to hold onto an elevated position such as the top of a coral or sponge. From such vantage points, they array their arms into a feeding position. Many reef-dwelling stalkless

Table 14.1 Characteristics of the crinoid subclasses and orders of two subclasses. Based on information from Ubaghs *et al.* (1978), Lane & Webster (1980), and Sprinkle in Boardman *et al.* (1987).

"Subclass Inadunata"	Calycal cup cylindrical to bowl-shaped with relatively few plates No interradial plates except anal plates Plates of cup firmly joined Tegmen simple, with few plates, bearing mouth No arm plates fused into cup Arms uniserial or biserial and pinnulate Stalk often bearing cirri Some crinoid specialists consider Disparida and Cladida each to constitute a separate subclass Lower Ordovician–Middle Triassic (one known genus survived into Triassic)
Order Disparida	Monocyclic Arms uniserial, rarely pinnulate One to four anal plates Many with unusual symmetry or plate arrangements
Order Cladida	Arms primitively uniserial but derived forms biserial and pinnulate Dicyclic
Subclass Camerata	Cup conical or globular, with firmly joined plates Interradials common Lower arm plates commonly fused into top of cup Heavily plated tegmen usually present, often bearing anal tube Mouth subtegmenal Biserial arms always pinnulate, nearly all arms biserial Stalk often bearing cirri Lower Ordovician–Upper Permian
Order Monobathrida	Monocyclic
Order Diplobathrida	Dicyclic
Subclass Flexibilia	Cup conical to bowl-shaped, dicyclic, plates loosely sutured Basal arm plates incorporated into cup Tegmen of many small, loosely joined plates, bearing mouth Arms uniserial, non-pinnulate Stalks lacking cirri Middle Ordovician–Upper Permian
Subclass Articulata	Cup reduced, dicyclic with infrabasals lost or fused during growth Anal plates and compound radial plates lacking in post-larval stage Tegmen flexible, skeletal plates thin or reduced to spicules Mouth and ambulacral grooves exposed on tegmen Arms generally uniserial, long, always pinnulate Muscular articulations between radials and base of arm and between some or all arm plates Radials and arm plates always perforated by a nerve canal Stalk bearing cirri, or in stalkless forms cirri extend from single plate at base of cup Lower Triassic–Recent

crinoids are nocturnal, spending the daylight hours in cryptic recesses. Deeper-water crinoids typically have stalks and therefore better resemble the majority of fossil species. Currents bend the stalk-bearing crinoids so that the calices trail downcurrent, and the arms are flared so widely that in many species the outer ends bend back upcurrent (Fig. 14.3). The ambulacra and tube feet are on the downcurrent side, where food particles are removed from the turbulent region caused by flow of the ambient current past the radial array of arms.

A greater diversity of life habits of fossil crinoids enabled them to occupy a wider array of environments during much of the Phanerozoic. An example may be found in Chesterian (upper Lower Carboniferous) carbonate platform deposits of south-central Kentucky. Commonly, carbonate platforms have three broadly defined zones in which sediments are deposited: open shelf, high-energy belts, and lagoons or other back-barrier protected regions (Table 14.2). In Chesterian rocks of south-central Kentucky, a vast array of crinoids and some other echinoderms occur locally in deposits interpreted as back-barrier lagoonal (Fig. 14.4). More specifically, the crinoids formed dense stands on washover lobes on the landward side of the barrier and on local shoals within the lagoon. (The lagoon must have had good circulation with the open sea so that normal marine salinity was maintained, otherwise the echinoderms could not have lived there.) The crinoids and associated blastoids growing to various heights on the shoals

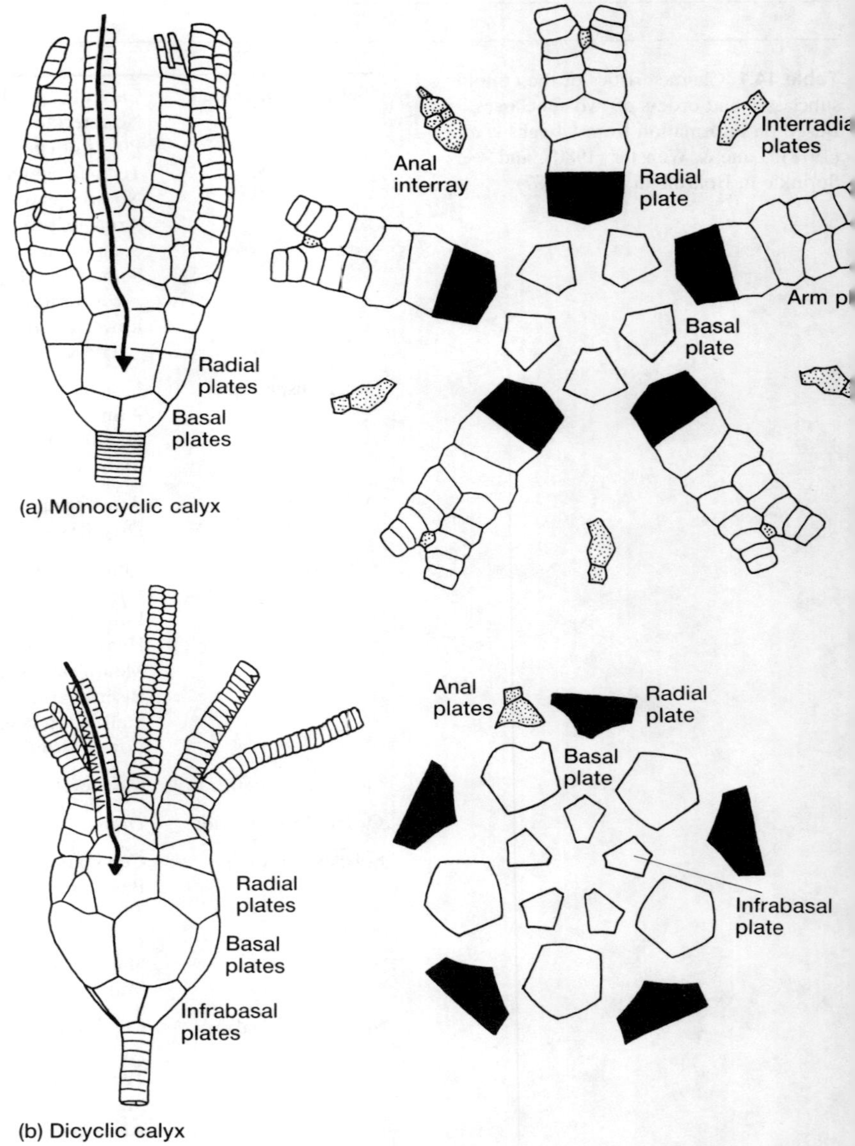

Fig. 14.2 Examples of crinoids with (a) a single circlet of plates below the radial plates and (b) two circlets of plates below the radial plates in the cup, with diagrams for comparison of plate numbers, shapes, types, and distributions. The arrows in the lateral views trace down an arm system to the radial plate at its base. Based on examples in Ubaghs *et al.* (1978).

Fig. 14.3 Stalked crinoids living on a lithified carbonate mound 600 m deep in the Straits of Florida, northwest of Little Bahama Bank; flow is toward camera. Courtesy of A. C. Neumann, made from DSRV *Alvin*.

Table 14.2 Characteristics of shore-parallel sedimentary zones in shallow carbonate platforms. Adapted from Heckel (1972) and McKinney (1979).

Open-ocean waters below normal wave	Barrier where wave action is vigorous	Back-barrier regions including lagoons
1 Biota diverse	**1** Biota diverse in reefs, otherwise intermediate	**1** Biota restricted except where open circulation with open ocean occurs
2 Sediment predominantly shelly lime mud	**2** Calcareous skeletal sands and gravels, much of which is fragmented and abraded	**2** Sediment commonly pelleted; lime mud with sparse shells or if open circulation then shells may be abundant
3 Frequently site of deposition of authigenic glauconite	**3** Oolite or skeletal sands; reefs may be developed locally	**3** Glauconite rarely or never formed; if evaporation high with respect to mixing, evaporites are precipitated
4 Inclusion of fine detrital organic matter forms dark-colored sediment and rock	**4** Relatively small amounts of detrital organic matter trapped in sediment because of continual winnowing of surface materials	**4** Relatively small amounts of detrital organic and rapid oxidation result in light-colored sediment and rock except where high organic input washed from land
5 Sediment lamination destroyed by bioturbation	**5** Cross-bedding, fairly well-sorted sediment typical	**5** Sediment lamination preserved due to small numbers of burrowers

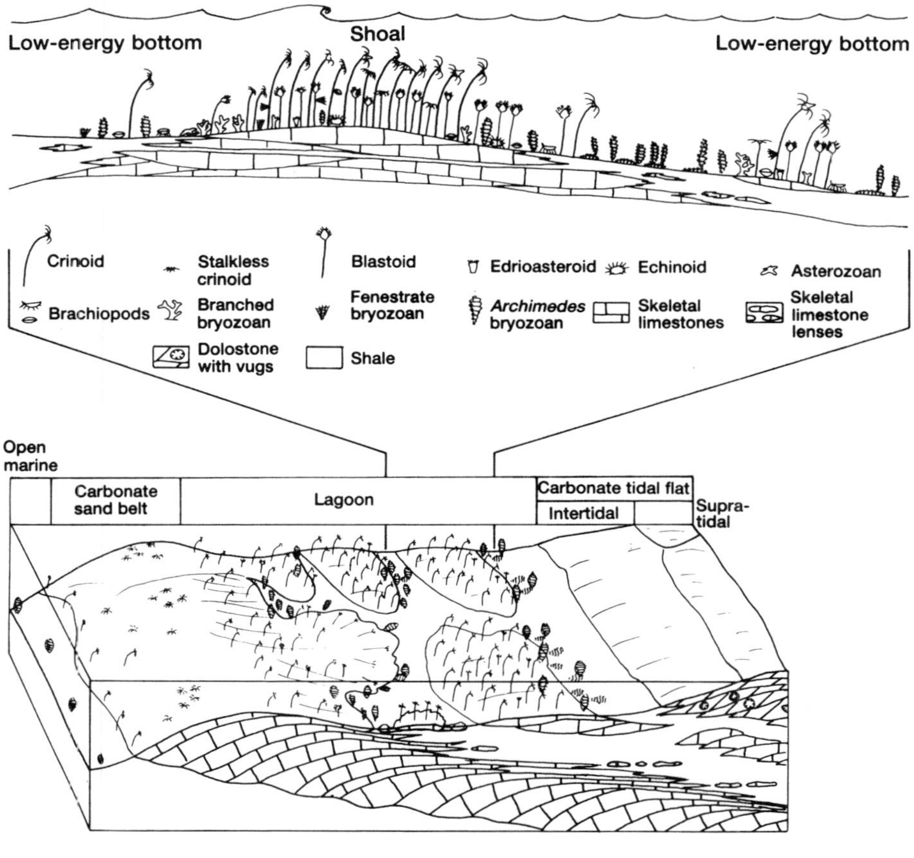

Fig. 14.4 Reconstruction of environments in south-central Kentucky during late Chesterian time, showing the sedimentary record left as environmental belts shifted seaward. Dense crinoid-dominated gardens inhabited shoals and other sandy shallows within the lagoon, as shown in the enlargement of a single shoal in the upper part of the figure. Modified from Chestnut & Ettensohn (1988).

apparently benefitted from an increase in velocity of flow — and therefore in rate of delivery of food particles — as water moved over the shallow shoals from the adjacent, somewhat deeper lagoonal floors, much as for living oysters that build shoal patches in lagoons around the Gulf of Mexico. However, the stalked crinoids apparently could not withstand the substantially more turbulent water that moved across the even shallower high-energy sand belt, for they are absent from these deposits.

The aberrant stalkless crinoid *Agassizocrinus* (Fig. 14.5) is an exception. It is found — often in dense clusters — in deposits of the carbonate sand belt not only in the Glen Dean–Bangor deposits shown in Fig. 14.4 but also in another, more ooid-rich deposit of carbonate sand below the underlying Hardinsburg–Hartselle Formation. Stalkless crinoids without cirri evolved several times, and are commonly found in relatively clean carbonate sands or are encrusting, whereas stalk-bearing crinoids more typically are found in mixed mud and sand.
(Question 4)

14.4 Class Blastoidea

Blastoids were the most diverse class among the Blastozoa, but their body plan was remarkably conservative. The oldest blastoids are known from Middle Ordovician rocks and clearly show well-defined characteristic features of the class: small number of test plates, superbly developed pentameral symmetry, well-developed ambulacra, and thin, folded sheets defining the respiratory structures (called **hydrospires**) along the row of pores at the margins of each ambulacrum. Aside from ambulacral plates, there are 13 or much less commonly 14 plates constituting the test: three (or very rarely four) basals, five radials with an ambulacrum centered in each, and five **deltoid** plates that are interradial and positioned between the upper ends of the ambulacra, above the sutures between adjacent radial plates (Fig. 14.6). The ambulacral plates consist of a single elongate **lancet plate** in each ambulacrum, covered on each side by two rows of tiny **side plates** that supported a row of brachioles along each side of the ambulacrum.

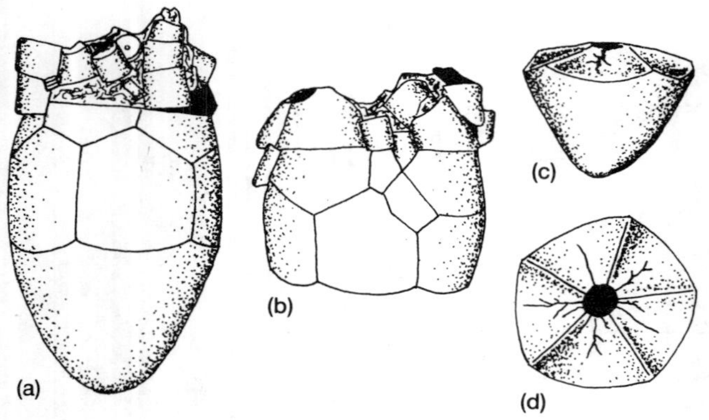

Fig. 14.5 The stalkless inadunate crinoid *Agassizocrinus conicus*, which occurs in Chesterian deposits of carbonate sands. (a) Dicyclic calyx and lowest plates of arms; (b) view of partial calyx showing anal plates between the radials; (c) and (d) lateral and upper views of fused infrabasal plates. Drawn from illustrations in Chestnut & Ettensohn (1988).

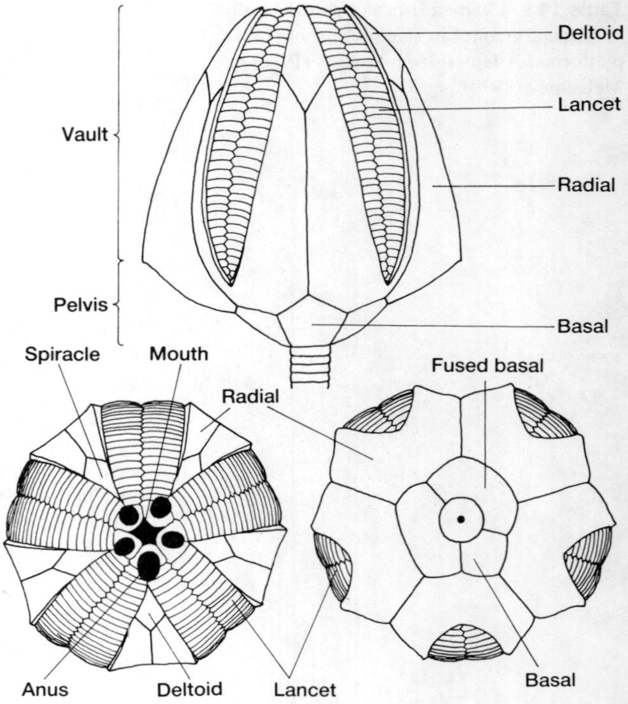

Fig. 14.6 External morphology of *Pentremites*.

The mouth was centered at the top of the test and in many blastoids was surrounded by **spiracles**, which were exhalent openings from the respiratory hydrospires. The spiracles may number 10 (two per ambulacrum), or neighboring spiracles from adjacent ambulacra may be united into a single opening so that only five spiracles occur. One spiracle usually is enlarged to include the anus.

Although about 65 genera of blastoids have been described, the most widely known is *Pentremites*, which is widespread in Upper Mississippian (Lower Carboniferous) marine deposits of the eastern United States and Pennsylvanian (Upper Carboniferous) marine deposits of the central United States. Blastoids were most successful and abundant during the Lower Carboniferous. *Pentremites* tests may be locally found by the hundreds where the animals formed dense stands, usually with abundant crinoids.

Among the several characteristics used to distinguish genera and species of blastoids are the morphology of deltoids, hydrospires, and ambulacra; shape of the test in lateral profile; shape of the test as seen from the oral surface; and size of plates.
(Questions 5–7)

14.5 Recognition of plates from crinozoan and blastozoan stalks in sedimentary rocks

Throughout each plate of an echinoderm (with a few known exceptions) there is uniform crystallographic alignment of its constituent calcite, although the microstructure consists essentially of a three-dimensional network (stereom) of rods

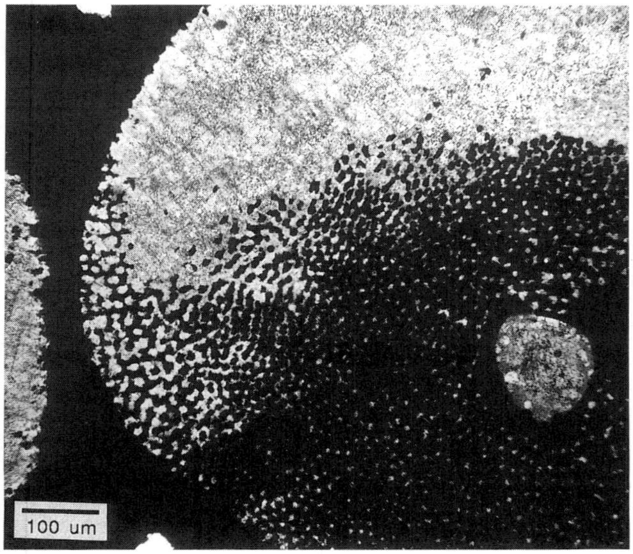

Fig. 14.7 Photomicrograph of thin section of echinoderm plate (Carboniferous; Alabama). Stroma in part of the plate are filled with dense hydrocarbon, which seeped into the spaces soon after death of the animal, before growth of calcite crystals had a chance to fill the spaces.

spaced so that there is a microscopic gallery system (stroma) extending among the rods (see Fig. 13.4). During diagenesis, the galleries are filled by calcite that is in crystallographic continuity with the rods, which results in each plate forming a single, solid calcite crystal. These crystals may grow along their outer surfaces to fill any original empty space that was beside them. Because the solid crystals break with highly lustrous cleavage faces, originally clean-washed echinodermal limestones commonly appear highly crystalline and are marketed as marbles. However, even in such "marbles" the original shapes of plates can usually be determined by close examination, and microscopic examination of the plates reveals a characteristic reticulated pattern because of a contrast in color between the original calcite of the skeleton and the calcite that filled the stromal galleries (Fig. 14.7).

Stalked echinoderms, predominantly crinozoans but also including blastozoans, locally were important or dominant components of Paleozoic limestones. The plates from stalks were most common, and in thin section can be recognized by the limited but characteristic shapes generated where the two-dimensional section cuts the three-dimensional cylinders in various orientations. These characteristic shapes, plus the reticulated pattern and unitary extinction seen in polarizing microscopes, because each plate has a single crystallographic orientation, make echinoderm plates perhaps the most easily recognized of all skeletal materials in thin-section.
(Questions 8 and 9)

FURTHER READING

Beaver, H. H. *et al.*, 1967. Echinodermata 1 in R. C. Moore (ed.), *Treatise on invertebrate paleontology*, Part S. Geological Society of America and University of Kansas: Lawrence, KS.

Chestnut, D. R., Jr & Ettensohn, F. R., 1988. Hombergian (Chesterian) echinoderm paleontology and paleoecology, south-central Kentucky. *Bulletins of American Paleontology* **95:**1–102.

Kammer, T. W. & Ausich, W. I., 1987. Aerosol suspension feeding and current velocities: distributional controls for late Osagian crinoids. *Paleobiology* **13:**379–395.

Sprinkle, J., 1973. *Morphology and evolution of blastozoan echinoderms*. Harvard University Museum of Comparative Zoology Special Publication. Cambridge, MA.

Ubaghs, G. *et al.*, 1978. Echinodermata 2 (Crinoidea; Vols 1–3), in R. C. Moore & C. Teichert (eds), *Treatise on invertebrate paleontology*, Part T. Geological Society of America and University of Kansas Press: Lawrence, KS.

EXERCISE 14

Worksheet

Name

QUESTION 1

Casts of several complete or partial crowns of representatives from various blastozoan and crinozoan classes are provided for study and identification to class, using the information in Table 13.1 (see the previous exercise). For each specimen, give the information required in the table on p. 218.

QUESTION 2

Casts of two complete or partial crowns of crinoids are provided for study of plate patterns. Carefully study each specimen and identify radial plates, basal plates, and, if present, arm plates, anal and other interradial plates, and infrabasal plates. Use the plate diagrams in Fig. 14.2 as a stylistic guide, and draw the plate patterns of the two crinoids on the two five-rayed spaces provided below. For each specimen, identify whether it is monocyclic or dicyclic.

Genus:__________ Monocyclic, or dicyclic?__________

Genus:__________ Monocyclic, or dicyclic?__________

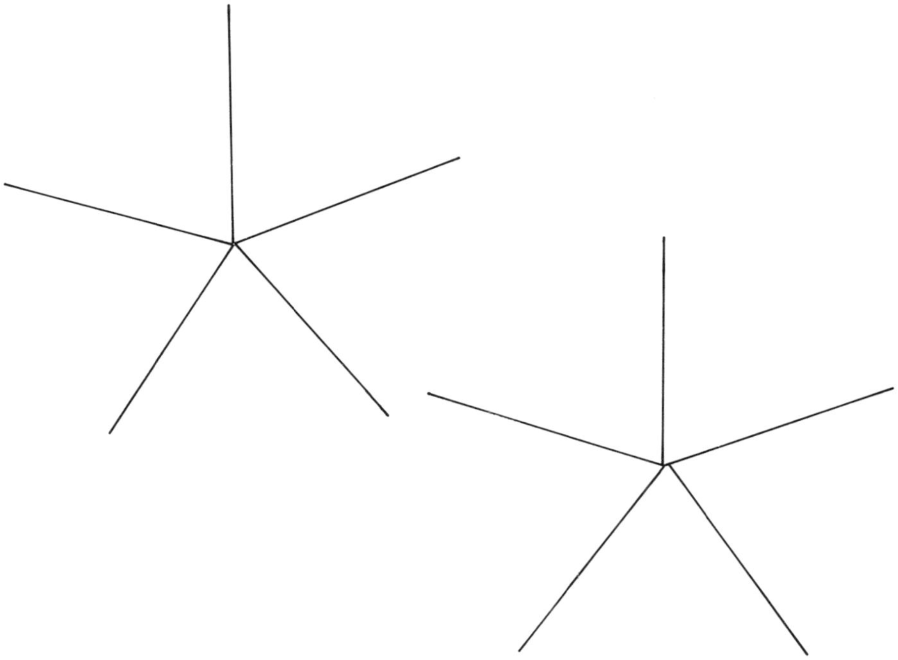

QUESTION 3

Casts of several complete or partial crowns of crinoids are provided for identification to subclass and, where possible, to order, using the information in Table 14.1. For each specimen, give the information required in the table on p. 219. Do the same for the two specimens used for Question 2.

QUESTION 4

How and where did *Agassizocrinus* live? Calyx plates of *Agassizocrinus* were substantially more dense than typical echinoderm plates (Fig. 13.4 and 14.8), what might the difference in density between *Agassizocrinus* and the more typical density of *Antedon* suggest about how *Agassizocrinus* lived where it did? *Agassizocrinus* was probably a suspension feeder, but was it necessarily so, or might it have been at least a part-time deposit feeder? Explain your answer.

Genus	*Symmetry*	*Plates*		*Main body of ambulacrum (recumbent on test, or erect)*	*Special respiratory structures in test (pore rhombs, sutural slits, diplopores, etc.)*	*Class*
		Arranged in regular circlets of repeated shapes?	*Approx. no. in test*			

Genus	*Monocyclic or dicyclic*	*Arms uniserial or biserial*	*Interradial plates other than anals?*	*Lower arm plates fused into cup?*	*Subclass*	*Order*

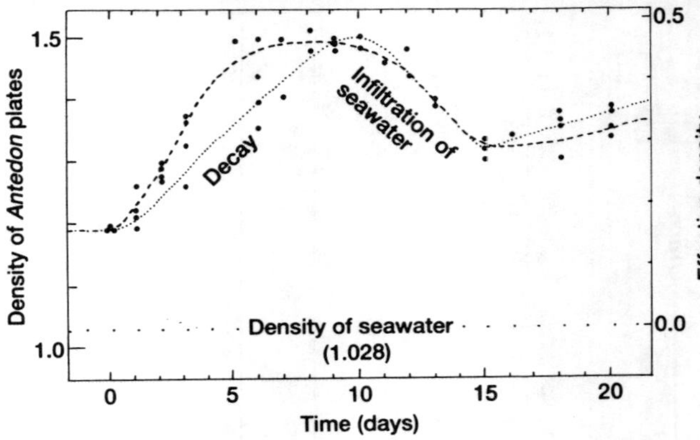

Fig. 14.8 Density of *Antedon bifida* plates before, during, and after decomposition. From Cain (1968).

____

____

____

____

____

____

QUESTION 5

In order to demonstrate your ability to recognize various parts of the blastoid specimen provided, make the following determinations.

Number of basal plates: ______________________

Number of spiracles: ______________________

Length of deltoid: ______________________

Length of ambulacrum: ______________________

Height of test: ______________________

Ratio of length, deltoid:ambulacrum: ______________________

Ratio, length of ambulacrum:height of test: ______________________

QUESTION 6

Data given below were determined from *Pentremites* collected from a single assemblage in the lower part of the Bangor Limestone (upper Lower Carboniferous), Blount County, Alabama. For each pair of measurements from a specimen, place a point at the correct place on the graph on p. 222 to make a scatter diagram, and determine the vault:pelvis ratio for each by dividing the vault measurement by the pelvic measurement (see Fig. 14.6).

Specimen number	*Vault height (mm)*	*Pelvis height (mm)*	*Vault:pelvis ratio*	*Specimen number*	*Vault height (mm)*	*Pelvis height (mm)*	*Vault:pelvis ratio*
1	10.1	13.0		26	8.2	12.5	
2	6.3	1.8		27	7.4	4.3	
3	7.8	1.5		28	9.1	13.0	
4	9.6	13.2		29	10.3	3.0	
5	9.4	4.3		30	8.1	2.4	
6	4.0	8.8		31	5.0	8.8	
7	9.8	2.4		32	7.8	2.1	
8	9.2	2.5		33	6.8	1.2	
9	7.8	3.2		34	6.0	10.3	
10	7.3	10.8		35	7.1	12.3	
11	5.0	1.8		36	5.1	8.7	
12	7.5	11.5		37	8.7	3.8	
13	9.3	2.2		38	6.0	9.1	
14	7.4	2.0		39	8.7	3.2	
15	6.5	3.1		40	5.5	9.0	
16	10.0	1.9		41	6.4	9.5	
17	7.6	11.5		42	5.2	10.1	
18	10.8	4.1		43	6.1	9.6	
19	10.7	4.0		44	8.3	3.1	
20	3.4	8.0		45	7.1	3.4	
21	7.0	3.7		46	7.5	4.0	
22	9.4	2.8		47	10.5	4.7	
23	4.7	9.2		48	9.3	2.5	
24	7.0	2.1		49	7.2	2.8	
25	6.0	9.2		50	8.9	11.9	

Is there one species represented, or might there be at least two species?

If you decide that there are two species, circle with a solid line the cloud of points that represents one of the species, and encircle the other set of points with a dashed line. [In order to persuade someone that two species are present in the sample, one could use the vault:pelvis ratios in a *t*-test (refer to Exercise 2) or the nonparametric Kolmogorov–Smirnov test, which does not assume a "normal" distribution of the values in the two groups. If computers and software are available, this could be easily done and the results written into the space immediately below.

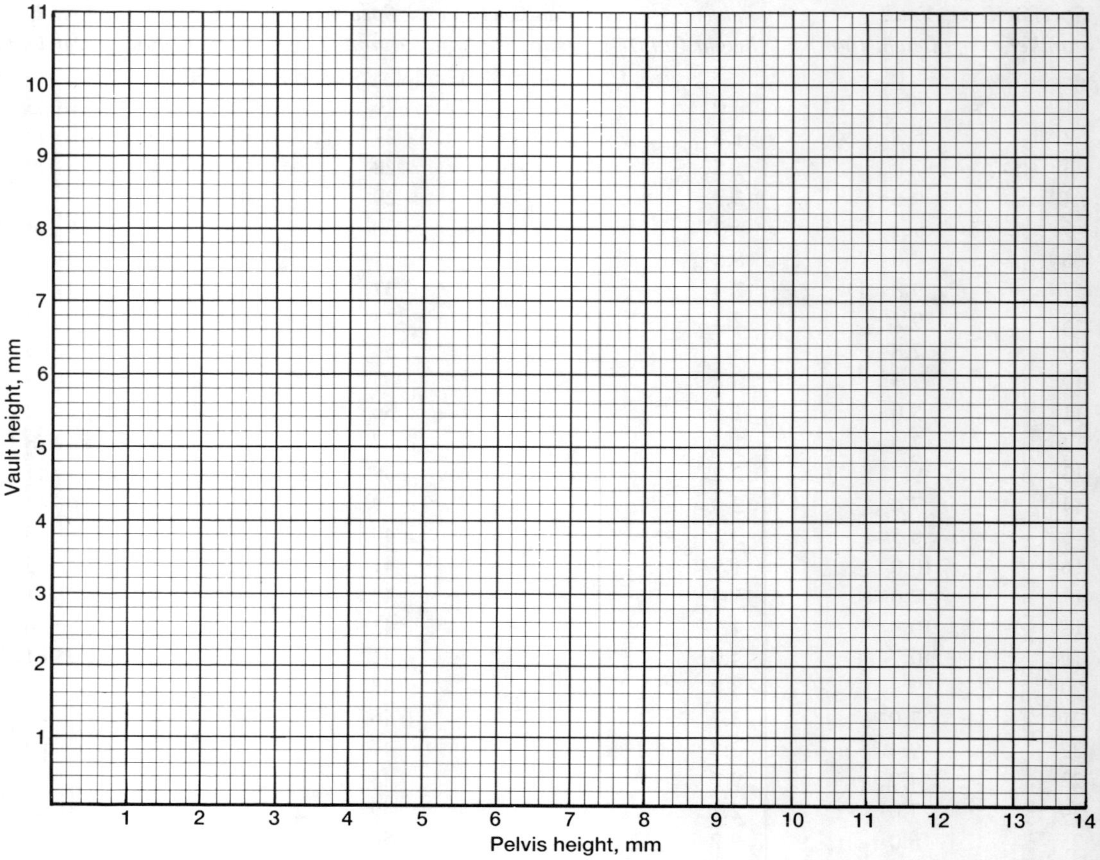

QUESTION 7

Refer back to Exercise 2 for the discussion of isometric and anisometric growth, including allometry. If only one species seems to be present in the sample for Question 6, which type growth does it show? If two species are present, do both show the same type growth, or different types? Which type(s) is or are shown?

____

Explain the basis for your answer.

____

____

____

____

____

____

____

____

QUESTION 8

In the space below, draw all possible different shapes that can be made with a two-dimensional cut through the perforated cylinder drawn at the top left. The perforated cylinder represents a typical plate from an echinoderm stalk. Simple changes in proportion (fatness, elongation, etc.) do not count as different shapes; however, you should draw all possible different outline shapes. There is a total of nine different shapes, two of which are drawn as examples.

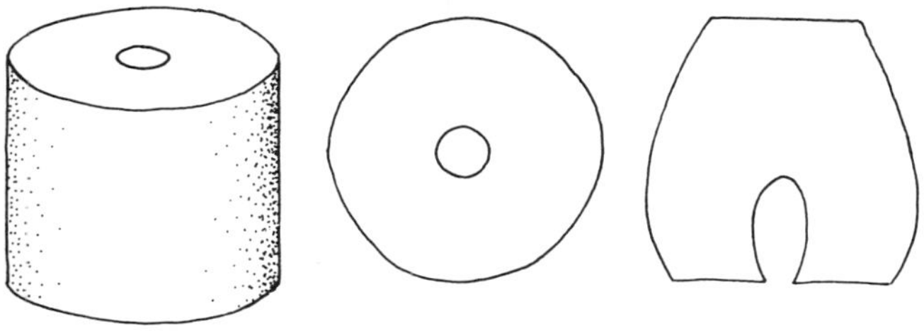

QUESTION 9

Fig. 14.9 is a photograph of a thin-section of carbonate rock. Using a clearly visible pen, circle each echinoderm plate included in the photograph.

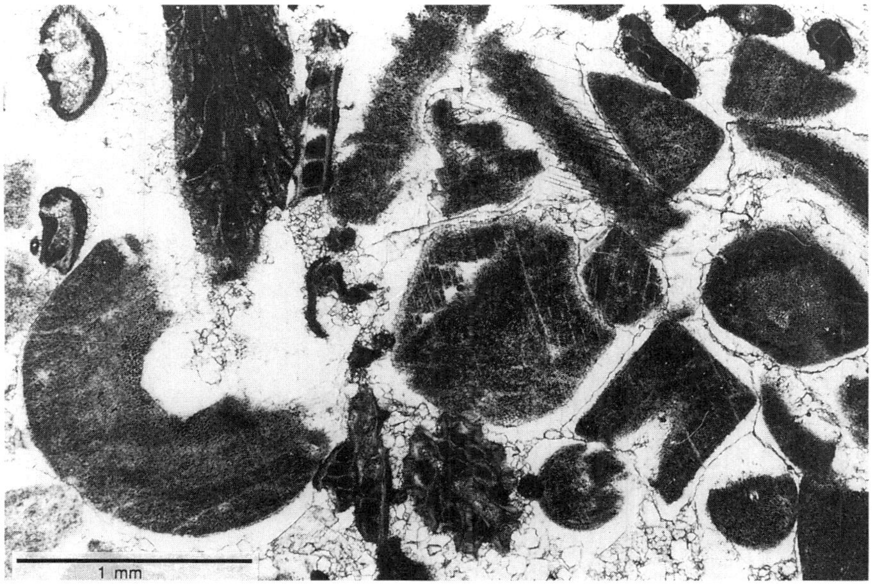

Fig. 14.9 A thin section of a skeletal grainstone, which is a limestone made of whole or broken carbonate fossils and has calcite crystals instead of mud filling the spaces between the fossils. The specimen is from the lowest beds of the Paragon Formation (Upper Mississippian, or Upper Lower Carboniferous), eastern Kentucky.

EXERCISE 15 Hemichordata

15.1 INTRODUCTION

Hemichordates include large solitary worms as well as clonal organisms consisting of small zooids. These divergent morphologies are related in their overall body plan, which is constituted of three parts: anterior shield, medial collar, and posterior trunk. Another feature that indicates the coherence of the group is an elongate pouch (**diverticulum**) that extends forward from near the anterior end of the digestive tract. This diverticulum is the structure that was earlier mistaken as a notochord, causing the hemichordates to be incorrectly grouped with the chordates. Developmental similarities, however, do suggest close relationship with the Echinodermata, which in turn are linked with the chordates.

The purposes of this exercise are:

1 to compare the characteristics of graptolites with living hemichordates;

2 to examine the morphology of graptolite colonies and zooids; and

3 to determine the age of an assemblage of graptolites by the concurrent range zone method in order to see their high biostratigraphic potential.

15.2 ENTEROPNEUSTA AND PTEROBRANCHIA

The class Enteropneusta contains the so-called "acorn worms," which may be up to 2.5 m long. These solitary, infaunal, generally burrowing animals are completely soft-bodied and have left no known body fossils, although they undoubtedly have contributed to disturbance (**bioturbation**) of sediments.

The class Pteriobranchia has a very sparse fossil record for both of its constituent orders, extending back to the Lower Ordovician (Tremadocian Stage). They are **clonal** organisms, which means that they reproduce by asexual budding of modules, and they are generally **colonial**, which means that the modules remain attached to one another. (See also Exercises 5 and 12.) Individual zooids may be about 0.5–7 mm long. Their three-part body organization consists of an anterior pre-oral disc, a medial lophophore, and a posterior stalk or peduncle that contains the digestive tract and by which the zooid may be attached (Fig. 15.1).

The order Rhabdopleurida are the pterobranchiates that are of greatest interest for comparison with the graptolites, an extinct group of fossils that apparently belong among the hemichordates. *Rhabdopleura* is colonial, with its zooids attached to a common stolon, each by its contractile stalk (Fig. 15.2). The common stolon is encrusting and gives rise laterally to the zooids; some lateral extensions develop into bifurcations of the stolon rather than into zooids. The stolon extends through the proximal, encrusting portion of each zooidal tube. Zooidal tubes are composed of proteinaceous material. In the encrusting parts of zooids, the tubes are composed of narrow half segments that are offset to left and right and are joined along a serrated suture, but in the erect distal portions of zooids the tube consists of complete, ring-shaped segments (Fig. 15.2).

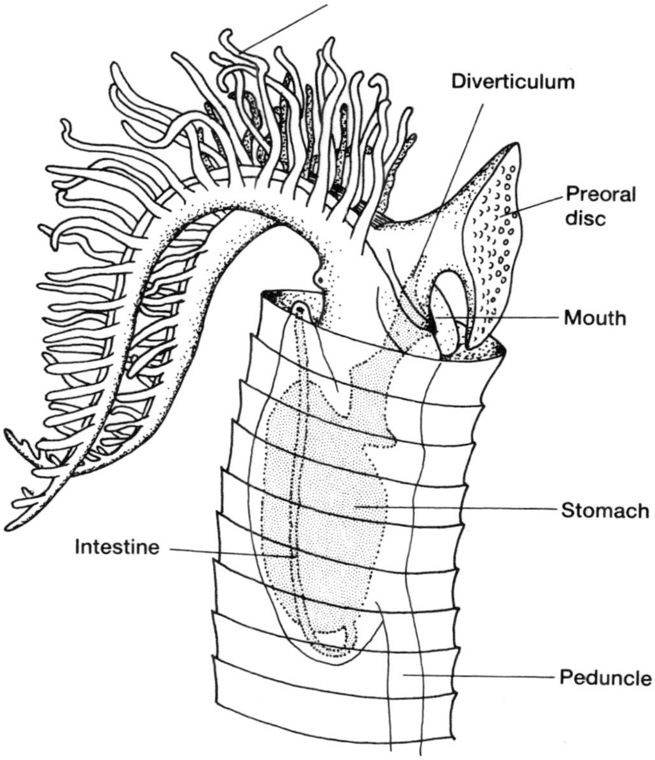

Fig. 15.1 Polypide of a single autozooid of the pterobranch *Rhabdopleura*, positioned in the outer end of its autotheca. The autotheca is about 0.3 mm in diameter. After Bulman (1970) with permission from the Geological Society of America.

15.3 GRAPTOLITHINA

The Graptolithina are considered to be a class of organisms within the Hemichordata. There is reservation about their assignment to the phylum, because of complete lack of knowledge of their embryology and zooidal soft tissues. However, organization of the colonies and skeletons is so similar between graptolites and rhabdopleurids that the two groups are inferred to be very closely related.

Graptolites were exclusively colonial and were either attached to a substratum or were planktic. Attached colonies were either encrusting or erect. Zooids were small, with generally conical tubes, a fraction to a few millimeters long. Skeletons of zooids are of a single type (**monomorphic**) in the most common order, the Graptoloidea (Fig. 15.3). This single type zooid is called the **autozooid**, and its skeleton is referred to as the **autotheca** (pl. **autothecae**). All other orders are characterized by polymorphic thecae, constituting up to three types of zooids (Table 15.1): autozooids, smaller **bizooids**

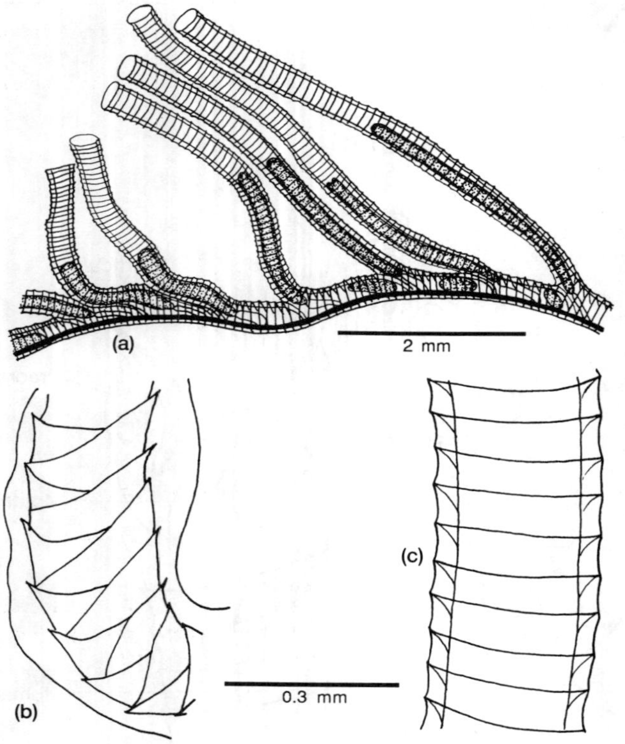

Fig. 15.2 Portion of a colony of *Rhabdopleura*. (a) Vaguely-seen polypides of autozooids retracted deep into their autothecae and connected by their peduncles to the black stolon; (b) detail of the half-ring structure of the encrusting portions of autothecae; (c) detail of the full-ring structure of the erect portions of autothecae. After Lopez Gappa (1987).

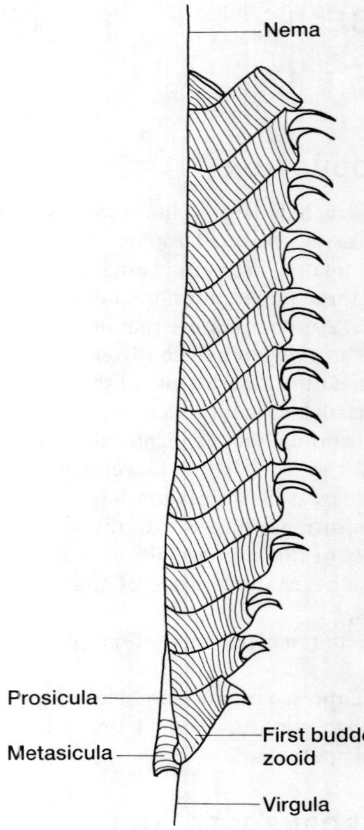

Fig. 15.3 Skeleton (rhabdosome) of a short graptoloid graptolite colony consisting of a single row of thecae of monomorphic zooids. The sources of variation among the thecae are position within the colony (astogeny), developmental stage of the individual zooids (ontogeny), and possible microenvironmental effects; there is no polymorphism in this group of graptolites.

(preserved as **bithecae**), and colony-extending **stolozooids** (preserved as **stolothecae**).
(Question 1)

15.3.1 Preservation of graptolites

Graptolites are most commonly preserved in sedimentary rocks that were originally deposited on the sea floor in oxygen-deficient environments, most commonly dark gray or black, basinal shales or limestones that accumulated on foul bottoms and often retain a sulfurous smell that can be detected on fresh breaks. This restriction in their occurrence is due to the original proteinaceous nature of their organic skeletons, which would have been — and presumably were — destroyed in oxygenated conditions. Refer back to Fig. 1.3. Graptolites that are preserved in limestones preserve their three-dimensional shape, but those preserved in dark shales are crushed flat and are preserved as carbonaceous films.

The carbonaceous films that constitute most graptolite fossils, as for most fossils with low relief, are best seen in light that is at a very low angle to the surface on which they are preserved. This is because the shadows generated by light almost parallel with the surface highlight even the smallest differences in relief, many of which are due to the shapes of the zooidal tubes.

15.3.2 Graptolite morphology

The following description of the structure of autothecae and colonial skeletons (the **rhabdosome**) is based on graptoloid graptolites. The founding autotheca of a colony is the **sicula**, a conical tube open at the larger end and extending as a hollow tube (the **nema**) from the opposite, apical end. The narrower, initially formed portion of the siluca is the **prosicula**, which has continuously spiralled structure and appears to have formed as a unit in a single growth phase. The widening, more distal portion of the sicula, the **metasicula**, appears to have grown incrementally because it consists of half rings that overlap alternately from side-to-side, producing a longitudinal serrated suture where this occurs, as in *Rhabdopleura*. The half ring structure is characteristic of the remainder of the inner (**fusellar**) layer of the entire rhabdosome, and it is covered exteriorly by the laminated **cortical** layer. Graptolite rhabdosomes were composed of proteinaceous material.

One or more small holes (**foramina**, singular **foramen**) were formed in the side of the sicula, either as a gap in the wall when the sicula was growing, or by resorption later. From the foramina, the first generation of asexually budded

Table 15.1 Characteristics of the class Graptolithina and its constituent orders.

Class Graptolithina	Colonial, typically of branches that extended free from substratum or were planktic, less commonly completely encrusting Zooids in linear series, in series of clusters, or infrequently in irregular aggregates Autothecae and, where present, bithecae connected by stolons (stolothecae) Black thread extending through stolons of some Skeleton originally proteinaceous Middle Cambrian–Upper Carboniferous
Order Dendroidea	Rhabdosomes attached at base, branched, typically with many branches that may or may not be laterally connected Branches consisting of autothecae, bithecae, and stolothecae Black thread in stolons of most Middle Cambrian–Upper Carboniferous.
Order Camaroidea	Rhabdosomes encrusting, with autothecae, bithecae, and stolothecae Autothecae consisting of proximal inflated portion and distal, narrower, tubular neck Bithecae small, tubular, distributed irregularly throughout rhabdosome Stolothecae hard, black Apparently endemic to a region in Poland Upper Cambrian–Lower Ordovician
Order Crustoidea	Rhabdosomes encrusting, with autothecae, bithecae, and stolothecae Autothecae inflated proximally, constricted at base of distal portion, but with flared, complexly infolded apertures at distal end Apparently endemic to a region in Poland Lower Ordovician–Upper Silurian
Order Stolonoidea	Encrusting or sessile, with extensively developed stolothecae divided irregularly into irregular clusters Autothecae questionably known Apparently endemic to a region in Poland Upper Cambrian–?Ordovician
Order Tuboidea	Benthic, predominantly encrusting but some with erect stipes Autothecae and bithecae in clusters, irregularly budded Distal portions of autothecae or of both auto- and bithecae erect Stolothecae apparently absent or confined to encrusting disc Upper Cambrian–Upper Silurian
Order Graptoloidea	Rhabdosomes planktic, with one to several stipes not laterally linked Stipes consisting of autothecae in one to four series Presumed stolons without durable, proteinaceous skeletons Lower Ordovician–Lower Devonian

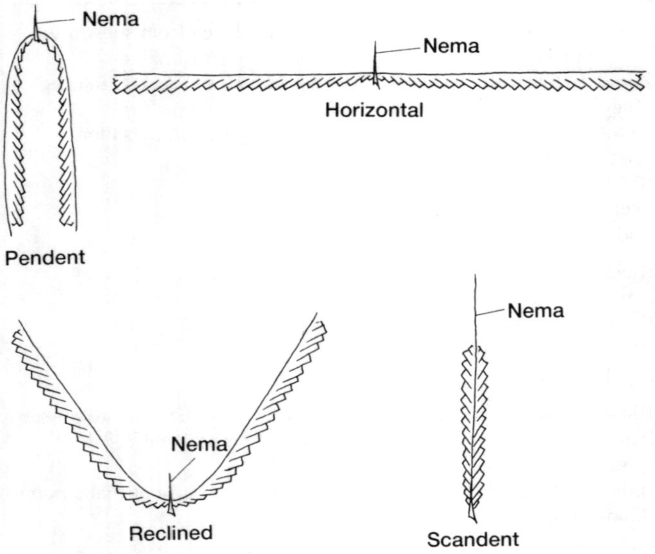

Fig. 15.4 Names of the basic rhabdosomal types for graptoloids. The names are based on the geometrical relationships of the stipes to the nema, which extends from the apical end of the sicula.

zooids grew, generally either extending out at high angles, or doubled back so that their distal, wider ends were directed oppositely to that of the sicula and lay adjacent to the nema. In colonies of the latter type, the nema led the growth of the colony as a lengthening thread, as one or more series of zooids budded end-to-end along it. Graptoloid colonies apparently floated, and some have been preserved with thin, sac-shaped objects interpreted as floatation bladders at the ends of the nemas.

Asexually budded graptoloid autothecae may be simple tubes with constant diameter, minimal flexures, and plain, circular apertures. Others variously pinch and swell, are unusually short or elongated, are highly flexed, are reflexed at the distal end, have expanded or constricted apertures, or have elaborate apertural sculptures. These morphological features are most useful in identification of genus and species.

Rhabdosomes of graptoloids are composed of one or more branches, termed **stipes** (singular, **stipe**). These may consist of one, two or four rows of autothecae, hence are said to be uniserial, biserial, or quadriserial, and this character is important in genus- and higher-level graptoloid taxonomy. Stipes may be straight or curved, and they may number one, two, four, or more within a colony; this too is important in taxonomy above the species level.

Rhabdosomes of graptoloids with one to four branches are further described by the relationship between the sicula and the stipe(s) (Fig. 15.4). If the stipes extend in the same direction as the sicula, so that the aperture of the sicula and apertures of asexually budded zooids open in one direction, the rhabdosome is **pendent**. If stipes extend out roughly perpendicularly to the axis of the sicula, the rhabdosome is **horizontal**. Where stipes are recurved so that they extend oppositely to the aperture of the sicula, that is, more-or-less in the same direction as the nema, they are **reclined** or **scandent**. The difference between these latter colony types is whether the nema is free (reclined) or has the stipe growing along it (scandent) where two rows of zooids are back to back.

(Questions 2 and 3)

The dendroid graptolites are moderately common fossils but were not as diverse and generally do not occur in as large numbers locally as do graptoloids. As for graptoloids, they are most commonly poorly preserved. Dendroids were apparently attached throughout life of the colony and consisted of many subparallel branches that in some species were linked laterally by rod-shaped clusters of zooids. Branches consisted of autozooids, bizooids, and stolozooids. Dendroids superficially resemble fenestrate bryozoans but can generally be recognized because of their typical preservation as carbon films of low relief, whereas fenestrates are preserved as three-dimensional, originally calcareous skeletons or, if the skeleton has been leached away, by the distinctly three-dimensional molds.

(Question 4)

15.3.3 Geologic distribution of graptolites

Wherever there are lower to middle Paleozoic marine sedimentary facies that accumulated as anoxic substrates, graptolites may be present and are potentially important for stratigraphic correlation and for determination of geologic age. Planktic graptoloids have played a particularly important role in stratigraphy.

Vertical stratigraphic distributions of particular species have been used for recognition of major time-stratigraphic boundaries, such as the internationally recognized stratotype of the Silurian–Devonian boundary in black, basinal shales and limestones in the Prague Basin, Czechoslovakia (Fig. 15.5), where the boundary is based on the lowest occurrence of *Monograptus uniformis*. In addition, concurrent ranges of readily recognized genera of graptolites permit recognition of particular series, stages, and portions of stages in Upper Ordovician strata where diverse assemblages occur (Fig. 15.6).

Even in the absence of preservation that permits identification to genus, graptoloids have general biostratigraphic usefulness. Many-branched rhabdosomes appeared in Lower Ordovician rocks, but rhabdosomes with one or two stripes dominate Upper Ordovician assemblages, giving way to an overwhelming preponderance of single stipe rhabdosomes in Lower Silurian assemblages. In addition, in mid-Ordovician, pendent and horizontal rhabdosomes were replaced by reclined and scandent rhabdosomes; and Ordovician and lowest Lower Silurian graptoloid faunas are dominated by taxa with biserial stipes, whereas the remaining Silurian and Devonian faunas contain taxa with uniserial stipes (Fig. 15.6)

(Question 5)

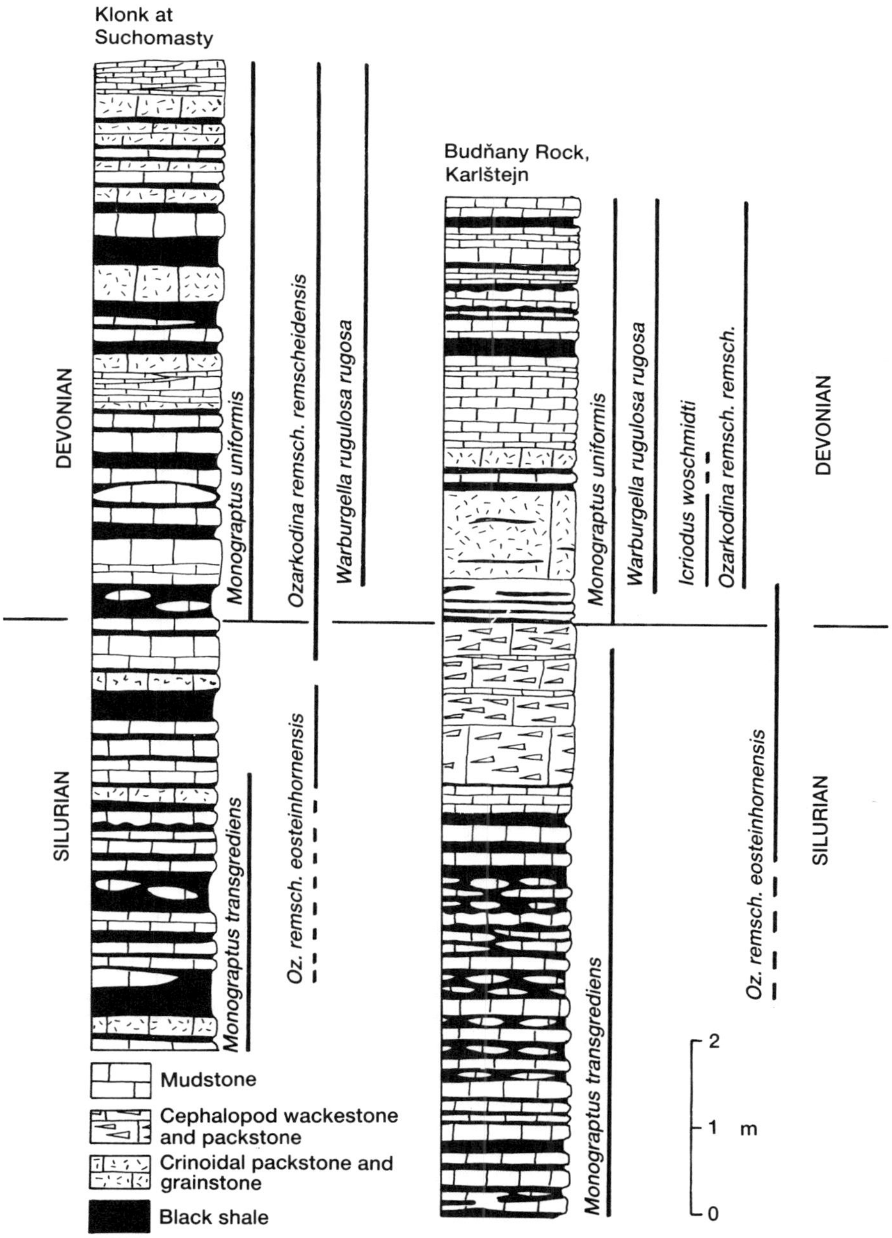

Fig. 15.5 Columnar representation of portions of the rock (stratigraphic) sequence through the Silurian–Devonian boundary at two localities in the Prague Basin, Czechoslovakia. The vertical ranges of species of the graptoloid *Monograptus* and several conodonts are shown for both of the sequences. There is an international agreement that the Silurian–Devonian boundary is set at the lowest occurrence of *M. uniformis* in the dark-colored shales and limestones that were deposited in a deep basin and are now exposed in a steep hillside called Klonk, by the village of Suchomasty. The lowest occurrence of *M. uniformis* at Klonk is within limestone bed 21, which correlates several kilometers away at Budnany Rock with the shale between limestone bed 41 (a cephalopod-rich limestone) and bed 42 (fine-grained limestone at the base of a thin reefal buildup), deposited nearer the edge of the basin. After Chlupać *et al.* (1980).

	Name	Diagnostic characters	Other forms with which this might be confused	Time-range
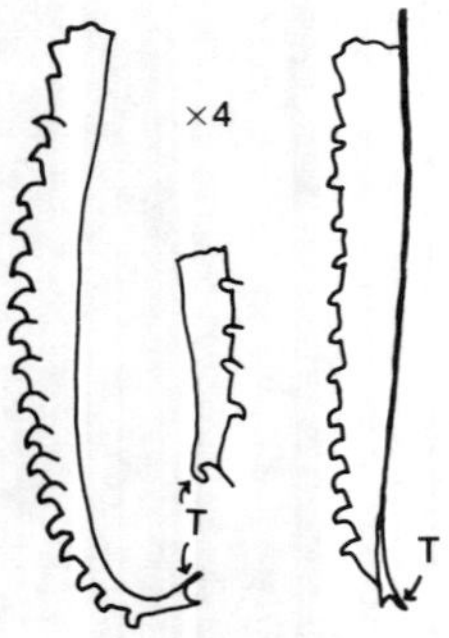	*Monograptus hercynicus* group	Straight or slightly curved rhabdosome. Some with strongly curved initial portion (J-shaped). A small tongue-like structure (T) developed on dorsal side of sicula. Initial thecae strongly hooked and generally changing shape along rhabdosome.	Difficult to distinguish from some Early Silurian curved species. Look for conspicuous tongue (T) on sicula. Earlier monograptids, particularly Late Silurian (Ludlovian) species, also have dorsal tongues but these are much less pronounced. Sometimes associated with *Linograptus* and *Abiesgraptus*.	Early Devonian
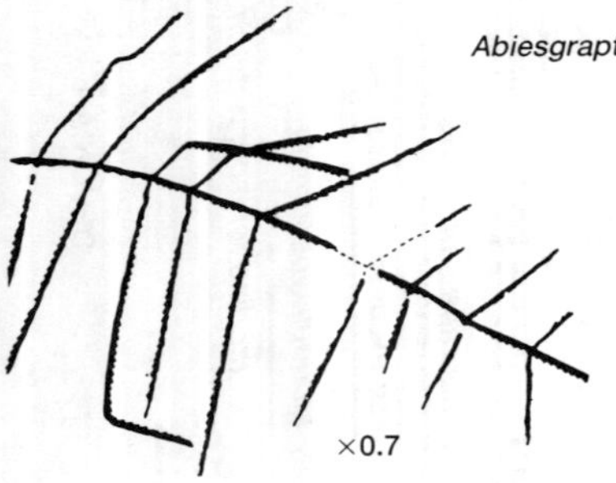	*Abiesgraptus*	Main stipe with regularly spaced, paired lateral branches. Dorsal rod visible in well preserved specimens.	Very rare, can be confused with *Pleurograptus*, which differs by having unpaired irregularly spaced lateral branches without dorsal rod. *Abiesgraptus* occurs with monograptids rather than biserial genera. Stipe fragments of *Abiesgraptus* and *Linograptus* may be confused with fragments of the Ordovician genera *Didymograptus*, *Tetragraptus*, etc., but the Ordovician graptolites lack a dorsal rod along their axes.	Early Devonian
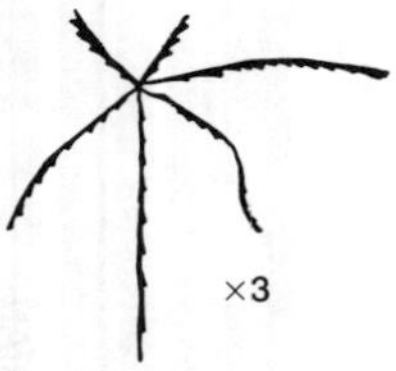	*Linograptus*	Rhabdosome composed of two or more branches radiating from a central point. No secondary branching.	*Linograptus*, like *Abiesgraptus*, occurs with monograptids.	Late Silurian and Early Devonian
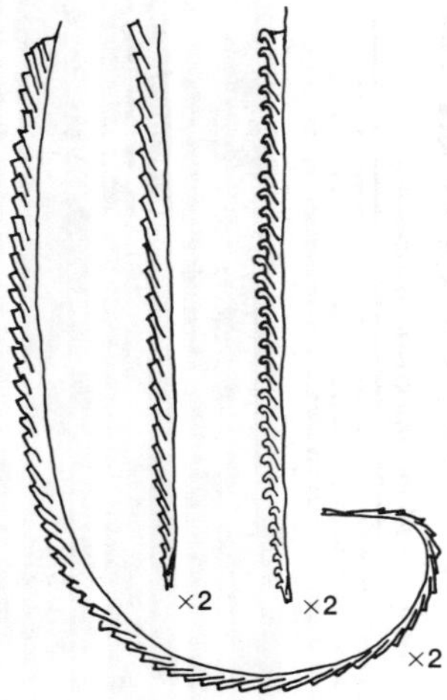	Monograptidae	Uniserial with dorsal rod. Shapes of thecae and of entire rhabdosome determine species. Some easily recognized species with restricted range are illustrated below.	Can be confused with broken stipes of *Didymograptus*. Look for sicula and dorsal rod in *Monograptus*. Look for symmetrical join of 2 stipes at sicula in *Didymograptus*.	Silurian and Early Devonian. Late Silurian and Devonian monograptids are generally straight; curved species are few. Early and Middle Silurian faunas contain many strongly curved species. Association of monograptids and biserial graptolites indicates Early Silurian (Llandoverian).

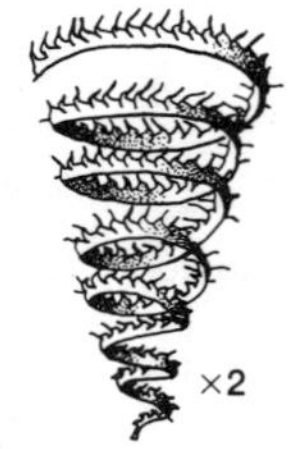

M. turriculatus
Early Silurian

Saetograptus
Late Silurian

M. discus
Early Silurian

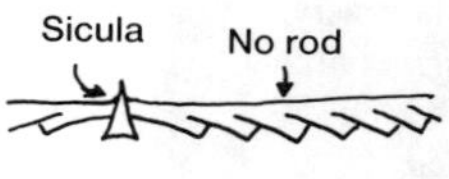

Didymograptus

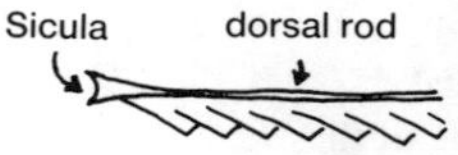

Monograptus

Fig. 15.6 (pp. 230–235) Characteristics and time ranges of some of the most stratigraphically useful graptolites. From Churkin & Carter (1972).

	Name	Diagnostic characters	Other forms with which this might be confused	Time-range
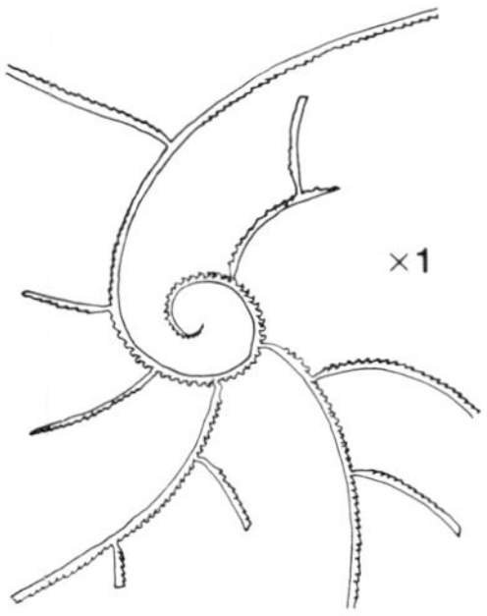 ×1	*Cyrtograptus*	Rhabdosome spirally coiled with secondary branches developed like spokes on a wheel.	None	Middle Silurian
 ×2	*Retiolites*	Lacy secondary mesh within basic network of alternating thecae.	*Reteograptus*, which lacks a secondary mesh.	Early and Middle Silurian
 ×2	*Monograptus spiralis* group	Spirally coiled rhabdosome.	None	Late Early Silurian
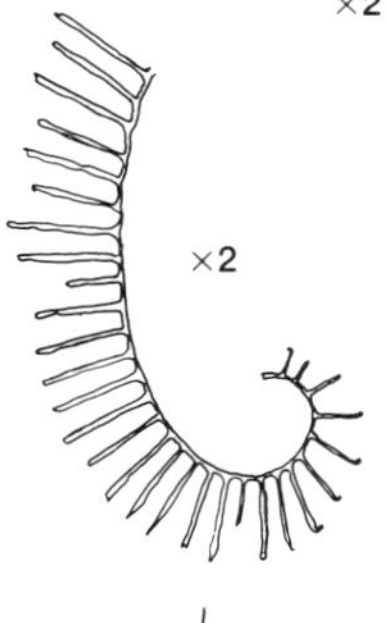 ×2	*Rastrites*	Curved rhabdosome. Thecae widely spaced, very long, narrow and straight. Thecae attached to extremely thin thread-like rod. Hooked ends of thecae sometimes preserved.	A few species of *Monograptus*, but thecae in *Rastrites* are completely isolated from one another and are more nearly perpendicular to the stipe.	Early Silurian
 ×2	*Petalograptus*	Leaf-shaped rhabdosome. Thecae long, straight, or gently curving, and greatly overlapping.	*Phyllograptus* has similar outline but can be distinguished by presence of thecae along central axis.	Early Silurian
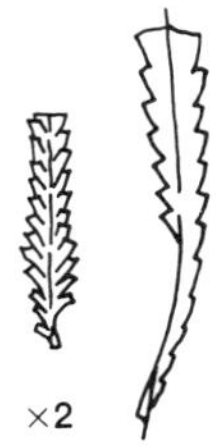 ×2	*Dimorphograptus*	Initial portion uniserial, changing distally to biserial. Uniserial protion often gently curved.	None	Early Silurian

Fig. 15.6 *(Continued)*

Name	Diagnostic characters	Other forms with which this might be confused	Time-range
Diplograptidae	Straight single stipe with thecae on both sides (biserial). Proximal (narrow) end often bearing distinctive spines. Some of these spinous species (see below) are very useful for precise age determination.	This is a long-ranging group consisting of many genera and species. Genera are determined on basis of thecal shape.	Middle Ordovician Early Silurian

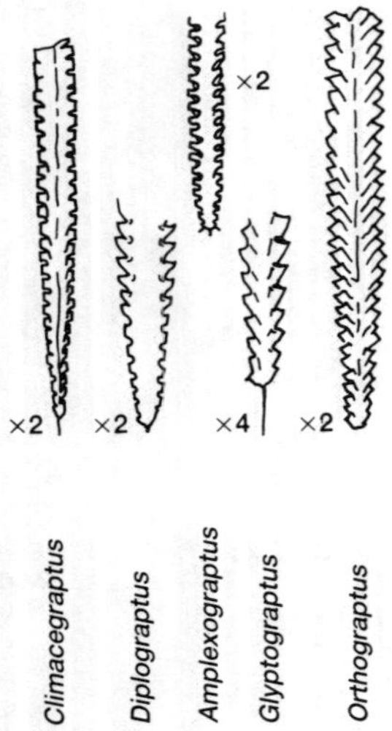

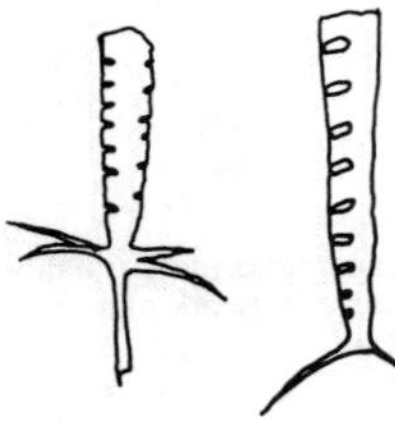

Climacograptus hastatus Late Ordovician

C. bicornis Middle Ordovician

Orthograptus Thecae straight.

Glyptograptus Thecae with gently sigmoidal curvature.

Climacograptus-Amplexograptus Thecae strongly sigmoidal with aperatures in semicircular excavations.

Diplograptus Thecae are amplexograptid types at start and more gently sigmoid (glyptograptid) type distally.

Name	Diagnostic characters	Other forms with which this might be confused	Time-range
Dicellograptus	V-shaped rhabdosome. Very strong curvature of thecae.	See *Leptograptus*.	Middle and Late Ordovician

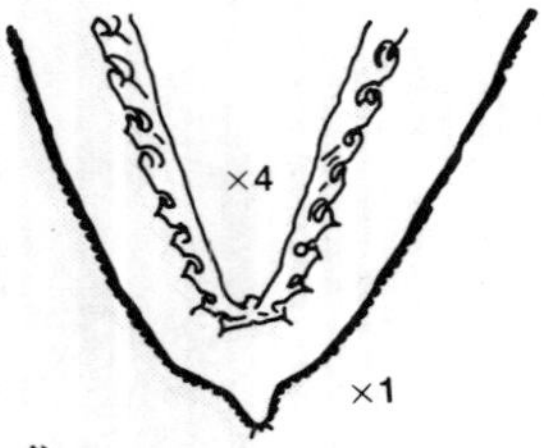

Name	Diagnostic characters	Other forms with which this might be confused	Time-range
Pleurograptus	Rhabdosome branched. Main stipe slightly curved and with irregularly spaced secondary branches on one or both sides. Secondary branches may be branched.	Could be confused with *Abiesgraptus*, which differs by not having branched secondary branches. *Pleurograptus* occurs with biserial genera, whereas *Abiesgraptus* occurs with monograptids. *Pleurograptus* is very rare in North America.	Early Late Ordovician

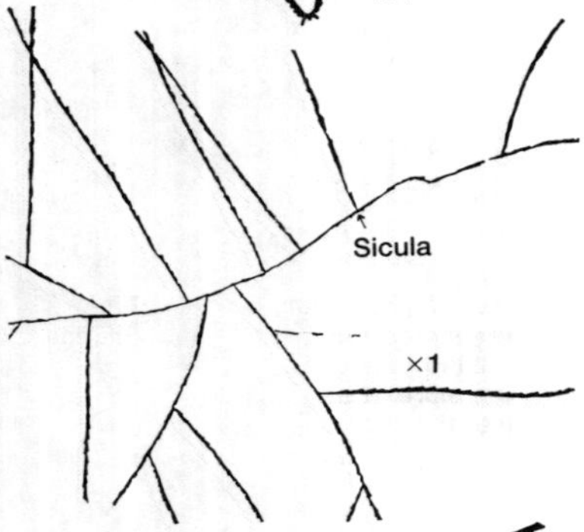

Name	Diagnostic characters	Other forms with which this might be confused	Time-range
Leptograptus	Two slender stipes forming an angle of divergence of nearly 180° (slightly reclined). Serrated margin of stipe difficult to see because thecae are inclined at low angle to stipe.	*Dicellograptus* has more complexly curved thecae, and the angle of divergence of its branches is generally less than 90°.	Middle and Late Ordovician

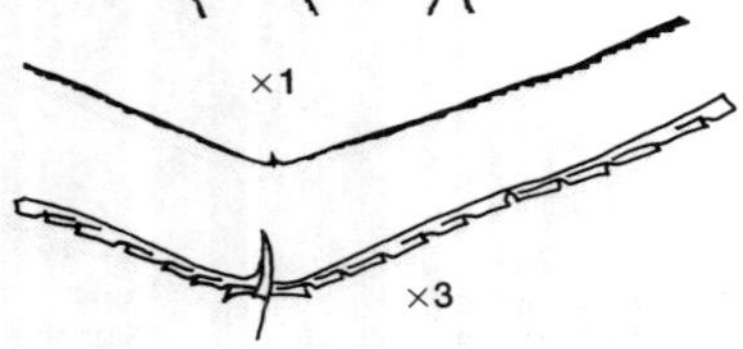

Name	Diagnostic characters	Other forms with which this might be confused	Time-range
Reteograptus	Orderly, boxlike network of alternating thecae.	See *Retiolites*.	Middle Ordovician

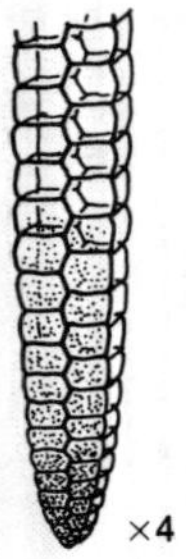

Fig. 15.6 *(Continued)*

	Name	Diagnostic characters	Other forms with which this might be confused	Time-range
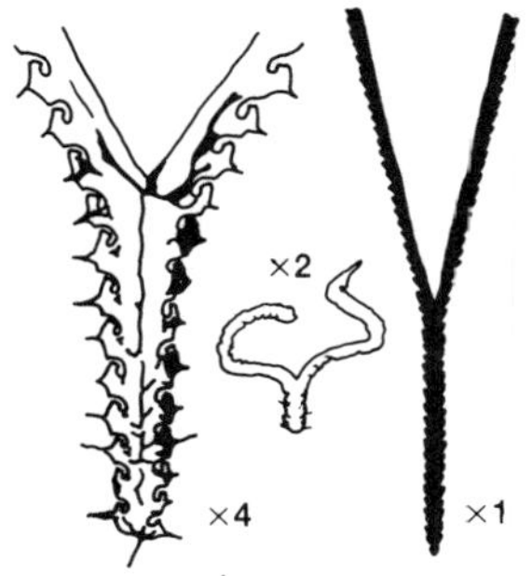	*Dicranograptus*	Y-shaped rhabdosome. Biserial initially and then dividing into 2 uniserial branches. Curved thecae, often with small spines.	Broken uniserial stipes can be confused with monograptids or didymograptids. However, dicranograptids lack the dorsal rod of monograptids and have curved thecae instead of the straight thecae of didymograptids.	Middle Ordovician
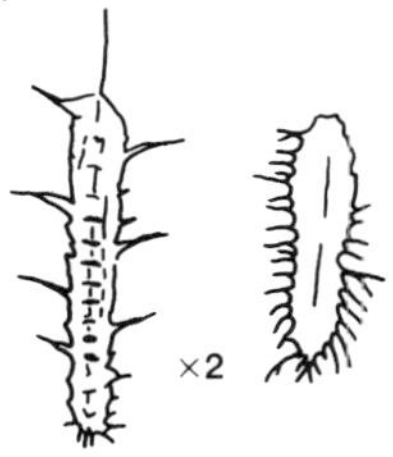	*Glossograptus*	Biserial with large spines on both sides of the rhabdosome.	Could be confused with spined species of Diplograptidae, which generally have shorter and straighter spines.	Middle Ordovician
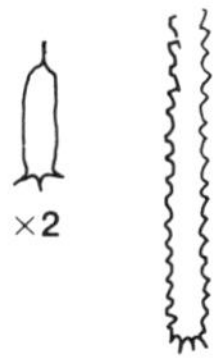	*Cryptograptus*	Straight biserial stipe with spines on proximal end. Thecae may appear *Orthograptus*-like to nearly *Climacograptus*-like. Entire rhabdosome parallel-sided.	Scalloped margins (a series of semicircular curves) generally distinguish *Cryptograptus* from most Diplograptidae. Can be oriented so that thecae are not visible and rhabdosome margins appear straight. In this orientation, *C. tricornis*, a very diagnostic species, is characterized by 3 proximal spines.	Early and Middle Ordovician
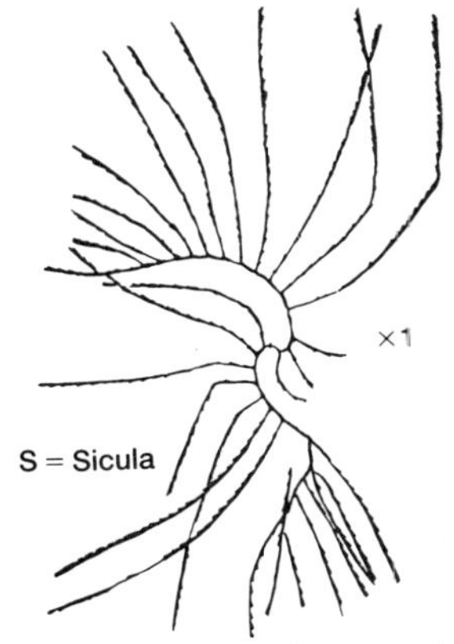	*Nemagraptus*	Rhabdosome branched and bilaterally symmetrical. Two main stipes reclined or more usually curved into S-shape. Evenly spaced secondary branches on convex side of each main stipe.	Fragments can be confused with other branched graptolites. Look for diagnostic associates *Didymograptus* and *Glossograptus*.	Middle Ordovician
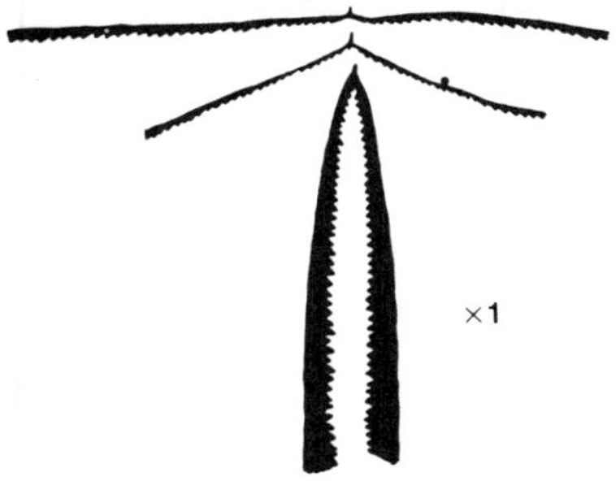	*Didymograptus*	Two-branched, pendent, horizontal or reclined. Thecae simple, straight or slightly curved. 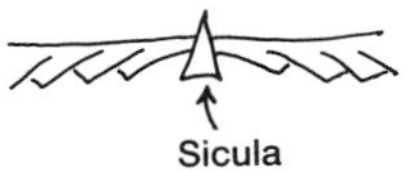	See Monograptidae	Early and Middle Ordovician

Fig. 15.6 *(Continued)*

	Name	Diagnostic characters	Other forms with which this might be confused	Time-range
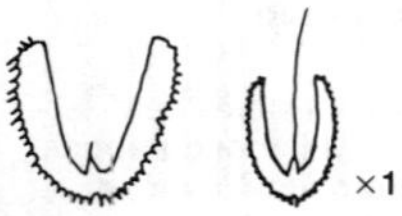	*Isograptus*	Two wide stipes, reclined; sicula and thread-like extension usually prominent.	Could be confused with wide-stiped *Tetragraptus*, which usually will show portions of more than 2 stipes.	Early Ordovician and early Middle Ordovician
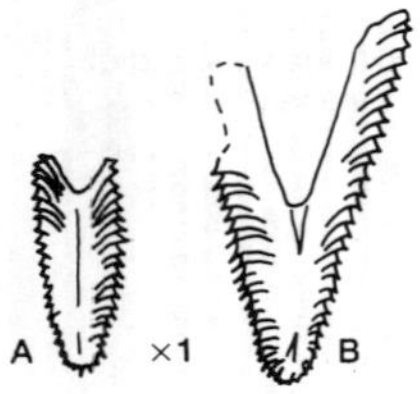	*Cardiograptus* (A) and *Oncograptus* (B)	*Cardiograptus* has two stipes back to back (biserial); *Oncograptus* is also biserial initially; later the stipes diverge.	Could be confused with flattened *Phyllograptus*, but does not have any sign of a row of thecae along central axis	Early Ordovician earliest Middle Ordovician
 Axial view of unflattened specimen 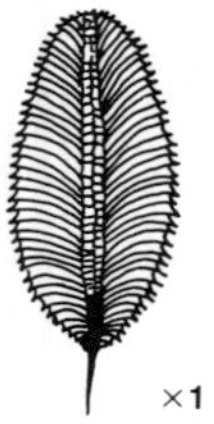	*Phyllograptus*	Four stipes joined along their dorsal margins at right angles to each other. Usually flattened so only 2 stipes appear in profile and a third appears in ventral view along the central axis.	See *Cardiograptus* and *Petalograptus*.	Early Ordovician earliest Middle Ordovician
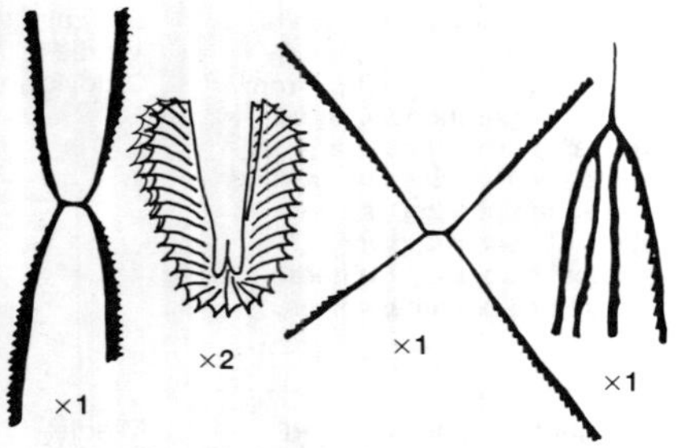	*Tetragraptus*	Four-branched, bilaterally symmetrical. Thecae simple, straight or slightly curved.	None when four branches are visible. See *Isograptus*.	Early Ordovician earliest Middle Ordovician
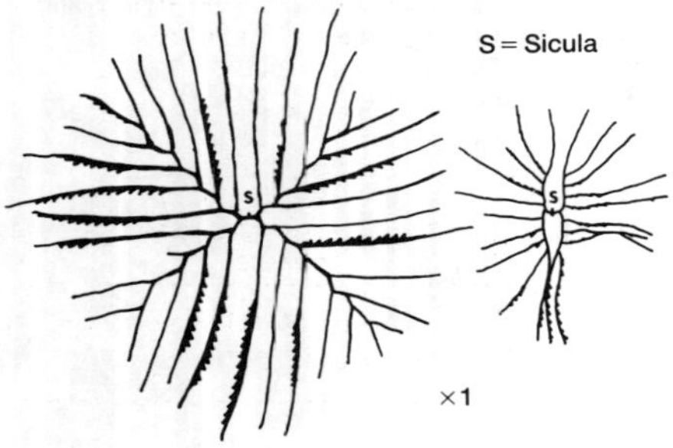 S = Sicula	Sparsely branched genera with bilateral symmetry	Bilateral symmetry of rhabdosome. Stipes uniserial. Thecae straight.	Single stipes can be confused with *Monograptus*. Look for sicula.	Mainly Early Ordovician but rarely into Middle Ordovician
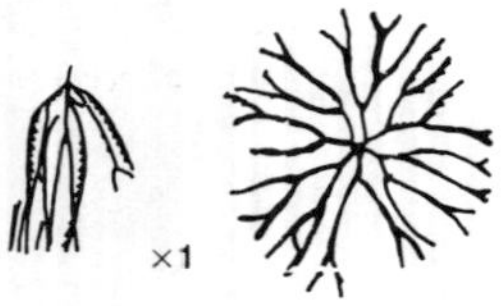	Sparsely branched genera with radial symmetry	Dichotomous (paired) branching. Sicula present.	Could be confused with plants. Look for thecae on sides of branches.	Mainly Early Ordovician but rarely into Middle Ordovician
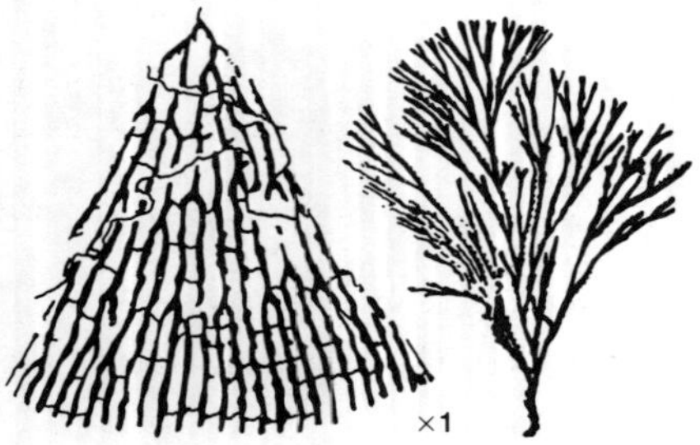	Abundantly branched genera	Rhabdosome with many branches, sometimes anastomosing or connected by transverse branches. "Stem" and root-like processes at one end. Many fan-shaped.	Could be confused with plants. Look for thecae on sides of branches.	Late Cambrian-Mississippian. Abundant specimens without associated different types of graptolites, generally Cambrian. Common with other types of graptolites in the Early Ordovician. Rare in Middle Ordovician-Missippian

Fig. 15.6 *(Continued)*

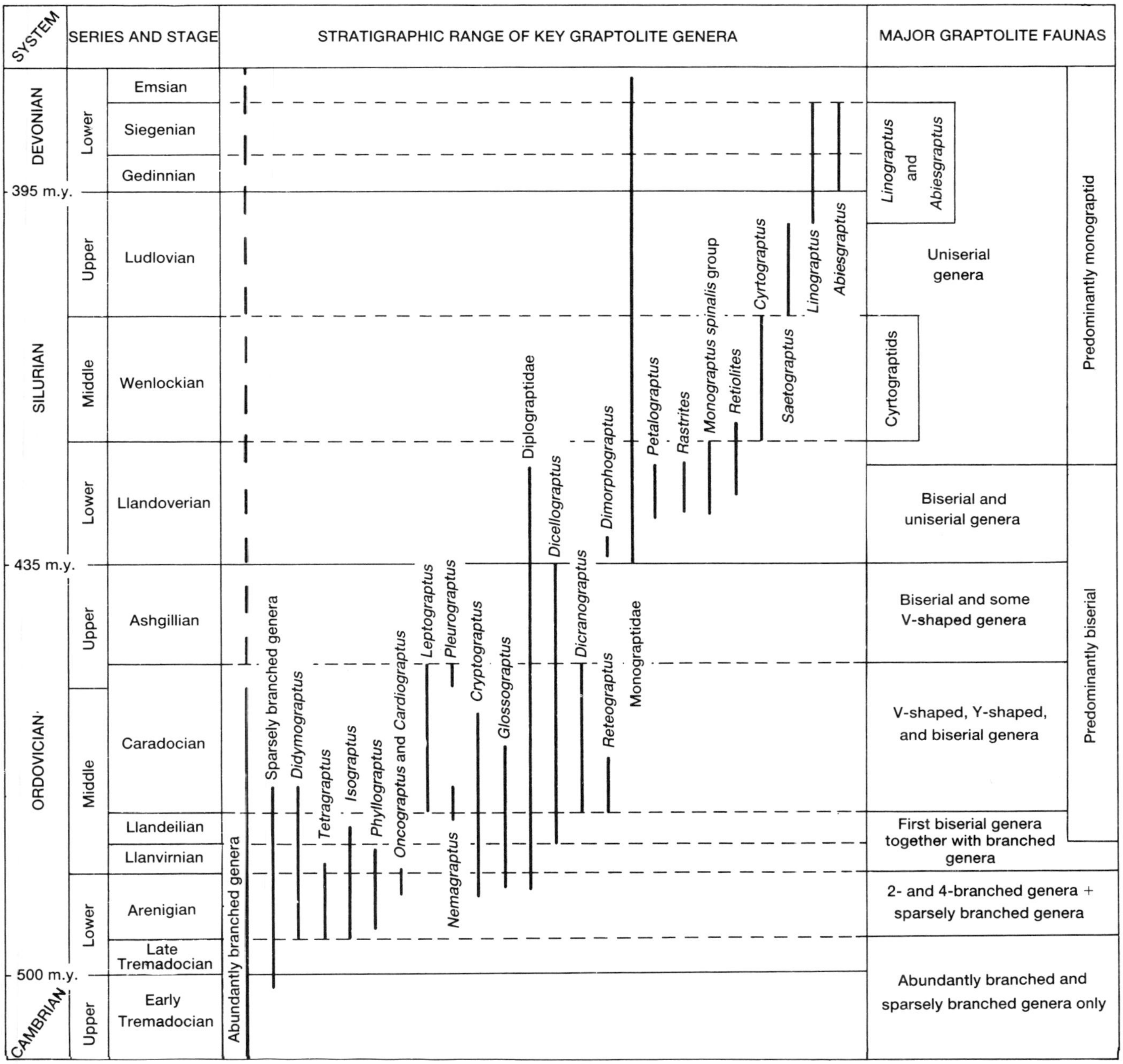

Fig. 15.6 *(Continued)*

FURTHER READING

Berry, W. B. N., 1977. Graptolite biostratigraphy: a wedding of classical principles and current concepts, in E. Kauffman & J. E. Hazes (eds), *Concepts and methods of biostratigraphy*, pp. 321–338. Dowden, Hutchinson & Ross: Stroudsburg, PA.

Bulman, O. M. B., 1970. *Graptolithina*, in C. Teichert (ed.), *Treatise on invertebrate paleontology*, Part V (Revised). Geological Society of America and University of Kansas Press: Lawrence, KS.

Urbanek, A., 1973. Organization and evolution of graptolite colonies, in R. S. Boardman, A. H. Cheetham & W. A. Oliver Jr (eds), *Animal colonies; their development and function through time*, pp. 441–514. Dowden, Hutchinson & Ross: Stroudsburg, PA.

EXERCISE 15 Worksheet

Name

QUESTION 1

The chart below lists some characteristics of the pterobranch *Rhabdopleura* on the left side. Based on information in the text of this exercise, plus any supplemental information from a paleontology textbook, compare graptolites with *Rhabdopleura* by writing in the corresponding information about graptolites for each characteristic of *Rhabdopleura*. If no corresponding information is available for any particular characteristic of graptolites, write NA in the space.

Comparison of Rhabdopleura (Class Pterobranchia) with the Graptolithina:

Rhabdopleura	*Graptolithina*
1 Colonial, organized into zooids	
2 Encrusting, marine	
3 Zooids 1–3 mm long	
4 Zooids encased in proteinaceous tubes	
5 Half-ring growth segments present in encrusting portion	
6 Feeding zooids bud from an elongate stolon	
7 Stolon is distinctly darker in color than is outer proteinaceous tube (and is termed the black stolon)	
8 Zooids have a shield-like, discoidal proboscis with a few tentaculate, curved arms	
9 Zooidal alimentary tract is U-shaped	
10 Zooids have two or more gill slits	

QUESTION 2

In Fig. 15.6, circle pendent rhabdosomes with a solid line, and circle scandent rhabdosomes with a dashed line. If colored pencils are available, use different-colored pencil lines for solid and dashed. Use the same types of lines to circle the ranges of the pendent and scandent taxa on the range chart within Fig. 15.6. Do not circle horizontal or reclined rhabdosomes.

Name four graptolites from Fig. 15.6 that have reclined rhabdosomes:

__________ __________ __________ __________

Note the ranges of the four genera, and compare them with ranges of scandent rhabdosomes given in Fig. 15.6. What would you predict to be the approximate stratigraphic position of graptolite faunas containing:

1 Common reclined rhabdosomes? ____________________;

2 Common scandent rhabdosomes? ____________________

QUESTION 3

Several species of graptolites with pendent and with reclined rhabdosomes are illustrated in Fig. 15.7. Circle each reclined rhabdosome. If both arms of the rhabdosome are preserved, but their junction is obscured, how would a rhabdosome be recognized as pendent or as reclined, based on contrasts observed in Fig. 15.7?

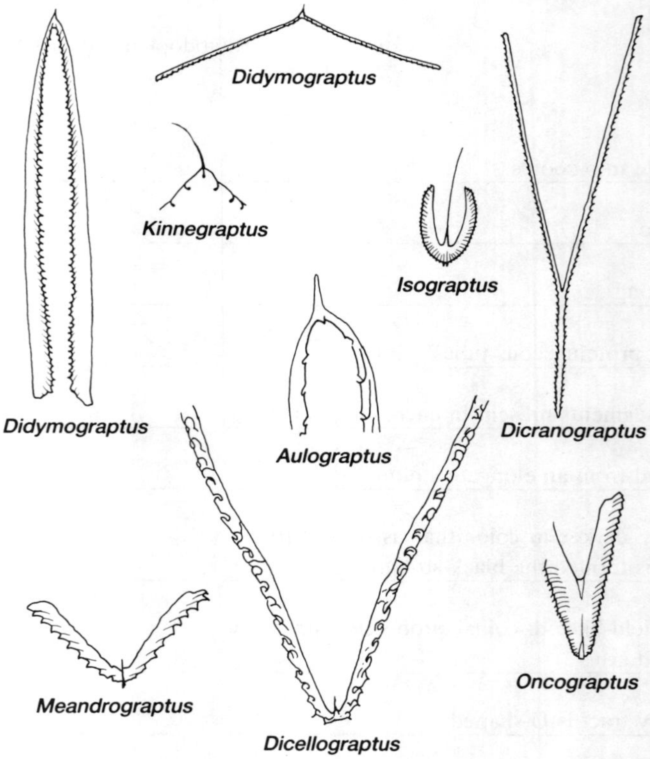

Fig. 15.7 Various pendent and reclined rhabdosomes of graptoloids. After Bulman (1970).

QUESTION 4

Examine representative graptolites provided, which are identified to genus. Record the available information for each on p. 239.

Genus	*Geologic age of specimen*	*Rock type*	*Preservation*	*Graptoloid or dendroid*	*No. of stipes*	*Scandent, pendent, etc.*	*No. of rows of zooids*	*Zooid profile*

QUESTION 5

Examine the assemblage provided, take note of the characteristics of rhabdosomes and stipes, and identify specimens to genus where possible, using the information in Fig. 15.6. Determine the most probable age of the rock that contains the assemblage (period, epoch, or age as appropriate), using the common overlap of generic ranges or geologic distribution of characteristics seen in the specimens. (This procedure determines age by concurrent range.) Be as precise and specific as possible in age assignment of the assemblage, but be careful not to be more specific than the actual fossils warrant.

Generic assignment of specimens: ______________________

____

____

____

____

____

Age assignment of assemblage: ______________________

EXERCISE 16 Conodonta

16.1 INTRODUCTION

Conodonts are an extinct group of organisms that have long been known from their isolated or commonly grouped sets of phosphatic tooth-like **elements**. Discovery of the impression of soft parts of the conodont animals (Fig. 16.1) was reported only in the 1980s. These discoveries plus the previous findings of symmetrically arranged elements in natural assemblages helped to narrow the possibilities for a taxonomic home for conodonts. Although there is still no universal agreement among specialists on assignment, the majority consider them to belong among the agnath fishes or to constitute an independent phylum.

The conodont animal was bilaterally symmetrical, with paired extensions of the anterior part of the cephalic lobe, behind which sets of paired, mineralized elements were located. The trunk consisted of a series of chevron-shaped muscles, with a posterior and a caudal fin. Many functions had been previously suggested for the conodont elements, but their position within the animal suggests that they likely served as teeth or possibly as tentacle supports.

Conodonts ranged from the Late Proterozoic — if one accepts the questionably included group known as paraconodonts — through the Triassic Period (Table 16.1), and they apparently were exclusively marine dwellers. Conodont elements are small, generally 0.1–1 mm in length, so are considered to be microfossils. They are generally freed from carbonate rocks by dissolving the rock in a dilute solution of acetic or some other organic acid, which does not affect the phosphatic conodont elements. Commonly, they are studied by searching the surface of black shales or other fissile marine rocks; such surfaces may yield individual elements or the multi-element sets termed **assemblages** or **apparatuses**.

Although multi-element assemblages were known almost from the beginning of investigations of conodonts, species names were applied to individual elements for over 100 years of study. In the late 1960s, however, statistical procedures provided a way to recognize linked associations of isolated elements, and studies since then increasingly have emphasized assemblages as the basis for taxonomy (Fig. 16.2). As a result of change from single-element to multi-element species and generic names, conodont taxonomy had undergone a colossal reorganization, which still continues in substantial degree.

Abundance of conodont elements ranges from near zero to

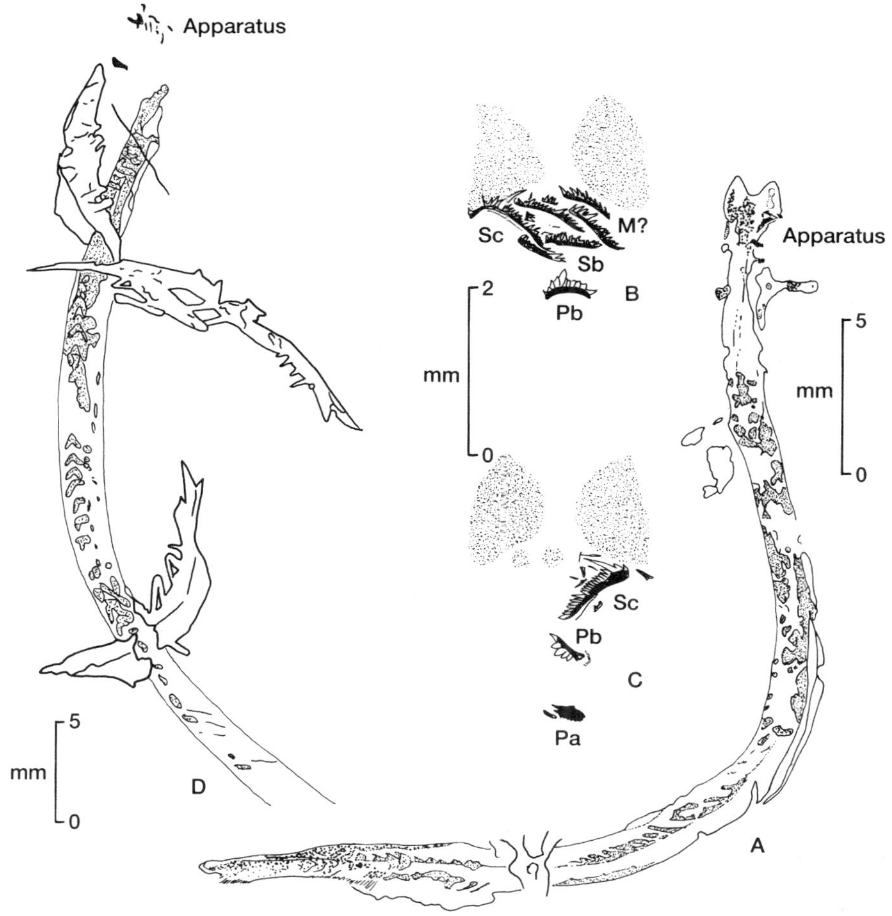

Fig. 16.1 Impressions of the body of a conodont animal from the "Shrimp band" (some shrimp lie across part D of the figure) of the Lower Carboniferous (Dinantian) Granton Sandstones near Edinburgh, Scotland. The originally discovered specimen is drawn in part A, and B and C show the conodont elements that adhere to opposite surfaces of the break that exposed the fossil. Letters in B and C identify positions of elements within the assemblage; see Fig. 16.2. Part D illustrates a subsequently discovered specimen. The two specimens clearly show that the conodont elements are located at the anterior end of an elongate, bilaterally symmetrical body composed of chevron-shaped muscles and possessing fins. From Briggs *et al.* (1983) and Aldridge *et al.* (1986) with permission from Norwegian University Press (Universitetsforlaget AS), Oslo.

Table 16.1 Simplified classification and characteristics of conodonts as given by Clark in Boardman *et al.* (1987). Sweet (1988) provides a much more detailed classification.

Class Paraconodontida	Characterized by predominantly coniform elements with large, deep basal cavity Lamellae few and widely spaced Relatively high proportion of organic matter within skeleton Assemblages typically of one type coniform element, less commonly of multiple types of elements Late Proterozoic–Middle Ordovician
Class Conodontophorida	Characterized by coniform, ramiform, or pectiniform elements with generally shallow basal cavity or with specialized basal pit Lamellae numerous and closely spaced Surface of elements commonly highly textured Assemblages of one to seven different types of elements Cambrian–Upper Triassic
Order Distacodontina	Assemblages of single type or of multiple types of coniform or ramiform elements Some assemblages may contain coniform, ramiform, and primitive pectiniform elements Cambrian–Devonian
Order Hibbardellina	Assemblages may contain several types of ramiform elements, or both ramiform and pectiniform elements, or only a single type of pectiniform element Middle Ordovician–Late Triassic

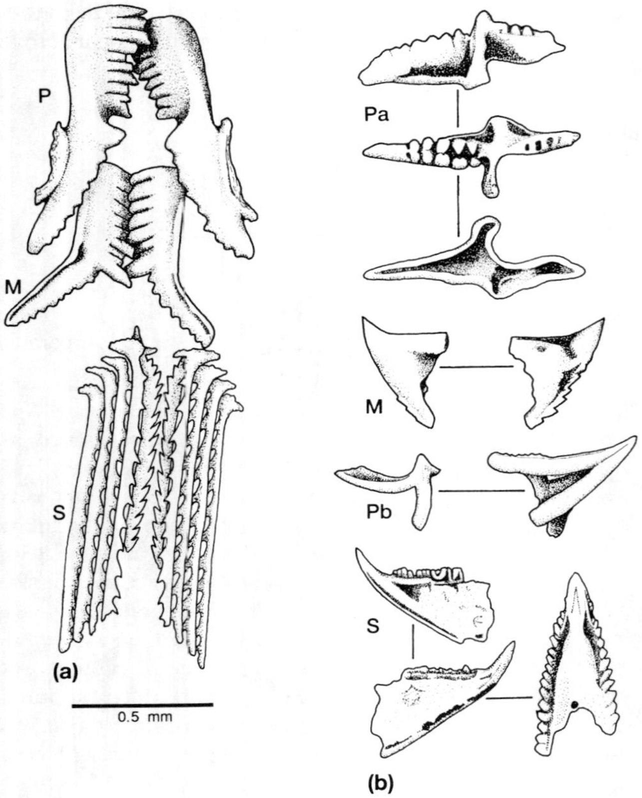

Fig. 16.2 Assemblages of conodont elements. (a) Natural assemblage of *Scottognathus typicus* (Rhodes) from Pennsylvanian (Missourian) rocks in Illinois; (b) statistically determined assemblage of *Icriodella* from Middle Ordovician rocks of Kentucky. Primary (P), medial (M), and symmetry (S) elements are marked in both assemblages; where used, lower case letters indicate the presumed sequence of subsets. Drawn from specimens illustrated by Rhodes (1952) and Clark in Boardman *et al.* (1987).

over several thousand per kilogram of Paleozoic to early Mesozoic marine rock. They obviously were widespread in the Paleozoic and Triassic seas. The apatite composition of elements resisted destruction on the sea floor, and rate of background sedimentation seems to have been the primary determinant of their abundance.

In this exercise, three aspects of conodont paleontology will be introduced:

1 the general morphology of elements and assemblages;
2 the value of conodonts as biostratigraphic tools; and
3 the use of conodonts as guides to thermal alteration of their host rocks.

16.2 MORPHOLOGY OF CONODONTS

16.2.1 Individual elements

There are three basic groups of conodont element types. All are composed of superposed laminae of apatite. In some, laminae extend up into a tapered cusp and are overlapped slightly at the base so that a broadly conical cavity exists in the underside. In others with more complex surface sculpture, edges of laminae on the basal side are overlapped only in the center so that a small pit is produced, but the surrounding laminae may become progressively shorter so that a relatively flat or slightly convex surface is produced. Some specimens retain on the undersurface a basal filling which is also composed of superposed laminae, although these laminae are coarser and less densely mineralized than those of the upper part of the element. Most collections of conodonts consist of specimens without the basal filling, which was probably quite weakly mineralized in most taxa.

The three basic forms and descriptive terminology of elements are illustrated in Fig. 16.3. **Coniform elements** ("cones") consist of a more-or-less well developed base surmounted by a single tapered cusp that may have a smooth

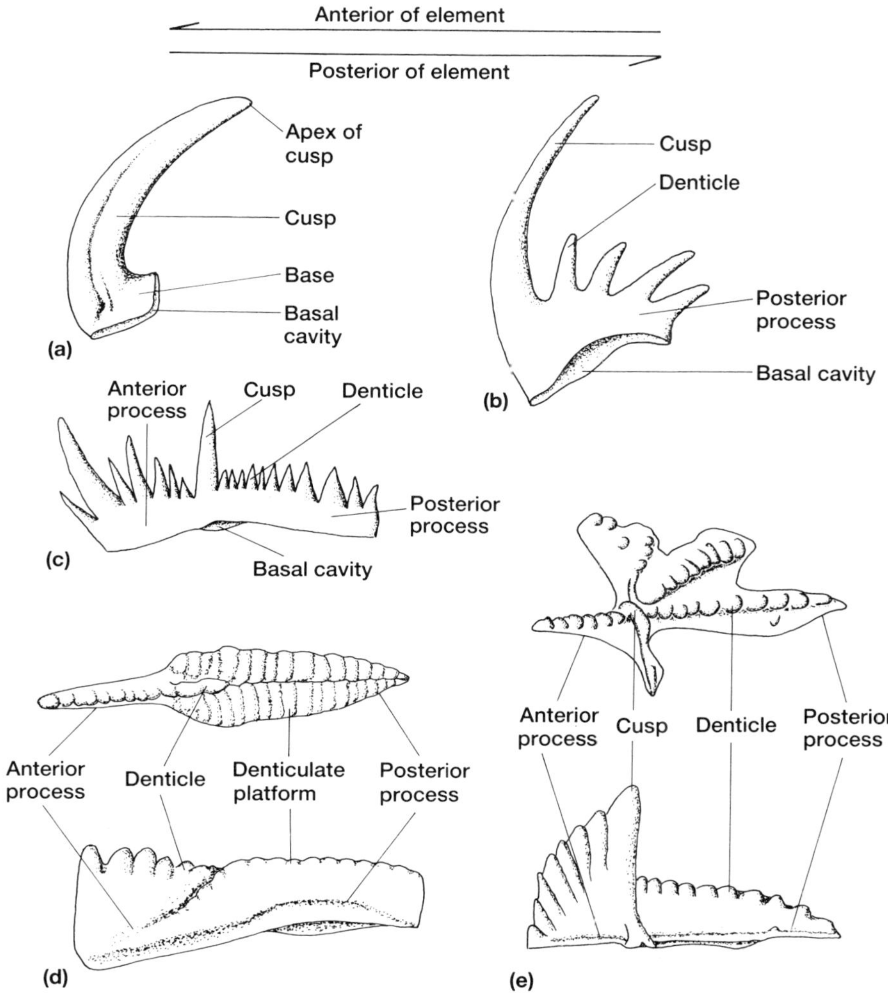

Fig. 16.3 The three basic forms of conodont elements, with some of the descriptive terminology. "Anterior" and "posterior" are given as standard orientations for description of individual elements and are not meant to designate orientation within the conodont animal. (a) Coniform, (b) and (c) ramiform, (d) and (e) pectiniform. Parts (a) and (d) are modified from Hass (1962); (b) and (c) are modified from Clark *et al.* (1981) with permission from the Geological Society of America; (e) is modified from Sweet (1988).

surface or possess lateral ridges termed costae. A few coniform elements have a small expansion of the posterior side that is subdivided into a few denticles on the upper edge. These are transitional into the form described next.

Ramiform elements ("bars") have an elongated ridge (a **process**) on the posterior or on both posterior and anterior sides of the cusp. In almost all cases the upper edges of the processes are subdivided into denticles. Some ramiform elements also have laterally extended processes.

Pectiniform elements ("platforms") include diverse forms in which there are posterior and usually anterior denticulate processes extending from the region of the basal cavity. The cusp is reduced in size and in many indistinguishable from the denticles, except for its position over the innermost point of the basal cavity. In addition, the base is narrowly or broadly expanded laterally as a platform that bears denticles on its upper surface.

A large number of polysyllabic terms and combinations of terms are applied to the numerous types of conodont elements. Elaborate terminology is, of course, highly useful for communicating concepts of morphology in the shortest possible way, but luckily in this case it is beyond what one needs to know in a brief introduction to conodonts.

16.2.2 Assemblages

Natural assemblages seem to have the elements arranged into three regions. From anterior to posterior, these regions consist of "principal" (P), "medial" (M), and "symmetry" (S) elements (Fig. 16.2a). Previously, P was considered to be anterior and S to be posterior, but impressions of the Carboniferous conodont animal from Scotland indicate that S was anterior and P was posterior. By analogy with natural assemblages, statistical assemblages also have elements designated P, M and S (Fig. 16.2b). All elements within an assemblage may be of one basic form, but more commonly elements within an assemblage are of two or more different forms. Pectiniform or ramiform elements most commonly occupy P positions, and ramiform or coniform elements generally occur in M and S positions.
(Question 1)

16.3 BIOSTRATIGRAPHY

Some species of conodonts may be distributed through several sedimentary facies, within a range of rocks inferred to have been deposited from the shallowest environments to deep ocean basins. Others are confined to a range of facies that indicate shallow to moderate depths, while others are confined to sediments that were deposited at depth, including black shales deposited on oxygen-deficient floors of basins that were stagnant in their deepest parts. This distribution pattern suggests that most, if not all, conodonts were pelagic (living up within the water column), so that they were relatively independent of the nature of the sea floor over which they lived.

The trans-facies distribution of conodonts, together with their rapid evolution, distinctive morphologies, and commonly high abundance in rocks, make them pre-eminent fossils for Paleozoic and Triassic biostratigraphy. Cambrian through Triassic rocks have been subdivided into approximately 150 conodont-based zones, although biogeographic restrictions to some extent interfere with and complicate world-wide correlation of these zones.

Proportions of coniform, ramiform, and pectiniform elements changed through time, although convergent evolution and the waxing and waning of various conodont stocks at different times cause complications in the pattern. Coniform elements appeared in the Proterozoic and were the only form present through the Cambrian. Lower Ordovician rocks contain both coniform and ramiform elements, with rare pectiniform elements. In Silurian rocks, coniform elements generally are least common, ramiform elements dominate, and pectiniform elements are common. During the Devonian, coniform elements dropped out entirely (except for a brief reappearance during the Permian), while ramiform- and pectiniform-bearing species underwent a major evolutionary radiation. In some conodont groups, ramiform elements were more prevalent within apparatuses during the Devonian, while in others there were relatively more pectiniform elements. Pectiniform elements commonly dominate Devonian and Carboniferous faunas because of the preponderance of pectiniform-rich taxa such as the superfamily Polygnathacea. Pectiniform elements continued to constitute a progressively larger proportion within individual apparatuses until they were dominant in all conodont groups during the Permian and Triassic (Fig. 16.4). *(Questions 2 and 3)*

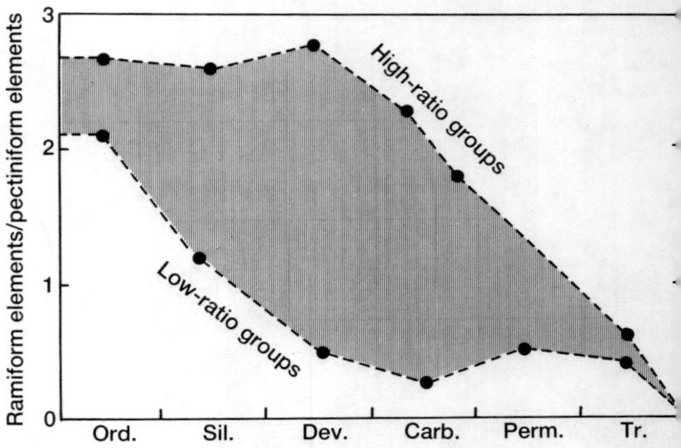

Fig. 16.4 Trends in ratio between ramiform and pectiniform elements within conodont apparatuses in Ordovician through Triassic conodonts. After Sweet (1988).

16.4 THERMAL ALTERATION

Conodont elements are composed of carbonate apatite containing trace amounts of organic matter. In an unaltered, unweathered state, their color is basically pale yellow. The color is more intense in the base and may grade into white within the cusps; in addition, the color intensity corresponds roughly with the conodont's thickness. Where conodonts have been subjected to increased temperatures and progressively extended burial, carbonization of the organic matter occurs, and a series of progressively darker colors is generated until the temperature is so high that the carbon is driven off and the elements become clear (Epstein *et al.*, 1977). Beginning with the pale yellow color, eight color index levels are recognized up to the darkest levels, and the range of the temperatures that produce them have been determined (Table 16.2).

The progressive alteration of color in conodont elements is useful as a rough guide to maturation of organic materials within their enclosing rocks. Where more than approximately 62–65% of carbon has been fixed so that it is no longer mobile, oil cannot be commercially extracted from rocks; this corresponds with a conodont color alteration index of between 1.5 and 2. If 95% of the carbon has been fixed, dry gas is no longer available in commercial quantities; this corresponds with color alteration index of about 4.5. Therefore, careful determination of color alteration of conodont elements for any particular stratigraphic level at an appropriate number of control points across a region should give a clear idea of

Table 16.2 Chart relating color alteration in conodont elements to temperature range to which they have been exposed and to percentage of carbon fixed in place (i.e. no longer mobile). After information in Epstein *et al.* (1977).

Color alteration index	Actual color	Temperature range (°C)	Fixed carbon range (%)
1	Pale yellow	<50–80	<60
1.5	Very pale brown	50–90	55–70
2	Brown to dark brown	60–140	55–70
3	Very dark grayish brown to dark reddish brown	110–200	70–80
4	Brownish black	190–300	80–95
5	Black	>300	>95
6	Gray		>99
7	Opaque white		>99
8	Colorless, transparent	~950	–

where oil and gas may possibly occur and where the thermal history has eliminated the possibility of commercial hydrocarbon production. It should be kept in mind that younger beds occurring in the same region may have experienced less severe thermal histories, so that the potential for oil and gas in relatively young rocks may occur over a larger area than that determined for older rocks.
(Question 4)

FURTHER READING

Aldridge, R. J. (ed.), 1987. *Palaeobiology of conodonts*. Ellis Horwood Ltd: Chichester.

Aldridge, R. J., Briggs, D. E. G., Clarkson, E. N. K. & Smith, M. P., 1986. The affinities of conodonts — new evidence from the Carboniferous of Edinburgh, Scotland. *Lethaia* **19:**279–291.

Briggs, D. E. G., Clarkson, E. N. K. & Aldridge, R. J., 1983. The conodont animal. *Lethaia* **16:**1–14.

Clark, D. L., Sweet, W. C., Bergstrom, S. M., Klapper, C., Austin, R. L., Rhodes, F. H. T., Muller, K. J., Ziegler, W., Lindstrom, M., Miller J. F. & Harris, A. G., 1981. Conodonta, in R. A. Robinson (ed.), *Treatise on invertebrate paleontology*, Part W (Suppl. 2). Geological Society of America and University of Kansas: Lawrence, KS.

Epstein, A. G., Epstein J. B. & Harris, L. D., 1977. *Conodont color alteration — an index to organic metamorphism*. U.S. Geological Survey Professional Paper **995:**1–27.

Sweet, W. C., 1988. *The Conodonta*. Oxford University Press: New York.

EXERCISE 16 Worksheet

Name

QUESTION 1

If samples of conodont elements are available for study, carefully draw a representative of each designated form. If the specimen is coniform, ramiform, or narrow-platform pectiniform, draw it in lateral view. For broad-platform elements, draw as viewed from above the denticulate upper surface. For each drawing, label anterior direction, posterior direction, upper and basal side if drawn in profile, and the various morphologic features as given in Fig. 16.3.

Form:__________ Genus:__________	Form:__________ Genus:__________
Form:__________ Genus:__________	Form:__________ Genus:__________

QUESTION 2

Figures 16.5–16.7 illustrate three different conodont faunas. For each, draw a box around each coniform element, draw a circle around each ramiform element, and leave the pectiniform elements untouched. For each fauna, enter on the table on p. 249 the number of different elements that represent each basic form, read again Section 16.3, determine the most likely time interval during which the fauna lived, and write in the characteristics of the fauna on which you base its probable age range.

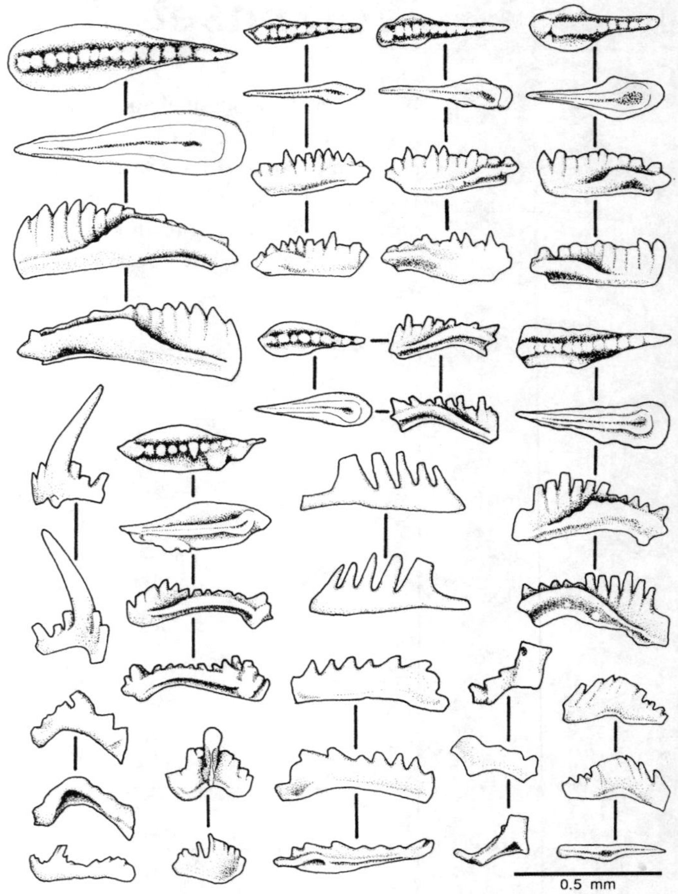

Fig. 16.5 Conodont elements from Limestone Point on San Juan Island, Washington. Lines connect different views of a single species. From Savage (1983).

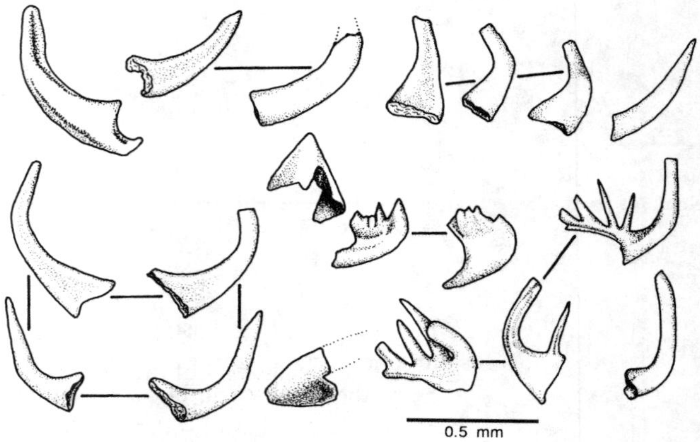

Fig. 16.6 Conodont elements from the Cass Fjord Formation of Devon Island, part of the Canadian Arctic Islands. From Nowlan (1985).

	Number of each element type			*Probable age range*	*Reason for decision on probable age*
	Coniform	*Ramiform*	*Pectiniform*		
Fauna #1 (Fig. 16.5)					
Fauna #2 (Fig. 16.6)					
Fauna #3 (Fig. 16.7)					

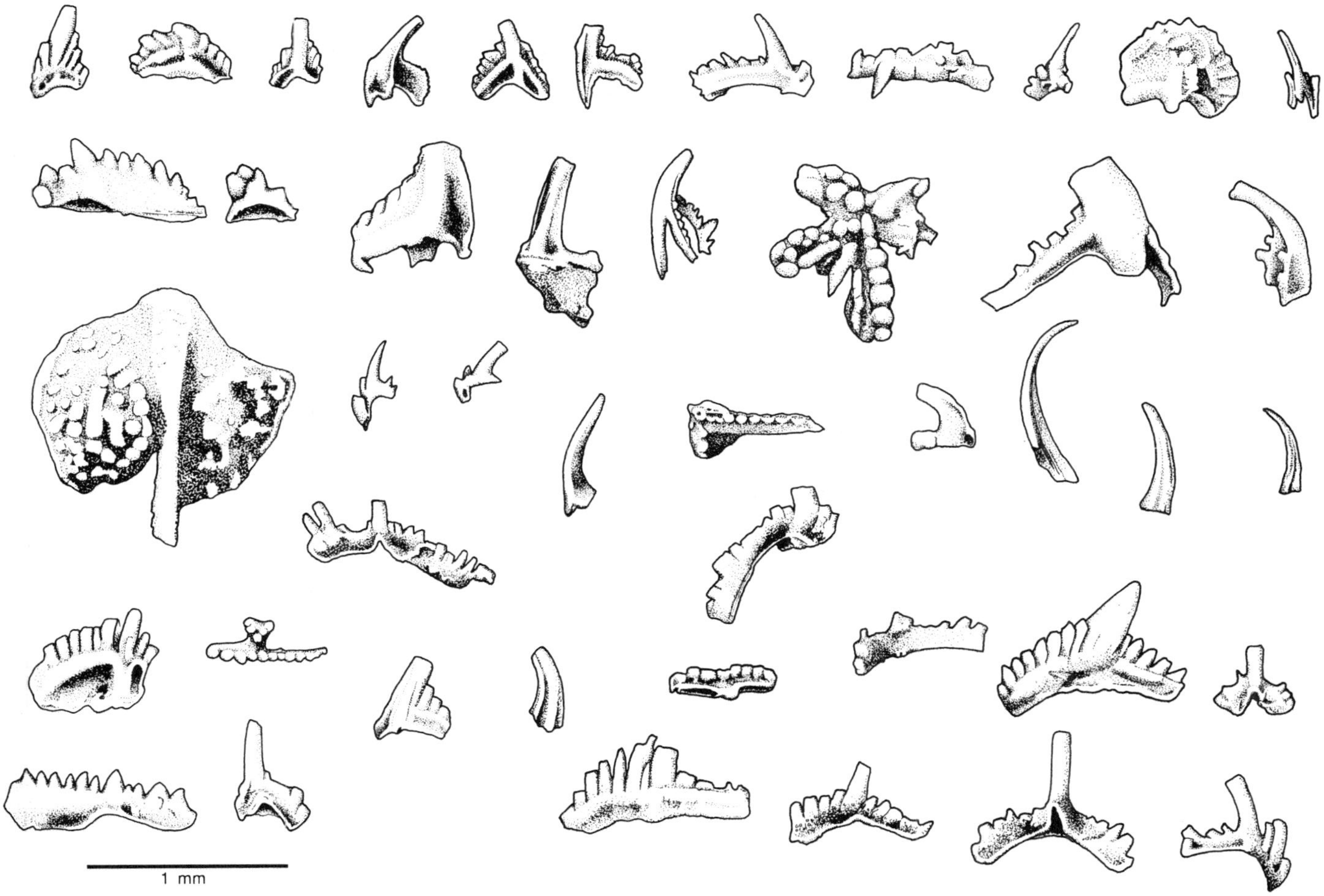

Fig. 16.7 Conodont elements from adjacent portions of the Rose Hill and Mifflintown Formations, central Appalachians. From Helfrich (1980).

QUESTION 3

The Late Mississippian strata in the Illinois Basin (located in southern Illinois, southern Indiana, and northwestern Kentucky) constitute the basis for definition of the North American Chesterian Epoch, which is a partial equivalent of the Visean and Namurian Ages of the Lower Carboniferous Epoch in Europe. The sequence of sandstones, shales, and limestones deposited during Chesterian time in the Illinois Basin record multiple cycles of progradation of a delta and marine innundation across the region. Distribution of some conodont elements through the stratigraphically important Chesterian Series and underlying beds is shown in Fig. 16.8. Figure 16.9 illustrates a portion of the Wymps Gap Limestone in southwestern Pennsylvania, including a list of conodonts obtained from it. Megafossils in the Wymps Gap Limestone suggest but do not demonstrate assignment to the Chesterian Epoch. Are any of the conodont elements from the Wymps Gap Limestone restricted to the Chesterian Series within the Illinois Basin? If so, list them on p. 251.

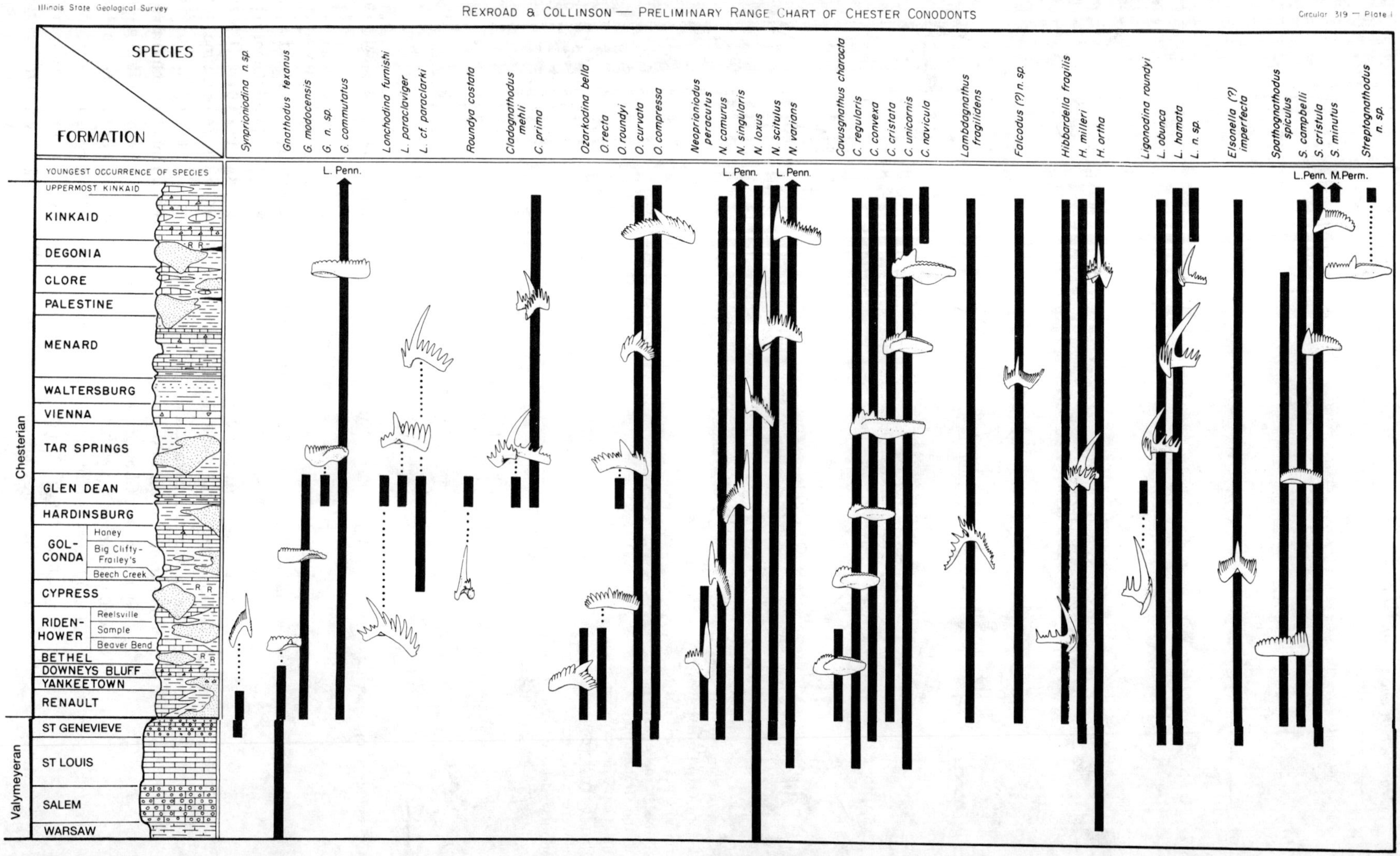

Fig. 16.8 Distribution of some Valmeyeran (Middle Mississippian) and Chesterian (Late Mississippian) conodonts in the Illinois Basin. Based on information in Rexroad & Collinson (1961, 1963, 1965).

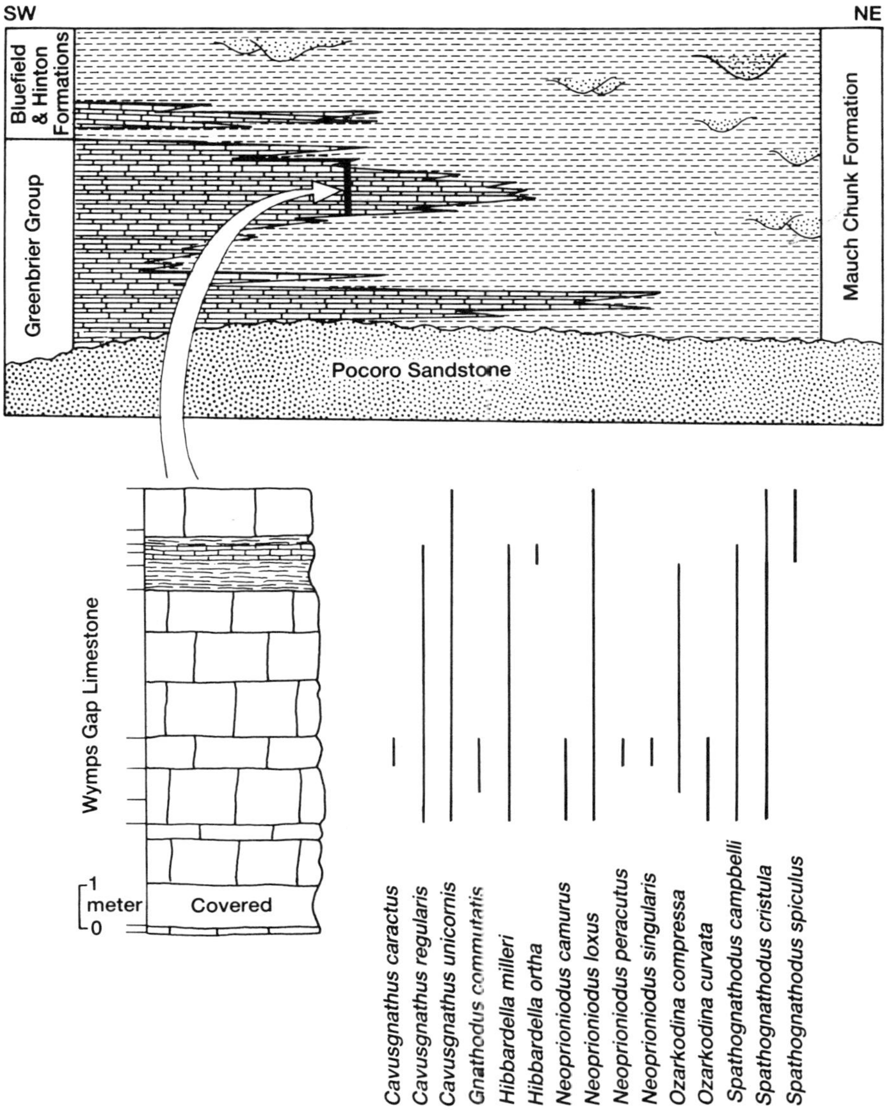

Fig. 16.9 Generalized Mississippian (Lower Carboniferous) facies from southwest to northeast across the central Appalachians, and vertical distribution of some of the conodonts in the Wymps Gap Limestone in an abandoned quarry in Chestnut Ridge, Fayette County, southwestern Pennsylvania. Dashes to the left of the stratigraphic section of the Wymps Gap Limestone refer to the intervals within the rock sequence from which samples were collected and processed for conodonts. Modified from Horowitz & Rexroad (1972) and Brezinski (1984).

____

____

____

____

____

____

Do any of the conodonts suggest correlation with any particular part of the Chesterian Series in the Illinois Basin? If so, list them and write the part of the sequence that is suggested.

____

Based on the conodonts, which is more firm, correlation with the Chesterian Series, or correlation with a particular part of the series? Explain your answer.

QUESTION 4

Color alteration levels of conodont elements from Paleozoic localities in southern Arizona are recorded in Fig. 16.10. Draw a line **(isograd)** that traces the approximately 2.0 color alteration level and another that traces the approximately 4.5 color alteration level. Based on the information given in Section 16.4, lightly color in blue the area where potential for oil as well as gas exists in the Paleozoic rocks. Using a different color, shade in the area where dry gas but not oil potentially may occur. Label the three areas (including the uncolored area) indicating their potential for producing the two types of hydrocarbons (oil and gas). Actual occurrence of hydrocarbons in commercial quantities depends on a complex of favorable conditions, especially subsurface presence of source and reservoir rocks in critical geometrical relationships.

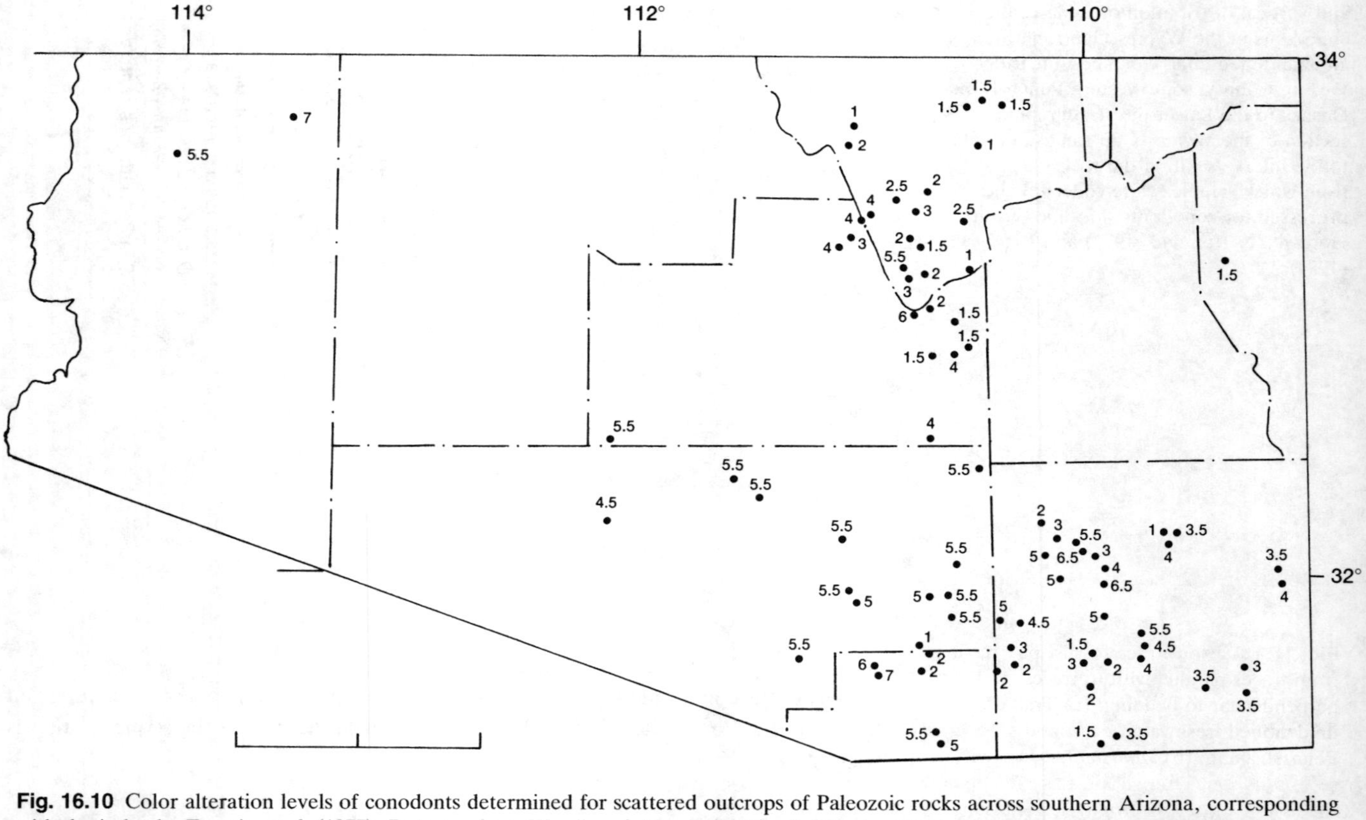

Fig. 16.10 Color alteration levels of conodonts determined for scattered outcrops of Paleozoic rocks across southern Arizona, corresponding with the index by Epstein *et al.* (1977). Data are from Wardlaw & Harris (1984), slightly simplified, with permission from the American Association of Petroleum Geologists.

FURTHER READING

Curran, H. A. (ed.), 1985. *Biogenic structures: their use in interpreting depositional environments*. SEPM Special Publication 35. Society of Economic Paleontologists and Mineralogists: Tulsa, OK.

Ekdale, A. A., 1988. Pitfalls of paleobathymetric interpretations based on trace fossil assemblages. *Palaios* **3:** 464–472.

Ekdale, A. A., Bramley, B. G. & Pemberton, S. G., 1984. *Ichnology – trace fossils in sedimentology and stratigraphy*. SEPM Short No. 15. Society of Economic Paleontologists and Mineralogists: Tulsa, OK

Frey, R. W. (ed.), 1975. *The study of trace fossils*. Springer-Verlag: New York.

Frey, R. W. & Pemberton, S. G., 1985. Biogenic structures in outcrops and cores. 1. Approaches to ichnology. *Bulletin of Canadian Petroleum Geology* **33:**72–115.

Hantzschel, W., 1975. *Trace fossils and problematica*, 2nd edition in C. Teichert (ed.), *Treatise on invertebrate paleontology*, Part W, Miscellanea (Suppl. 1). Geological Society of America and University of Kansas Press: Lawrence, KS.

Osgood, R. G. Jr, 1970. Trace fossils of the Cincinnati area. *Palaeontographica Americana* **6:**281–444.

EXERCISE 17

Worksheet

Name

QUESTION 1

Several named trace fossils are provided. Describe the traces, the sediment in which they are preserved, and assign each to the appropriate behavioral category.

1 Name: ______________________ Sketch:

Description: ______________________

____

Sediment: ______________________

____

Behavioral category: ______________________

2 Name: ______________________ Sketch:

Description: ______________________

____

Sediment: ______________________

____

Behavioral category: ______________________

3 Name: ______________________ Sketch:

Description: ______________________

____

Sediment: ______________________

____

Behavioral category: ______________________

4 Name: ______________________ Sketch:

Description: ______________________

____

Sediment: ______________________

____

5 Name: ______________________ Sketch:

Description: ______________________

____

Sediment: ______________________

____

Behavioral category: ______________________

6 Name: ______________________ Sketch:

Description: ______________________

____

Sediment: ______________________

____

Behavioral category: ______________________

7 Name: ______________________ Sketch:

Description: ______________________

____

Sediment: ______________________

____

Behavioral category: ______________________

8 Name: ______________________ Sketch:

Description: ______________________

____

Sediment: ______________________

____

Behavioral category: ______________________

QUESTION 2

Name the trace fossils in Figs 17.4 and 17.5, assign each illustrated group to it[s] proper assemblage (see Fig. 17.3), and determine the benthic marine environmen[t] in which the assemblage probably formed. The trace fossils may be identified b[y] comparison with those used for Question 1, Figs 17.1 and 17.3, and descriptions an[d] figures in the *Treatise on invertebrate paleontology*, Part W (Miscellanea).

Names of trace fossils in Fig. 17.4:

Skolithos | Diplocraterion

Glossifungites | Ophiomorphora

Trypanites | ______________________

Assemblage to which the group belongs: Domichnia (Dwelling Traces)

Probable environment: Shallow, high energy environment

Names of trace fossils in Fig. 17.5:

Nereites

Cruziana

Zoophycos

Assemblage to which the group belongs: Pascichnia, Fodichnia

Probable environment: Deep water, abyssal, flysh like

QUESTION 3

If a group of trace fossils is provided in the laboratory, identify the rock type in which each is preserved, name each type of trace, determine the category of trace to which each belongs, assign the group to its proper assemblage, and give the estimated environment in which it probably formed.

Rock type in which assemblage occurs: ____________________

Trace fossil name: ____________________ Category: ____________________

Trace fossil name: ____________________ Category: ____________________

Trace fossil name: ____________________ Category: ____________________

Trace fossil name: ____________________ Category: ____________________

Trace fossil name: ____________________ Category: ____________________

Trace fossil name: ____________________ Category: ____________________

Trace fossil name: ____________________ Category: ____________________

Assemblage to which the group belongs: ____________________

Probable environment: ____________________

References

Aldrige, R. J., Briggs, D. E. K., Clarkson, E. N. K. & Paul Smith, M. 1986. The affinities of conodonts — new evidence from the Carboniferous of Edinburgh, Scotland. *Lethaia* **19:**279–291.

Arkell, W. J. *et al.*, 1957. Mollusca 4. Cephalopoda Ammonoidea, in R. C. Moore (ed.), *Treatise on invertebrate paleontology*, Part L. Geological Society of America and University of Kansas Press: Lawrence, KS.

Barnes, R. D., 1963. *Invertebrate zoology*. W. B. Saunders: Philadelphia.

Barnes, R. S. K. (ed.), 1984. *A synoptic classification of living organisms*. Blackwell Scientific Publications: Oxford.

Barnes, R. S. K., Calow, P. & Olive, P. J. W., 1988. *The invertebrates: a new synthesis*. Blackwell Scientific Publications: Oxford.

Bayer, F. M. *et al.*, 1956. Coelenterata, in R. C. Moore (ed.), *Treatise on invertebrate paleontology*, Part F. Geological Society of America and University of Kansas Press: Lawrence, KS.

Benson, R. H. *et al.*, 1961. Arthropoda 3. Crustacea, Ostracoda, in R. C. Moore (ed.), *Treatise on invertebrate paleontology*, Part Q. Geological Society of America and University of Kansas Press: Lawrence, KS.

Boardman, R. S., Cheetham, A. H. & Rowell, A. J. (eds), 1987. *Fossil invertebrates*. Blackwell Scientific Publications: Palo Alto, CA.

Brasier, M. D., 1980. *Microfossils*. George Allen & Unwin: London.

Bretsky, P. W. Jr, 1970. Upper Ordovician ecology of the central Appalachians. *Peabody Museum of Natural History Yale University Bulletin* **34**, 150 p., 34 pls.

Brezinski, D. K., 1984. *Dynamic lithostratigraphy and paleoecology of Upper Mississippian strata of the northcentral Appalachian Basin*. Unpublished PhD. dissertation, University of Pittsburgh: Pittsburgh, Pennsylvania.

Briggs, D. E. G., Clarkson, E. N. K. & Aldridge, R. J., 1983. The conodont animal. *Lethaia* **16:**1–14.

Broadhead, T. W., 1980. *Blastozoa*. University of Tennessee Department of Geological Sciences Studies in Geology 3, pp. 118–132.

Buchsbaum, R., 1948. *Animals without backbones*, 2nd edition. University of Chicago Press: Chicago.

Bulman, O. M. B., 1970. Graptolithina, in C. Teichert (ed.), *Treatise on invertebrate paleontology*, Part V (Revised). Geological Society of America and University of Kansas Press: Lawrence, KS.

Butts, C., 1926. The Paleozoic rocks. *Geological Survey of Alabama Special Report* **14:**41–230.

Butts, C., 1941. Geology of the Appalachian Valley in Virginia. *Virginia Geological Survey Bulletin* **52.**

Cain, J. D. B., 1968. Aspects of the depositional environment and palaeoecology of crinoidal limestones. *Scottish Journal of Geology* **4:**191–208.

Casey, R. E., 1987. Radiolaria. *University of Tennessee Department of Geological Sciences Studies in Geology* **18**, 213–241.

Cheetham, A. H., 1968. Morphology and systematics of the bryozoan genus Metrarabdotos. *Smithsonian Miscellaneous Collections* **153:**1–121.

Chestnut, E. R., Jr & Ettensohn, F. R., 1988. Hombergian (Chesterian) echinoderm paleontology and paleoecology, south-central Kentucky. *Bulletins of American Paleontology* **95:**1–102.

Chlupać, I., Kříž, J. & Schönlaub, H. P., 1980. Silurian and Devonian conodont localities of the Barradian, in H. P. Schönlaub (ed.), *Second European Conodont Symposium — Ecos II, Guidebook. Abstracts*, pp. 146–180. Geologischen Bundesanstalt: Vienna.

Churkin, J. Jr & Carter, C., 1972. Graptolite identification chart for field determination of geologic age. *United States Geological Survey Oil and Gas Investigations Chart OC* 66.

Clark, D. L. *et al.*, 1981. Conodonta, in R. A. Robison (ed.), *Treatise on invertebrate paleontology*, Part W (Suppl. 2). Geological Society of America and University of Kansas Press: Lawrence, KS.

Clark, W. B., 1904. *Miocene*. Maryland Geological Survey. Johns Hopkins Press: Baltimore.

Clarkson, E. N. K., 1986. *Invertebrate palaeontology and evolution*, 2nd edition. Allen & Unwin: London.

Coates, A. G. & Oliver Jr, W. A., 1973. Coloniality in zoantharian corals, in R. S. Boardman, A. H. Cheetham & W. A. Oliver, Jr (eds), *Animal colonies–development and function through time*, pp. 3–27. Dowden, Hutchinson & Ross: Stroudsburg, PA.

Cooper, G. A., 1975. Brachiopods from West African waters with examples of collateral evolution. *Journal of Paleontology* **49:**911–927.

Copper, P., 1985. Fossilized polyps in 430-Myr-old *Favosites* corals. *Nature* **316:**142–144.

Cox, R. L. *et al.* 1969. Mollusca 6. Bivalvia, Vols 1 and 2, *in* R. C. Moore (ed.), *Treatise on invertebrate paleontology*, Part N. Geological Society of America and University of Kansas Press: Lawrence, KS.

Crimes, T. P., 1975. The stratigraphical significance of trace fossils, in R. W. Frey (ed.), *The study of trace fossils*, pp. 109–130. Springer-Verlag: New York.

Cronin, T. M., 1987. Evolution, biogeography, and systematics of *Puriana*: evolution and speciation in Ostracoda, III. *Paleontological Society Memoir* **21:**1–71.

Darwin, C., 1881. *The formation of vegetable mold through the action of worms, with observations on their habits*. John Murray: London.

Dawes, C. J., 1974. *Marine algae of the west coast of Florida*. University of Miami Press. Coral Gables, FL.

de Laubenfels, M. W., 1955. *Porifera*, in R. C. Moore (ed.), *Treatise on invertebrate paleontology*, Part E, pp. 21–112. Geological Society of America and University of Kansas Press: Lawrence, KS.

Ekdale, A. A., Bromley, R. G. & Pemberton, S. G., 1984. *Ichnology — trace fossils in sedimentology and stratigraphy*. SEPM Short Course No. 15. Society of Economic Paleontologists and Mineralogists: Tulsa, Oklahoma.

Epstein, A. G., Epstein, J. B. & Harris, L. D., 1977. Conodont color alteration — an index to organic metamorphism. *United States Geological Survey Professional Paper* **995:**1–27.

Feray, D. E., Heuer, E. & Hewatt, W. G., 1962. Biological, genetic, and utilitarian aspects of limestone classification.

American Association of Petroleum Geologists Memoir **1**:20–32.

Flower, R. H., 1962. Notes on the Michelinoceratida. *New Mexico State Bureau of Mines and Mineral Resources Memoir* **10**:21–40.

Flower, R. H., 1968. The first great expansion of the actinoceroids. *New Mexico State Bureau of Mines and Mineral Resources Memoir* **19**:1–16.

Frey, R. W., 1973. Concepts in the study of biogenic sedimentary structures. *Journal of Sedimentary Petrology* **43**:6–19.

Frey, R. W. & Pemberton, S. G., 1985. Biogenic structures in outcrops and cores. 1. Approaches to ichnology. *Bulletin of Canadian Petroleum Geology* **33**:72–115.

Frey, R. W. & Seilacher, A., 1980. Uniformity im marine invertebrate ichnology. *Lethaia* **13**:183–207.

Friedman, G. M. & Amiel, A. J., 1974. Submarine cementation in reefs: example from the Red Sea. *Journal of Sedimentary Petrology* **44**:816–825.

Galloway, J. J., 1933. *A manual of Foraminifera*. Principia Press: Bloomington, IN.

Gould, S. J. & Woodruff, D. S., 1987. Systematics and levels of covariation in *Cerion* from the Turks and Caicos Islands. *Bulletin of the Museum of Comparative Zoology* **151**:321–363.

Hall, J., 1847. *Palaeontology of New York*, Vol. 1. Albany.

Handfield, R. C., 1971. Archaeocyatha from the Mackenzie and Cassiar Mountains, Northwest Territories, Yukon Territory and British Columbia. *Geological Survey of Canada Bulletin* **201**:1–119.

Haq, B. U., 1978. Calcareous nannoplankton, in B. L. Haq & A. Boersma (eds), *Introduction to marine micropaleontology*, pp. 79–107. Elsevier: New York.

Harrington, H. J. *et al.*, 1959. Arthropoda l. Arthropoda — general features; Protarthropoda, Euarthropoda — general features; Trilobitomorpha, in R. C. Moore (ed.), *Treatise on invertebrate paleontology*, Part O. Geological Society of America and University of Kansas Press: Lawrence, KS.

Hass, W. H., 1962. Conodonta, in R. C. Moore (ed.), *Treatise on invertebrate paleontology*, Part W, pp. 3–69. Geological Society of America and University of Kansas Press: Lawrence KS.

Heckel, P. H., 1972. Recognition of ancient shallow marine environments, in J. K. Rigby & W. K. Hamblin (eds), *Recognition of ancient sedimentary environments*. pp. 226–286. Society of Economic Paleontologists and Mineralogists Special Publication 16.

Helfrich, C. T., 1980. Late Llandovery–early Wenlock conodonts from the upper part of the Rose Hill and the basal part of the Mifflintown formations, Virginia, West Virginia, and Maryland. *Journal of Paleontology* **54**:557–569.

Horowitz, A. S. & Rexroad, C. B., 1972. Conodont biostratigraphy of some United States Mississippian sites. *Journal of Paleontology* **46**:884–891.

Hyman, L. H., 1940. *The invertebrates: Protozoa through Ctenophora*. McGraw-Hill: New York.

Kazmierczask, J., 1989. Halysitid tabulates: sponges in corals' clothing. *Lethaia* **22**:195–205.

Kern, J. P., 1980. Origin of trace fossils in Polish Carpathian flysch. *Lethaia* **13**:347–362.

Kier, P. M., 1972. Upper Miocene echinoids from the Yorktown Formation of Virginia and their environmental significance. *Smithsonian Contributions to Paleobiology* **13**:1–41.

Kier, P. M., 1975. The echinoids of Carrie Bow Cay, Belize. *Smithsonian Contributions to Zoology* **206**:1–45.

Kier, P. M., 1980. The echinoids of the Middle Eocene Warley Hill Formation, Santee Limestone, and Castle Hayne Limestone of North and South Carolina. *Smithsonian Contributions to Paleobiology* **39**:1–102.

Kier, P. M., 1982. Rapid evolution in echinoids. *Palaeontology* **25**:1–9.

Kirkpatrick, R., 1911. On *Merlia normani*, a sponge with a siliceous and calcareous skeleton. *Quarterly Journal of Microscopical Science* **56** (new series): 657–701.

Krumbein, W. C. & Garrels, R. M., 1952. Origin and classification of chemical sediments in terms of pH and oxidation–reduction potentials. *Journal of Geology* **60**:1–33.

Lane, N. G. & Webster, G. D., 1980. Crinoidea. *University of Tennessee Department of Geological Sciences Studies in Geology* **3**:144–157.

Lawrence, D. R., 1968. Taphonomy and information losses in fossil communities. *Geological Society of America Bulletin* **79**:1315–1330.

Leonard, A. B. & Frye, J. C., 1960. Wisconsinan molluscan faunas of the Illinois Valley region. *Illinois State Geological Survey Circular* **304**.

Lidgard, S., 1981. Water flow, feeding, and colony form in an encrusting cheilostome, in G. P. Larwood & C. Nielsen (eds), *Recent and fossil Bryozoa*, pp. 135–142. Olsen & Olsen: Fredensborg, Denmark.

Loeblich, A. R. Jr & Tappan, H., 1964. Sarcodina chiefly "Thecamoebians" and Foraminiferida, In R. C. Moore (ed.), *Treatise on invertebrate paleontology, Part C*, Protista 2, Geological Society of America and University of Kansas Press: Lawrence, KS.

Loeblich, A. R. Jr & Tappan, H., 1988. *Forminiferal genera and their classification*, 2 Vols. Van Nostrand Reinhold: New York.

Lopez Gappa, J. J., 1987. Presencia del genero *Rhabdopleura* (Hemichordata, Pterobranchia) en la plataforma continental Argentina. *Physis (Buenos Aires)* **A45**:33–36.

Macintyre, I. G., 1984. Preburial and shallow-subsurface alteration of modern scleractinian corals. *Palaeontographica Americana* **54**:229.

McKinney, F. K., 1979. Some paleoenvironments of the coiled fenestrate bryozoan *Archimedes*, in G. P. Larwood & M. B. Abbott (eds), *Advances in bryozoology*, pp. 321–336. Academic Press: London.

McKinney, F. K. & Jackson, J. B. C., 1989. *Bryozoan evolution*. Unwin Hyman: Boston.

Meyer, D. L., 1980. Ecology and biogeography of living classes. *University of Tennessee Department of Geological Sciences Studies in Geology* **3**, 1–14.

Miner, R. W., 1950. *Field book of seashore life*. G. P. Putnam's Sons: New York.

Moore, R. C., Lalicker & Fischer, A. J., 1952. *Invertebrate fossils*. McGraw-Hill: New York.

Mori, K., 1987. Intraspecific morphological variations in a Pleistocene solitary coral, *Caryophyllia (Premocyathus) compressa* Yabe and Eguchi. *Journal of Paleontology* **61**:21–31.

Munk, W. H. & Riley, G. A., 1952. Absorption of nutrients

by aquatic plants. *Journal of Marine Research* **11**:215–240.

Murphy, M. A. & Cebecioglu, M. K., 1987. Morphometric study of the genus *Ancyrodelloides* (Lower Devonian, Conodonts), central Nevada. *Journal of Paleontology* **61**:583–594.

Nichols, D., 1969. *Echinoderms*, 4th edition. Hutchinson University Library: London.

Nicholson, H. A., 1872. *A manual of palaeontology*. William Blackwood & Sons: Edinburgh.

Nitecki, M. H., 1971. *Ischadites abbottae*, a new North American Silurian species (Dasycladales). *Phycologia* **10**:263–275.

Nitecki, M. H., 1972. North American Silurian receptaculitid algae. *Fieldiana Geology* **28**, 108 pp.

Nowlan, G. S., 1985. Late Cambrian and Early Ordovician conodonts from the Franklinian miogeosyncline, Canadian Arctic Islands. *Journal of Paleontology* **59**:96–122.

Palmer, A. R., 1954. The faunas of the Riley Formation in central Texas. *Journal of Paleontology* **28**:709–786.

Parsley, R. L. & Mintz, W., 1975. North American Paracrinoidea: (Ordovician: Paracrinozoa, new, Echinodermata). *Bulletins of American Paleontology* **68**:1–115.

Pemberton, S. G. & Jones, B., 1988. Ichnology of the Pleistocene Ironshore Formation, Grand Cayman Island, British West Indies. *Journal of Paleontology* **62**:495–505.

Petersen, M. S., 1975. Upper Devonian (Fammenian) ammonoids from the Canning Basin, Western Australia. *Paleontological Society Memoir* **8**:1–55.

Raup, D. M., 1962. Computer as aid in describing form in gastropod shells. *Science* **138**:150–152.

Raup, D. M., 1965. Theoretical morphology of the coiled shell. *Science* **147**:1294–1295.

Raup, D. M., 1966. Geometric analysis of shell coiling: general problems. *Journal of Paleontology* **40**:1178–1190.

Rexroad, C. B. & Collinson, C., 1961. Preliminary range chart of conodonts from the Chester Series (Mississippian) in the Illinois Basin. *Illinois State Geological Survey Circular* **319**:1–11.

Rexroad, C. B. & Collinson, C., 1963. Conodonts from the St. Louis Formation (Valmeyeran Series) of Illinois, Indiana, and Missouri. *Illinois State Geological Survey Circular* **355**:1–28.

Rexroad, C. B. & Collinson, C., 1965. Conodonts from the Keokuk, Warsaw, and Salem Formations (Mississippian) of Illinois. *Illinois State Geological Survey Circular* **388**:1–26.

Rhodes, F. H. T., 1952. A classification of Pennsylvanian conodont assemblages. *Journal of Paleontology* **26**:886–901.

Saunders, W. B. & Ward, P. D., 1987. Ecology, distribution, and population characteristics of Nautilus, in W. B. Saunders & N. H. Landman (eds), *Nautilus – The biology and paleobiology of a living fossil*, pp. 137–162. Plenum Press: New York.

Savage, N. M., 1983. Late Triassic (Karnian) conodonts from northern San Juan Island, Washington. *Journal of Paleontology* **57**:804–808.

Shattuck, G. B., 1906. *Pliocene and Pleistocene*. Maryland Geological Survey. Johns Hopkins Press: Baltimore.

Shaw, A. B., 1964. *Time in stratigraphy*. McGraw-Hill: New York.

Smith, A., 1984. *Echinoid palaeobiology*. Allen & Unwin: London.

Sprinkle, J., 1980. An overview of the fossil record. *University of Tennessee Department of Geological Sciences Studies in Geology* **3**:15–26.

Stanley, S. M., 1970. Relation of shell form to life habits of the Bivalvia (Mollusca). *Geological Society of America Memoir* **125**:1–296.

Stearn, C. W., 1972. The relationship of the stromatoporoids to the sclerosponges. *Lethaia* **5**:369–388.

Stearn, C. W., 1983. Stromatoporoids: classification. *University of Tennessee Department of Geological Sciences Studies in Geology* **7**:157–163.

Størmer, L., Petrunkevitch, A. & Hedgpeth, J. W., 1955. Arthropoda 2. Chelicerata with sections on Pycnogonida and Palaeoisopus, in R. C. Moore (ed.), *Treatise on invertebrate paleontology*, Part P. Geological Society of America and University of Kansas Press: Lawrence, KS.

Sverdrup, H. U., Johnson, M. W. & Fleming, R. H., 1942. *The oceans, their physics, chemistry, and general biology*. Prentice Hall: Englewood Cliffs, NJ.

Sweet, W. C., 1988. *The Conodonta*. Oxford University Press: New York.

Taylor, P. D., 1983. *Ailsacrinus* gen. nov., an aberrant millericrinid from the Middle Jurassic of Britain. *Bulletin of the British Museum (Natural History)*, Geology series **37**:37–77.

Teichert, C. *et al.*, 1964. Mollusca 3. Cephalopoda Nautiloidea, in R. C. Moore (ed.), *Treatise on invertebrate paleontology*, Part K. Geological Society of America and University of Kansas Press: Lawrence, KS.

Thayer, C. W., 1975. Morphologic adaptation of benthic invertebrates to soft substrata. *Journal of Marine Research* **33**:177–189.

Ubaghs, G. *et al.*, 1978. Echinodermata 2 (Crinoidea), Vols 1–3, in R. C. Moore & Teichert C. (eds), *Treatise on invertebrate paleontology* Part T. Geological Society of America and University of Kansas Press: Lawrence, KS.

von Zittel, K. A., 1900. *Text-book of palaeontology*. Macmillan: London.

Wade, B., 1926. The fauna of the Ripley Formation on Coon Creek, Tennessee. *United States Geological Survey Professional Paper* **137**:1–272.

Waller, T. R., 1969. The evolution of the *Argopecten gibbus* stock (Mollusca: Bivalvia), with emphasis on the Tertiary and Quaternary species of eastern North America. *Paleontological Society Memoir* **3**:1–125.

Ward, P. D., 1987. *The natural history of Nautilus*. Allen & Unwin: Boston, MA.

Wardlaw, B. R. & Harris, A. G., 1984. Conodont-based thermal maturation of Paleozoic rocks in Arizona. *American Association of Petroleum Geologists Bulletin* **68**:1101–1106.

Wells, H. W., 1961. The fauna of oyster beds, with special reference to the salinity factor. *Ecological Monographs* **31**:239–266.

Weyl, P. K., 1970. *Oceanography, an introduction to the marine environment*. John Wiley & Sons: New York.

Whittington. H. B., 1959. Silicified Middle Ordovician trilobites. *Harvard University Museum of Comparative Zoology Bulletin* **121**:371–496.

Williams, A. *et al.*, Brachiopoda, in R. C. Moore (ed.), *Treatise on invertebrate Paleontology*, Part H. Geological Society of America and University of Kansas Press: Lawrence, KS.

Wilson, J. L., 1975. *Carbonate facies in geologic history*. Springer-Verlag: New York.

Index

Page numbers in *italics* refer to tables; page numbers in **bold** refer to figures.